RESEARCH IN PHILOSOPHY & TECHNOLOGY

Volume 1 · 1978

RESEARCH IN PHILOSOPHY & TECHNOLOGY

An Annual Compilation of Research

Editor: PAUL T. DURBIN
        Philosophy Department and Center for the
        Culture of Biomedicine and Science
        University of Delaware

Bibliography Editor: CARL MITCHAM
        St. Catharine College
        St. Catharine, Kentucky

OFFICIAL PUBLICATION OF THE SOCIETY FOR
PHILOSOPHY & TECHNOLOGY

VOLUME 1 • 1978

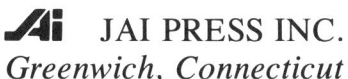 JAI PRESS INC.
Greenwich, Connecticut

Copyright © 1978 by JAI PRESS INC.
321 Greenwich Avenue
Greenwich, Connecticut 06830
All rights reserved. No part of the publication may be reproduced, stored on a retrieval system, or transmitted, in any form or by any means, electronic, mechanical, photocopying, recording or otherwise, without prior permission in writing from the publisher.

ISBN NUMBER: 0-89232-022-2

Manufactured in the United States of America

CONTENTS

INTRODUCTION TO THE SERIES	1
INTRODUCTION TO VOLUME ONE: APPROACHES TO PHILOSOPHY OF TECHNOLOGY OLD AND NEW	5

PART I: METHOD, DESCRIPTIVE FRAMEWORKS, AND A PRACTICAL PROGRAM FOR PHILOSOPHY OF TECHNOLOGY

CULTURE AND TECHNOLOGY *Joseph Margolis*	25
THE PROBLEM OF SCALE IN HUMAN LIFE: A FRAMEWORK FOR ANALYSIS *Robert E. McGinn*	39
TECHNOLOGY, MASS MOVEMENTS, AND RAPID SOCIAL CHANGE: A PROGRAM FOR THE FUTURE OF PHILOSOPHY OF TECHNOLOGY *Joseph Agassi*	53

PART II: THE UNIVERSITY OF DELAWARE CONFERENCE, 1975

TOWARD A SOCIAL PHILOSOPHY OF TECHNOLOGY *Paul T. Durbin*	67
THE EXPLANATION OF TECHNOLOGY *Albert Borgmann*	99
VALUES IN SCIENCE *Willis H. Truitt*	119
TECHNOLOGY AS IDEOLOGY *Kai Nielsen*	131

HUMANIZATION OF TECHNOLOGY: SLOGAN OR ETHICAL
IMPERATIVE?
   Edmund Byrne   149

WHAT IS TECHNOLOGY?
   Robert E. McGinn   179

SHIFTING FROM PHYSICAL TO SOCIAL TECHNOLOGY
   Jospeh Agassi   199

THE COGNITIVE DIMENSION OF TECHNOLOGICAL
CHANGE
   Stanley R. Carpenter   213

TYPES OF TECHNOLOGY
   Carl Mitcham   229

PART III: REVIEW AND BIBLIOGRAPHY

CURRENT BIBLIOGRAPHY IN THE PHILOSOPHY OF
TECHNOLOGY: 1972–1974
   Carl Mitchan and Jim Grote   313

INTRODUCTION TO THE SERIES

Rarely can the history of an intellectual movement be as easily delineated as in the case of the philosophy of technology movement in the United States. Even in the case of the movement elsewhere and at earlier times, the matter is clear. Carl Mitcham has outlined the preliminary stages of development, from the Greeks to the beginnings of philosophy of technology proper in the early twentieth century, in a paper that unfortunately cannot be included in this first volume of our series; it was the lead-off paper at a conference from which sprang both this volume and the whole series. We hope to include it in Volume Two. Mitcham, with his collaborator Robert Mackey, has also pinpointed the earliest works to bear the title "Philosophy of Technology" (all German and all but one published in the early years of the twentieth century), setting the pattern for the trend to follow. (Cf. Mitcham and Mackey, *Philosophy and Technology* [1972], introduction.)

As technology took on a recognizable form in the public consciousness in the United States, and particularly after the three big "technological events" of World War II and the decades following—the atom-bombing of Hiroshima and Nagasaki, Sputnik, and the American moon-landing adventure—the trickle of critiques of science and technology turned into a flood. The culmination of the antitechnology movement is perhaps epitomized in one book, Theodore Roszak's *The Making of a Counter Culture: Reflections on Technocratic Society and Its Youthful Opposition* (1969). Meanwhile, academic philosophers had also begun to examine the phenomenon.

In the early to mid-sixties three conferences took place in the United States which signal the beginnings of the new movement. The first, an Encyclopaedia Britannica conference held at the Center for the Study of Democratic Institutions, is reported in a volume, *The Technological Order* (1963), edited by Carl F. Stover. The second, the proceedings of which are printed as *Philosophy in a Technological Culture* (1964), edited by George F. McLean of Catholic University in Washington, D.C., represents the first step of American Roman Catholic Thomistic philosophers into the field. The third, a set of papers in the Summer 1966 issue of the history of technology journal *Technology and Culture* with the general title "Toward a Philosophy of Technology," is more commonly taken as the real beginning of philosophy of technology among academic philosophers in the United States. (Meanwhile, developments outside the United States and in other philosophical traditions continued; for instance, a symposium on Marxist-Leninist philosophy and the technical revolution appeared in *Deutsche Zeitschrift für Philosophie*, special issue, 1965.) The *Technology and Culture* symposium included contributions not only by one of the social critics, Lewis Mumford, but by a group of philosophers who have since come to be recognized almost as *the* American philosophers of technology: especially Joseph Agassi, Mario Bunge, and Henryk Skolimowski.

Mitcham and Mackey, in another of their landmark contributions to the growing field, *Bibliography of the Philosophy of Technology* (1973), note another indication of the growing movement (not only in the United States but worldwide) in a series of *Proceedings* volumes of the International Congress of Philosophy appearing at roughly five-year intervals: "These successive volumes of *Proceedings* [1953, 1958, 1964, 1968] indicate the progressive recognition of technology as a philosophical issue" (p. 13). Their evidence is: a single paper in 1953 in French; eight papers in 1958, three of which are in English and touch tangentially on technology; five papers in 1964, the majority in English but still only tangential to philosophy of technology; finally, the 1968 volume includes a veritable flood of papers on technology, engineering, cybernetics and automation, as well as the place of science in modern culture—with numerous Ameri-

Introduction to the Series 3

can contributors including names recognizable from the 1966 *Technology and Culture* symposium. Mitcham and Mackey add: "The XVth International Congress on Philosophy, to be in Varna, Bulgaria, in September 1973, will feature as a main theme, 'Science, Technology and Man'." Their advance notice was correct, and the proceedings of the XVth Congress are now a main source for philosophy of technology, with several contributions by Americans.

If at this point it needs to be noted separately, Mitcham and Mackey's two volumes, *Philosophy and Technology* (readings, 1972) and *Bibliography of the Philosophy of Technology* (1973), constitute one more indication of the growth of the movement. In the latter volume Mitcham and Mackey note: "We do hope to have listed the majority of works which are truly important and a good deal of those which (without proper annotation) might otherwise be thought to be important—since one main function of a bibliography of this type is to steer others away from deadwood" (p. xvi). This point will come up again shortly.

Finally we come to the present series, its genesis, and what it hopes to accomplish in terms of the movement. In 1975, building on what had gone before as well as the heroic championing of philosophy of technology by historian of technology Melvin Kranzberg, editor of *Technology and Culture,* a conference was held at the University of Delaware to attempt to coalesce the philosophy of technology movement. Out of that meeting came this series, the current first volume in it (comprising the proceedings of the Delaware conference and a follow-up symposium at the American Association for the Advancement of Science meeting in Boston, February 1976), plans for future conferences, and a newsletter. The function of the series of annual volumes will be described presently; the role of the newsletter relative to it is, basically, to maintain intellectual contacts between philosophers of technology, and to keep them abreast of developments in the field, between appearances of the Annual.

Before turning to the specific functions of the Annual, it is worth noting a theme that can be distilled from contributions to the philosophy of technology movement thus far. Though some philosophers remain skeptical (see the introduction to Volume One, below) those who see the movement as legitimate recognize two things: (1) *There are urgent problems connected with technology and our technological culture which require philosophical clarification,* and (2) *Much that has thus far been written on these problems is inadequate—making it all the more important for serious philosophers to get involved.*

As the general editor and his board of consulting editors see the matter, the Annual series should play a triple role:

1. Through the annual bibliographical update, under the general editorship of Carl Mitcham, it should become *the* place to look for bibliographi-

cal leads, documentation, and the like. (Between editions, the bibliography section of the newsletter, also under Mitcham's direction, will fill in the gap and keep workers in the field up on the latest developments.)

2. The Annual will serve as the outlet for the proceedings of as many philosophy of technology conferences and symposia as can be arranged. Plans are afoot to have several of these each year, with a major conference at least every other year.

3. Finally, in a sense, the Annual can take the place of a journal in the fledgling field. Contributions independent of conferences and symposia will be welcomed—though wherever possible authors will be encouraged to consider presenting them at conferences to guarantee peer criticism and to facilitate the development of dialogue within the field.

The decade from the mid-sixties to the mid-seventies has been an exciting one for philosophers concerned with problems of technology. We hope the appearance of this series will help maintain the excitement while contributing to the consolidation of the field and to the development and maintenance of the highest standards.

ACKNOWLEDGMENTS

In a cooperative series of this sort, based on conferences and symposia, personal acknowledgments might seem out of place—except for the numerous debts incurred by the general editor in the two years of organizational as well as intellectual effort that went into setting up the institutional framework for the series. At least a word of thanks must go to Frank Dilley, in 1975 running not only the Philosophy Department but also its Values Center at the University of Delaware; to Edward Lurie, Director of the Center for the Culture of Biomedicine and Science; to Kathy Krajewski, project assistant; to Bob Ross, formerly with Continuing Education, University of Delaware, who coordinated the Delaware Humanities Forum portion of the original conference, which forged our links to members of the scientific community; to the 1976 program committee of the American Association for the Advancement of Science; to a series of secretaries in our Philosophy Department and Center for the Culture of Biomedicine and Science; to numerous friends—one in particular—for advice and help at difficult times; to Fran Durbin for inspiration and forbearance. Two others deserve even fuller acknowledgment: Melvin Kranzberg, editor of *Technology and Culture,* and Herbert Johnson of JAI, Inc.; enough to say that without their vision there would be no series.

INTRODUCTION TO VOLUME ONE: APPROACHES TO PHILOSOPHY OF TECHNOLOGY OLD AND NEW

The proceedings of the 1975 Philosophy of Science Association biennial meeting include a symposium addressing the question, "Are there any interesting philosophical issues in technology?" In the planning stages some of the organizers of the symposium stated the background of the issue more bluntly: "Many philosophers of science doubt that there are any very interesting philosophical issues generated by technology as distinct from science." Stating the matter thus bluntly nicely poses the problem addressed in the first volume of this *Research in Philosophy & Technology Annual:* What would constitute an adequate philosophy of technology?

As has been noted above in the general introduction to the series, in the past ten years most academic philosophers in the United States who have addressed themselves to problems of technology have been concerned to identify what is philosophically worthy of notice in the recent flood of criticisms of technology and certain aspects of science. As these philosophers have begun to define their task in carving out a new field, they have had to face the obvious question of *clear methodological guidelines,* of criteria or standards for the evaluation of contributions to the new field. Thinking along these lines resulted in a decision to divide the preliminary task into two phases, with each to be treated in a separate conference or symposium. (Both are reported in this volume.)

The 1975 University of Delaware conference on philosophy of technology (Part II here) featured a variety of younger scholars in the emerging field representing a broad range of *different approaches* to philosophy of technology. It was decided to do this first for two reasons: (a) so that there would be fresh new material available for analysis in the process of seeking guidelines, and (b) to avoid any premature closure, too soon eliminating potentially fruitful candidates for an adequate philosophical approach to problems of technology.

Then in 1976, at the annual meeting of the American Association for the Advancement of Science, a follow-up symposium was held specifically on the question of *methodological guidelines* for an adequate philosophy of technology. That symposium is represented here as Part I.

Finally, the volume is rounded out, in Part III, with the first of an annual series of updates, by Carl Mitcham, of the *Bibliography of the Philosophy of Technology* (1973).

THE QUESTION OF THE LEGITIMACY OF PHILOSOPHY OF TECHNOLOGY AS DISTINCT FROM PHILOSOPHY OF SCIENCE

There are several ways of approaching the question of the legitimacy of philosophy of technology. One way would be to examine the rapidly developing literature in the field, especially over the past ten years. Another would be to look critically at the contributions to the 1975 University of Delaware conference—not all of which, admittedly, and as is usual with conference proceedings, are of equal merit. Still a third way would be to look at the question *a priori* in terms of recognizably legitimate general philosophical positions as these might be, or have been, brought to bear upon technology. We can begin with this third approach.

One of the giants among recent philosophers of science, Carl G. Hempel, makes a now-standard distinction that can get us started. Referring to a conception of Aristotle's (as interpreted by J. F. O'Brien) which claims "that the natural affinities underlying gravitational attraction are related to love," Hempel replies:

> This assertion is so elusive that it precludes the derivation of *any* test implications. . . . To illustrate this . . . , let us suppose someone were to offer the alternative thesis that physical bodies gravitationally attract each other and tend to move toward each other from a natural tendency akin to hatred. . . . The two verbally conflicting interpretations make no assertions at all. Hence, the question whether they are true or false makes no sense. . . . They are *pseudo-hypotheses* (*Philosophy of Natural Science* [1966], p. 31).

This is, of course, related to the old Logical Positivist criterion of (scientific) meaninglessness, and few would doubt that a Hempelian approach to philosophy of technology would be philosophically legitimate—if only in demonstrating the meaninglessness or vagueness or lack of empirical import of many assertions in the literature on technology. But Hempel goes on—drawing on certain conclusions of his own in *Aspects of Scientific Explanation* (1965)—to recognize that "it is not possible to draw a sharp dividing line between hypotheses and theories that are testable in principle and those that are not."

The spectrum thus engendered—from clearly untestable to questionable (shadowy or gray area) to clearly testable—can serve as a useful warning here. For one thing, it reopens the question whether or not there might be anything of interest to the philosopher of science (and technology) in Aristotle. More to the point: it suggests that what is philosophically interesting about technology may be a matter for some discussion, and specifically that *how distinct philosophy of technology is from philosophy of science will differ according to different philosophical persuasions.*

Let us look first at Hempel's object of scorn in the quotations above, the philosophy of Aristotle or his medieval or modern Thomistic followers. First off, these people would complain about Hempel's incredible misinterpretation of an Aristotelian theory of gravitational attraction in the quotes; any competent historian of Aristotle (cf., for instance, Friedrich Solmsen's *Aristotle's System of the Physical World* [1960]) would set Hempel right, that Aristotle's view depends on the notion of "natural inclination," not on any analogy with human love. This, however, only brings up the question of the legitimacy of Aristotle's "philosophy of nature"—although even there it can be maintained that the alternative theory Hempel ought to have brought up is not one based on hatred but on Aristotle's historical opponent, atomism; and the issue between the two is "philosophically interesting" enough for Hempel himself to discuss it in the same volume under its modern rubric of reductionism vs. antireductionism. We can here assume, however, that at least in many quarters an Aristotelian approach to philosophy of science would be considered legitimate.

What about an Aristotelian (or Thomistic) philosophy of technology? No widely recognized Aristotelian or Thomistic philosophy of technology currently exists. It would be possible to piece one together—strange as the thought might appear to anyone who reflects on the irony of applying Greek or medieval categories to problems peculiar to our technological culture—from elements of an Aristotelian-Thomistic philosophy of science, theory of craft-art, and natural law social ethic. In such an approach, philosophy of technology would clearly be distinct from what has been done in the twentieth century under the name of a Thomistic philosophy of science—by Jacques Maritain, for instance, among others.

Another traditional philosophical approach that might be thought relevant to technology is that of Marx; indeed, his approach is utilized in one of the papers in this volume (whereas Aristotle's is not). However, some philosophers, using criteria not unlike those of Hempel above, would question the validity of Marx's concepts. Consider the following:

> What I call "orienting statements" include some of the most famous statements of social science. One is Marx's statement that the organization of the means of production determines the other features of a society. . . . Whereas Boyle's Law says that, if pressure goes up, volume will assuredly go down, what Marx's Law says is that, if there is some, any, change in the means of production, there will be some unspecified change or changes in the other features of society. . . . I cannot grant that law the status of a real proposition. (George C. Homans, *The Nature of Social Science* [1967] p. 14.)

Aside from obvious corrections that a follower of Marx would make with respect to Homans's reading of "Marx's Law," our concern here is with the legitimacy of a Marxist approach to philosophy generally and more specifically with a Marxist critique of technology distinct from Marxist philosophy of science. Marxists and neo-Marxists of all sorts can speak for themselves on the first point. As to the second, it can plausibly be maintained—presumably on the basis of Marx's thesis of philosophy as *praxis*—that in Marxist theory philosophy of technology has priority over philosophy of science (if it is even legitimate to distinguish the latter from the former!). At any rate, in the hands of a neo-Marxist philosopher such as Herbert Marcuse, the tables can be turned and a critique of (capitalist or bureaucratic socialist) technology can be utilized to question the legitimacy of analytical philosophy of science of the Hempelian sort.

Introduction to Volume One

It might be objected at this point that Hempel's views on philosophy of science no longer represent the dominant view—and that non-Hempelians or anti-Hempelians would find philosophy of technology equally suspect. Here it may be enough to note three general trends in recent American philosophy of science: contributions influenced by Thomas Kuhn's "revolutionary" theory of scientific progress, the views of Karl Popper and others loosely labeled Popperians, and those who have moved logical empiricism in a vague "pragmatic" direction.

As formulated in *The Structure of Scientific Revolutions* (1962) and subsequent articles, Kuhn's approach to philosophy of science sticks more or less exclusively to "pure" science and major historical revolutions. Nevertheless, Kuhn's emphasis on group standards in particular scientific communities might suggest that if sociologists of science detect significant differences between pure-science communities, applied scientists, and exclusively technological technical communities (whatever that might mean), these differences would show up in different ways in the analysis of the scientific (or technical) standards of the various communities. In fact, this sort of thing may already be apparent in sociological studies based in part on Kuhn's model. At any rate, Kuhnians might be expected to be more tolerant of a distinct philosophy of technology than traditional logical empiricists.

Popperianism, at least in a modified form, has been represented throughout the past decade in philosophy of technology discussions, in the work of Joseph Agassi. His efforts continue in this volume.

Finally in this connection, "pragmatism" in philosophy of science can be given a less vague formulation by linking it explicitly to American pragmatism. Depending on the version, this could mean a greater or less distinction between philosophy of science and philosophy of technology. One pragmatist's views (those of G. H. Mead) are used in this volume to formulate a distinctive social philosophy of technology which shares many theses with a Marxist approach combining philosophy of technology with philosophy of science—but with both through-and-through social in a way standard philosophy of science usually is not.

This brief survey might well be rounded out simply by listing a number of other approaches that ought to be recognized simultaneously as legitimate and as distinct from the work customarily found in philosophy of science journals. First would come the openly metaphysical approaches—neo-Kantian, Hegelian, etc.—especially of the European authors Mitcham and Mackey credit with initiating philosophy of technology in the twentieth century. Next would come Martin Heidegger's work on technology, as well as that of other philosophers influenced by him. (One such view is presented here.) While hardnosed philosophers of science might be suspicious of much in both these traditions, it would be difficult to maintain that the work is not distinctively philosophical and, at least as commentary on our technological culture, worth examining. Finally, it seems long past due for theoretical ethicists such as John Rawls and those influenced by him, who have been applying their theories to contemporary normative concerns, to do the same for problems of technological society. There have been some attempts in the biomedical area and, to a lesser extent, in legal and ecological areas, but not enough.

PART II: THE UNIVERSITY OF DELAWARE CONFERENCE, 1975

In July 1975, several units at the University of Delaware—the Philosophy Department; the Center for the Study of Values; the Science, Technology, and Society Committee; and the Division of Continuing Education, with part of the program funded by the Delaware Humanities Forum— co-hosted a dual conference on philosophy of technology. One part was the philosophers' conference whose proceedings make up about half of this volume and which will be introduced presently. The other, under the general heading Scientists and Social Responsibility, gave members of the large Delaware technical community a chance to reply to their philosopher-critics. This latter portion of the program, many contributions to which were off-the-cuff, is not printed here in full—though most of the discussion is available on tape. It should be worthwhile, however, to report reasonable representative samples of the scientists' remarks, partly for the record, partly to dramatize the social context in which the philosophy of technology movement has had its birth.

One of the participants, Dr. Herman Skolnik of Hercules Research, editor of *The Journal of Chemical Information and Computer Sciences*, later polished up some of his remarks and comments and published them as an editorial in the September 1975 issue. Some representative comments:

> In talking about technology, the philosophers tended to ignore or confuse the continuum of science and technology and to allude to the work of engineers and inventors and the effect of Madison Avenue on selling their products to society. They viewed technology as ravaging nature, degrading humans, and impoverishing civilizations. . . . One philosopher's example for the dehumanizing effect of science and technology was the automobile assembly line. . . .
>
> I am all for developing a philosophy of technology. But I am not sure I am willing to let the philosophers do it for me. Certainly not the political and moral philosophers. Possibly those we should seek for this study are those who appear as philosophers to scientists when among scientists and as scientists to philosophers when among

philosophers. . . . If we take the holistic approach, which I favor, then the study needs to consider not only science and technology in terms of social responsibility but the social responsibility of every factor, political, educational, religious, economical, sociological, psychological.

The mentions of holism and social responsibility are references to some comments of Dr. Russell W. Peterson—former chemist, DuPont Company research administrator, governor of Delaware, and chairman of the Council on Environmental Quality—with which the Scientists and Social Responsibility portion of the program began:

In past decades political issues dominated the world scene. But today's issues of excessive population growth, shortages of the basics of life such as food, water, energy and shelter, threats to our health from malnutrition and pollution, and the unbalancing of our ecological life support systems lend themselves to attack by research and development.

But not just the traditional research and development which concerns itself with breaking down problems into smaller and smaller parts. What is needed in addition is a holistic, integrated approach. . . .

And here is where the scientist can make his greatest contribution to society—how he can most effectively fulfill his social responsibility.

One of the more perceptive comments especially pinpointing the disagreements between scientists and philosophers was made by Dr. P. Burke Welldon, personnel director of Hercules Research, who led off the Critique of the Critics session:

[Earlier] I referred to what is perhaps the philosophers' strongest criticism of scientists and technologists: that they are neutralists without a responsibility to humanity basic to their work. I hope that in this philosophical criticism there is no hint, no implication that sometimes it is better *not* to discover the facts of nature. Because this seems to me diametrically opposed to the credo of physical scientists—indeed of us all. If we take the position [for instance] that the secrets of nuclear energy are so threatening that it would be better they had never been discovered, we advocate a return to the Dark Ages.

Perhaps the most optimistic comment was made by Dr. S. T. Putnam:

I went from an initial feeling of antagonism to the philosophers' group (primarily because of the obvious great differences between my political philosophy and theirs) to a state of détente . . . to the realization near the end. . . that I could establish some basis for a common understanding [at least] with the more moderate members of the philosophers' group. I think that further discussions concentrated on defining the goals we need to achieve . . . would be worthwhile.

In addition to the members of the technical community, other participants in the conference included historians of science and technology, George Basalla, John Beer, and Melvin Kranzberg; and other philosophers of technology—Ian Barbour, Stuart Brown, Robert Mackey, and Joseph Margolis. Future conferences are planned, and it is hoped again to include scientists, engineers, physicians and biomedical researchers, and social scientists, as well as historians.

Now for the proceedings of the part of the program devoted specifically to philosophy of technology.*

The first paper laid the foundation of the conference by providing a straightforward history-of-ideas background to the philosophy of technology. In "The Philosophy of Technology: Origins and Issues," Carl Mitcham (St. Catharine College, Kentucky) described the philosophy of technology as the attempt, by means of reason and insight, to answer fundamental questions about the nature and meaning of technology, i.e., the making and using of artifacts. According to Mitcham, the philosophy of technology arose when its actual realization in processes and objects became present to consciousness in such a way as to "invite" questioning. Thus the philosophy of technology began with the Industrial Revolution and the resulting problems presented by technology.

Mitcham proceeded to sketch the history of technology from the Classical Period through the nineteenth century, presenting only a hasty summary of twentieth-century developments. After detailing the reasons for the Greek and Christian attitude of distrust toward technological activity, Mitcham asserted that not until Francis Bacon and the Renaissance was there approval of technological production. Bacon initiated a transvaluation of values when he proceeded to outline the methodology for a science of nature directed toward practice rather than theory.

In the conclusion of his paper, Mitcham called for a philosophy of technology which would include a metaphysical approach to clarify such concepts as science, art, technology, and design. What is also needed, said Mitcham, is a philosophical anthropology of technology to reveal how technological activity is grounded in human creativity. Finally, Mitcham stated the need for an ontology of technology to determine the kind of being technological objects possess.

*This summary was prepared, for the most part, by Ms. Kathleen Krajewski, assistant for the conference.

Unfortunately, Mitcham's paper cannot be included here; we hope that it will appear in a future volume of the Annual. Following Mitcham's paper, Paul Durbin (University of Delaware) extended to the conference participants an invitation to dialogue on the meaning of the philosophy of technology.

In his paper, "Toward a Social Philosophy of Technology," the first in Part II below, Durbin offers what he terms a "Mead-like" approach to the social philosophy of technology, strongly dependent on the pragmatist G. H. Mead. The working definition of the paper focuses on the communication of meaning in the technical community. Social philosophy generally, for Durbin, includes a theoretical perspective which would allow criticism of standard social science theories about social structure by means of alternative theories, a nonreductionist social epistemology and, finally, a concrete statement taking a stand on social problems. In his paper, Durbin aims to present a normative social philosophy of technology based on the concept or metaphor of deliberate meaning-community establishment and maintenance with respect to the technical community in the context of the larger cultural community.

Durbin discusses four alternatives to the metaphor/definition of communication as meaning-community establishment and maintenance, which include a Platonic/Idealistic philosophy of technology, behaviorism, a division of labor between philosophers and social scientists popular today in academia, and contemporary Marxist-oriented critiques of technology.

In conclusion, Durbin stresses the need for a continuation of vocational education and the broadening of the training of scientists, engineers, etc., in the humanities and social sciences. Durbin hopes that the scientific and technical professions will recognize their political functions and become more involved in liberal causes such as civil rights.

In "The Explanation of Technology," Albert Borgmann (University of Montana) presents a Heidegger-like attempt to get at the essence of technology. He endeavors to place technological explanation midway between two modes of explanation: a Hempel-like "deductive-nomological" explanation of the sciences ("apodeictic") which yields transformative but not guiding power, and the "deictic" explanation of the arts which sets standards of action and conduct but does not provide any effective means of compliance. Technological explanation, the middle ground, he calls "paradeictic."

Borgmann's thesis is that technology now articulates the world but the articulating power of technology has no focal point, so that it must itself be articulated. According to Borgmann, paradeictic or paradigmatic explanation would reveal the pattern by which the concrete workings of technology could be uncovered. Citing the technological device as the paradigm of the contemporary world, he proceeds to examine the notion of technology as the procurement of devices. The remainder of Borgmann's paper concentrates on points such as the limits to the progress of procurement, the possibility of guidance in the expansion of procurement and an analysis of a critique of the device paradigm. Borgmann concludes that, since "deictic" explanation is a part of "paradeictic" explanation, a critique of technology remains possible.

In his paper, "Values in Science," Willis H. Truitt (University of South Florida) promotes a Marcusean, neo-Marxist attack on the alleged "value-neutrality" of science, arguing against liberal-pragmatist philosophies of science and technology that do not categorically repudiate instrumentalism and optimism with respect to technology. Such approaches to the philosophy of technology proclaim value neutrality in science and technology where, Truitt maintains, a concealed ethic already exists.

Truitt notes particularly the areas where Durbin and he reached similar and different conclusions. With Durbin, Truitt concurs that science and technology cannot be isolated, for they are continuous, and that a philosophy of technology must be normative. He further agrees with Durbin that science and technology are not neutral and that class structure in the United States allows technology to support the rich and powerful and hinder the interests of the weak and poor. However, Truitt disagrees with Durbin on instrumentalism and espouses a Marxist critique of science and technology as a guarantee that they will not be subjected to class and undemocratic institutional exploitation.

Kai Nielsen's paper, "Technology as Ideology," is based on comments at the conference aimed primarily at Truitt (and to a lesser extent at Joseph Agassi, see below) but was only fully developed later. Nielsen argues that there was a general lack of awareness at the conference of the political and ideological dimensions of the problem. He claims that there are obvious evils associated with the uses of technology in capitalist and bureaucratic socialist countries the eliminating of which is primarily a political problem. The bulk of the paper is an elaboration of the Marx-based but novel theories of Jürgen Habermas.

Nielsen ends with a strongly normative claim that we ought to look primarily at whose interests the current utilizations of technology reflect. By understanding these interests we can assess the morality of these uses. Finally, the task is not scientific but social and political, requiring informed moral reflection for a rational resolution of the problem.

In general, these papers, presented on the first day of the two-day philosophers' conference, emphasize pre-twentieth and early twentieth-

century approaches or models for the philosophy of technology. (Nielsen's paper is a partial exception.)

In the first paper of the second day, "Humanization of Technology: Slogan or Ethical Imperative?," Edmund Byrne (Indiana University–Purdue University at Indianapolis) offers an approach to the philosophy of technology closely tied to contemporary legal problem-solving. His paper is an attempt to support the claim that man-machine teleology is not restricted to the level of machine design and operation but necessarily includes organizational policy and planning where ethical considerations are engendered on both levels. Byrne suggests a phenomenological comparison between the concepts of "cyborg" and "prosthesis" to show that considerations of design are "subservient to and dependent upon" those of management goals. He asserts that the teleology of man-machine relationships is a question of managerial responsibility.

Byrne proposes a philosophy of man which a given technologization presupposes as a basis for distinguishing between good and bad technologizations. A defensive philosophy of man yields a technology of despair grounded in the cyborg whereas a supporting philosophy of man brings forth a technology of hope, giving priority to the prosthesis model of man-machine relationships. Byrne concludes by calling for "appropriate governmental and citizen interest adequately complemented with duly issue-oriented professional responsibility on the part of scientists and engineers" so that personal and professional responsibility will not be destroyed by a technology based on a cyborg model of man-machine relationships.

Robert E. McGinn (Stanford University) puts forward another "systems" approach, this time to the description (not definition) of technology. In "What is Technology?," McGinn sets forth an analysis of the structure and content of technology as one form of human activity. He describes the structure of technology by isolating seven aspects of the activity form. Those factors include: material product-making or object-transforming, purpose, resources, knowledge, method, the socio-cultural-environment context within which the activity occurs, and, finally, the practitioner's mental set which relates to any of the preceding six aspects as well as to the meaning and importance of his activity. He elaborates his analysis of the structure of technology as a form of truly human activity by relating an example of a man now living who has mastered the old Indian art of birch-bark canoe technology.

McGinn describes the content of technology as the "set-theoretical union or sum of all particular (kinds of) technologies, each of which is a complex of . . . knowledge, resources, and methods involved in the production (transformation) of that kind." Using this differentiated concept of technology, McGinn proceeds to clarify the argument whether technology is value-neutral or value-laden and to discuss several specifically modern elements in modern technology.

In "Shifting from Physical to Social Technology," Joseph Agassi (Boston University and Tel Aviv University) offers a muted Popperian approach to the philosophy of technology which is basically a "participatory-democracy" extension of Popper's social-democracy critique of modern political systems—now turned against technocracy.

Agassi claims that, even though technology has many aspects, a particular technology is classified according to the most immediately problematic aspect of it. Thus, where the nature of the problem is evidently social, the specification social technology would apply. In explaining why physical technology is dominant, Agassi cites materialism as the chief cause. Furthermore, when technology is equated with physical technology, social and political technology is overlooked. Agassi calls for a reallocation of our resources of research technology to take into account social and political aspects of technology. The starting point for the reform movement, he holds, must be the refutation of all materialism. Agassi then discusses three views of rapid social change: the functionalist, the Hegelian, and the Weberian. He ends his paper by stating that, although the movement of concern for social technology is weak, the revolution must start in the Western industrialized countries "with the democratization of education and of the quality of working life" where rapid progress toward a social and political technology can be brought about.

The paper, "The Cognitive Dimension of Technological Change," by Stanley R. Carpenter (Georgia Tech) proposes that an adequate philosophy of technology should include among its concerns an elaboration of the process by which practical action itself becomes an object of reflection and control. The paper examines the rationalization of technological practice which brings action under the control of conscious organization and transforms man from reactor to willing agent. Action becomes an object of decision.

Rational action is contrasted with other forms of behavior including spontaneous, erotic, and tradition-bound forms. It is characterized as species-adaptive rather than an ideal logical type. It is thus contingent, containing within itself the potential for becoming maladaptive. The development of the rationalization of practice is traced from artisanal skills, technical maxims, and empirical generalizations through systematic technologies to its highest embodiment in modern science. Two essential properties of the scientific method are noted which have contributed to the success of modern technology—manipulability of nature as a road to knowledge and abstraction from sensible experience as embodied in high-level theory. The former has made applied science possible and the latter has generated surprising technologies. Additionally, substantive science is shown to rationalize action by furnishing constitutive knowledge of material properties, forestalling futile action, explaining past failures, and guiding the search for improvements. Cybernetics, game theory, operations research, econometrics are cited as applications of scientific methodology which result in new levels of practical mastery.

The potential of these new instruments for benevolence as well as for domination over human populations is enormous. While granting that the scale of human interference in natural processes has crossed a threshold, calling into question the adequacy of reason itself, it is argued that reason is not yet maladaptive. Courageous acceptance of responsibility for the stewardship of the planet and for perpetuation of life forms, including human life as presently constituted, represents a rational strategy for the future.

Carpenter, a former engineer, has been influenced in his thinking by, among others, Hans Jonas.

In the final paper of the conference, "Types of Technology," Carl Mitcham provides another attempt at isolation of the relevant factors for the definition of technology. His approach shares features of both McGinn's presentation (many similar relevant factors, although differently organized) and Borgmann's paper. Mitcham's paper is part of a larger project, with the "essence" of technology to come in a later paper.

Mitcham suggests that the term "technology" refers to the human making and using of material artifacts in all forms and aspects and should be differentiated modally and generically. Focusing on the functional or structural distinctions among technologies, Mitcham presents a detailed analysis of technology as object, technology as process (including invention, design, making and use), technology as knowledge, technology as volition (described in terms of will to power, to survival, to freedom, etc.). Although the generic differences in types of technology are found in the presence and organization of the elements under each of these four modes, Mitcham admits that this does not yet give us a definition of the essence of a technology. He concludes his paper with a diagram of the generic differences between ancient and modern technology, revealing that one of the characteristic features of the modern form—unlike ancient technology—is the tendency to remain highly unified. This will presumably serve as the transition to his later attempt to get at the essence of technology.

PART I: METHOD, DESCRIPTIVE FRAMEWORKS, AND A PRACTICAL PROGRAM FOR PHILOSOPHY OF TECHNOLOGY

As noted earlier, although general methodology, the business of criteria or standards, ought naturally to come at the beginning of a venture, in fact the historical sequence of our proceedings was the reverse. We first held the Delaware conference, laying out a variety of current approaches to the philosophy of technology, and only then, later, turned to the matter of presuppositions. The latter took place in a symposium at the 1976 American Association for the Advancement of Science meeting in Boston, with the title: "Philosophy of Technology: Methodological Guidelines."

The charge given to the participating philosophers, all of whom had been at the Delaware conference the preceding summer, was to reflect on those proceedings with the thought of providing some methodological guidelines or standards for the philosophy of technology movement. As things turned out, each interpreted this in a very distinctive way. One interpreted "method" as a set of minimal conditions for the rational discussion of problems of technology. (This happens to coincide with his general approach to philosophical problems.) The second took the task to be one of providing a descriptive framework for defining and controlling technology. Finally, the third of the panelists to be included here interpreted "method" as the providing of a practical program for the philosophy of technology movement. (Though some questioned it at the time, this approach is certainly legitimate in certain philosophical traditions.)

Joseph Margolis (Temple University), in "Culture and Technology," gives the "minimal constraints" interpretation. The technological is identified as a species of the cultural, not to be confused with practical knowledge or applied science. The cultural domain is functionally specified in terms of linguistic mastery or of Intentional or rulelike phenomena produced or exhibited by creatures that possess linguistic mastery. The technological, then, may be characterized as the intersection of practical knowledge and ideology, that is, of knowledge of means and ends—which need not be culturally informed—directed to the achievement of ideologically qualified objectives. The argument depends on construing persons, as distinct from members of *Homo sapiens,* as culturally emergent entities, specifically, as creatures that have mastered language and are, for that reason, capable of self-reference. Language itself is pretechnological though subject to technological manipulation; not natural, in the sense of not being explicable solely in terms of the lawlike regularities of physical nature. The admission of cultural emergence entails some form of relativism regarding values, normative objectives, ideologies. Hence, a philosophy of technology requires an account of the minimal constraints that may be nontendentiously imposed on any technological program. We are, in this sense, partisans, necessarily, but aspire to be rational partisans.

Robert McGinn (Stanford University), in "The Problem of Scale in Human Life: A Framework for Analysis," addresses himself indirectly to the "limits to growth" question, now reformulated in terms of turning back the clock on technology *versus* growth-as-progress. The means he chooses to employ is what he calls "scalar analysis," relating it to notions of "diseconomies of scale" in economic analysis. The bulk of his paper is devoted to describing appropriate and inappropriate scales (especially sizes) for human beings in various situations. He concludes with the claim that transformations of scale radically alter our environments, thus affecting the degree to which they facilitate or thwart our values and goals.

Finally, Joseph Agassi (Boston University and Tel Aviv University) proposes, in "Technology, Mass Movements, and Radical Social Change," a positive program for future philosophy of technology. He first describes recent mass movements and the way they operate. This is in the context of the ecology movement, which Agassi thinks the best of them and the one most likely to help us in the face of the "technological apocalypse." What Agassi finds typical of mass movements—that they quickly get dissipated either because a limited objective is accomplished or because the time is not ripe socially or historically—he hopes can be overcome by learning a generally successful technique that can be abstracted from the mass movements of the sixties. He thinks the mass movement offers the best hope of avoiding a population and pollution catastrophe that seems ever more threatening. Agassi concludes with a nod of recognition to those who might think he has gotten too deeply into politics for academic good taste, but he insists that a viable program for philosophy of technology must be political—and that in any case these are the items that will be on the agenda of philosophy of technology in the near future.

PART III: BIBLIOGRAPHICAL UPDATE AND REVIEWS: PHILOSOPHY OF TECHNOLOGY, 1973-1974

Compiled by Carl Mitcham, with the Assistance of Jim Grote

The review and bibliography section of this research annual builds on and continues the *Bibliography of Philosophy of Technology* (1973).

The existence of that comprehensive bibliography of the philosophy of technology grew out of the particular needs of two graduate students who, in the late 1960s, were trying to understand the world around them. In the past this effort might plausibly have terminated in the philosophy of nature, political philosophy, or even theology. In the present it could not help but lead to philosophy of technology—that is, to the attempt, by means of reason and insight, to grasp the nature and meaning of the making and using of artifacts.

To step back from the immediacy of our technological environment into words and arguments about it might seem at first to be a step away from the original intention. When the bibliographical project began ten years ago, however, it was predicated on the belief that we simply could not have been the first or only persons to try to think seriously about something which was a pervasive feature of our time. The bibliography was, as it were, an attempt to ask for guidance, to find friends—to stand back from an otherwise overpowering reality and thereby open up room for reflection. Paradoxically, this was facilitated by the fact that relatively few philosophers had given technology sustained consideration. As the number of discussions of technology has grown over the last decade, the danger of being overwhelmed by the world of technology through its own philosophical bibliography has correspondingly increased. And the bibliography, when published, was forced to take on a prophylactic as well as exploratory character.

In the tension between these two intentions can be found the germ of the particular features of the *Bibliography of the Philosophy of Technology*. On the one hand, not wanting unduly to restrict an emerging field, it tried to include analyses of technology from the points of view of all major branches of philosophy, as well as theoretical interpretations from a historical, sociological, or technical perspective; there was even a desire to make selective reference to the kind of concrete studies in engineering and the social sciences which could be a helpful stimulus to realistic philosophical work. On the other hand, the bibliography needed to be critical in annotating materials which did not bear directly on the question of technology and in evaluating an amorphous mass of popular literature. Thus the use of traditional distinctions between the branches of philosophy as a structural framework for the bibliography, in conjunction with the evaluative (although slightly ambiguous) subcategories of primary and secondary sources. There are weaknesses with this particular system, of course; still, the basic structure has held up well enough to serve as a reasonable foundation upon which to continue.

If the original bibliography arose out of personal need and upon the principles just outlined, it is nevertheless modified and continued because of the generally favorable reception accorded to that initial publication.

Only strong encouragement from respected scholars could persuade one to continue such an undertaking on a public scale. Not only have reviews of the bibliography—which have run the course from *Chemtech* (February 1976) to *Review for Religion* (May 1974) and *Technikgeschichte* (1974)—been uniformly favorable, but in 1974 the work was awarded the Abbott Payson Usher Prize as the best work published by the Society for the History of Technology (SHOT) in the previous three years. The only general criticism has been that the bibliography lacked an index. This defect is in the process of being remedied; an index will be published, along with some corrections and supplementary material, in a future issue of *Technology and Culture*.

At one point the possibility was raised of making this future index and supplement the beginning of a yearly bibliography of the philosophy of technology, perhaps in conjunction with Jack Goodwin's annual "Current Bibliography in the History of Technology." But while the initial encouragement and support of Melvin Kranzberg and SHOT, without which the original bibliography would not have appeared, might point in this direction, it did not seem right to undertake something of this magnitude in the pages of a journal devoted primarily to the history of technology. The needs of the historian and the philosopher are just different enough to make this slightly incongruous, not to mention the anomaly of having historians subsidize work of primary interest to philosophers. For his 1973 bibliography (*Technology and Culture* XVI, No. 2 [April 1975]) Goodwin did expand one category explicitly to include not only "general relationships between technology and culture" but also "the philosophy of technology." This conjunction parallels that in the *Répertoire bibliographique de la philosophie* between "philosophie de la culture et de la technique." Nevertheless, this association tends to confuse two distinct issues and misses the opportunity to help articulate either; moreover, neither Goodwin nor *Répertoire bibliographique* makes any attempt to do a comprehensive critical analysis of the field. Thus, with the inauguration of the *Research in Philosophy & Technology Annual* it was natural to consider an ongoing bibliography a desirable feature of this venture.

From the philosophical perspective a good bibliography ought to be more than a reference list—even a highly critical one. In its elaboration it ought to be able to provoke thought and develop reflection. As such it needs to suggest new relationships and keep some questions open, while not being so diffuse as to miss articulating special topics. To serve its part in coalescing the philosophy of technology movement it also needs to be conceptually critical in steering others away from a greater amount of the literary equivalent of processed food than ever before. In light of such needs, and in consideration of limitations otherwise imposed by a simple yearly collation, it has been decided to expand the bibliography to include short reviews and surveys of selected authors and issues. Reviews will be used when it seems desirable to have more extensive notice than annotation can provide. Bibliographical surveys will be able to bring together items from the original *Bibliography of the Philosophy of Technology,* various current bibliographies, as well as some outside the range of either, in order to pay detailed attention to some important author or topic. Over the years these may build toward a more philosophical introduction to the field. Although it has not been possible to develop these two new categories to the full in this initial issue of the Annual, still examples are provided by way of a review of Friedrich Rapp's *Contributions to a Philosophy of Technology,* Hans Jonas's *Philosophical Essays,* and Philip Hanson's bibliographical survey on the Canadian philosopher George Grant. Thus the bibliographical section of the *Research in Philosophy & Technology Annual* will normally consist of three parts: (1) Reviews, (2) Bibliographical Surveys of Authors and Issues, and (3) a Current Bibliography of the Philosophy of Technology. With index, it seems best to follow the example of Goodwin's "Current Bibliography in the History of Technology" and to do a yearly author index, supplemented by a subject index only once every two or three years.

In conclusion, readers are invited to comment on this arrangement and to suggest modifications they might think useful. Readers are also strongly encouraged to contribute, especially to the review and bibliographical survey departments. Authors are encouraged to send off-prints of their publications to ensure proper notice.

PART I: METHOD, DESCRIPTIVE FRAMEWORKS, AND A PRACTICAL PROGRAM FOR PHILOSOPHY OF TECHNOLOGY

CULTURE AND TECHNOLOGY

Joseph Margolis TEMPLE UNIVERSITY

If language is a technological achievement, it is a peculiar one since there is no recognizable phase of human history in which language is lacking. Not that the biological species *Homo sapiens* is defined by reference to linguistic capacity; only that no known human society is without language and that it is inconceivable that even the most prehistoric of human accomplishments were prelinguistic. The ubiquity of language, the grammatical convergence of all known natural languages, the incredible speed with which language is accurately mastered in infancy and childhood from imperfect and fragmentary samples have led some theorists—notably Noam Chomsky—to claim in effect that the human mind or the human brain (the accounts waver here) is programmed for language.[1] If that were true, of course, language could not be construed as a technological achievement. There are difficulties with the notion, particularly because language is essentially a rule-governed activity or at least an activity whose features are best approximated by somewhat idealized rules: the problem is, very simply, that we have absolutely no plausible model for the genetic transmission of rulelike regularities.[2] Chomsky's argument is intended to show that language, at any rate the deep structures of, or associated with, grammar that he calls linguistic universals, are a part of human nature. That thesis, so-called rationalism, is dubious for at least the reason just given; but it musters what in effect is a very strong argument against construing language as a technological accomplishment—which is not to deny that invented languages and even the refinement and development of language in new and specialized ways are technological achievements. The fact remains, however, that if language is, as it appears to be, essentially rulelike in nature, then it is impossible to reduce the phenomenon of language to any phenomenon that is explicable solely in terms of the laws of physical nature. Not merely the known laws of physical nature—any laws of physical nature, however comprehensive they may be.

The reason is an important one. Rules or rulelike regularities are explicable only in terms of a peculiar set of conditions that include at least the following. Rules must be articulated in some recognizably social way; must be capable of being replaced in a viable way by alternative systems of rules; must be capable of being followed and violated by beings themselves capable of recognizing that rules obtain and that rules are followed or violated; must be capable of being conformed to and reformed or revised for reasons to which the beings affected subscribe.[3] To insist on constraints of these sorts is, it will be readily seen, to insist on the paradigmatic features of culture itself. More than this, although language can hardly be supposed to be natural to *Homo sapiens,* it is essential to the nature of human persons. There is no more promising way to define what a person is—by accident only, what only human persons are—than by reference to the mastery of language and through that mastery, the capacity for a developed self-reference. Human animals, we may say, emerge as human persons insofar as they master language. But in that sense, though human animals are natural creatures subject to the laws of nature, human persons are creatures of culture, creatures whose distinctive mode of behavior must be specified in terms of the concepts of rules and norms and not merely of laws.

These distinctions provoke a fundamental philosophical query, possibly the central theme of all human inquiry: What is the relationship between *Homo sapiens* and human persons? Or, What is the relationship between physical nature and human culture? Reductionism faces the difficulty of accounting for the properties of language; dualism is regarded as a scandalous abdication of explanatory responsibility; idealism is a form either of theological mystery or of human arrogance; neutral monism is a name for a negative doctrine that cannot in principle be further specified. The puzzles of such alternatives are well known. But, apart from any attempt to solve them, in effect Schopenhauer's world-knot, we can see the sense in which technology cannot be the whole of culture even if, in our own time, all cultures tend to be increasingly technologically self-conscious.

To speak of technology is, at least *per accidens,* to exclude the agency of the nonhuman world, the world of (nonhuman) nonpersons. One senses at once the possibility of a quarrel. In a splendid book, *Animal Architecture*—the title alone is compelling—Karl von Frisch introduces his topic with the following words:

> When we stand before great churches, temples, pyramids, and other works of architecture built hundreds, if not thousands, of years ago, our minds are filled with awe and admiration. Yet there have been architects millions of years before that. Their work, it is true, owes its existence not to the inspired genius of great artists, but to the unconscious, unremitting activity of the force of life itself. Without tools, indeed without anything that could be called action, the coral polyps of the warm seas erected their limestone piles—edifices that can reach the size of mighty mountains—and they go on building today. . . . But mainly this book will be devoted to the activities of animals that actually build structures of the greatest diversity from extraneous materials or from substances they produce within their own bodies—using techniques akin to those that humans employ in masonry, weaving, plaiting, digging, and so on.[4]

Whatever may be the charm of an anthropomorphic idiom, the fact remains that the marvels of animal "techniques" in building, tools and the like utterly fail to exhibit the sort of cultural sensitivity minimally entailed in adherence to linguistic rules. Man, the human person, is, in some sense, a culturally emergent entity, thus distinguished and constituted by his mastery of language. Animals, we may say, if once we grant them knowledge of any sort, perceptual knowledge for instance, must be accorded practical knowledge as well: the lion that sees that there is an eland before it knows how to stalk the eland and, knowing that, knows that there is an effective way to stalk the eland. Even such ascriptions are obviously parasitic on the model of a language. But human technology is, precisely, the practical capacity of a creature that has mastered language and that can consider alternative ways of acting and making—in accord, to be sure, with some measured grasp of the laws of nature—that conform with linguistically informed rulelike considerations. Such are the norm-governed considerations of economy, efficiency, speed of construction, and the like.[5] Very simply put, then, culture is both the context of technology and the genus of which the technological cannot be more than a determinate species. Insofar as the cultural cannot be reduced to the merely physical, the technological also must be seen to concern matters beyond merely what, regarding the laws of nature, can be expressed in terms of the idiom of means and ends. In that sense, technology cannot be merely "applied science" or practical knowledge: the human "application" of science involves, precisely, the use of natural forces and materials, however modified, for culturally specified ends. Assuming the lion's perceptual knowledge, the lion's stalking technique may be read as a sort of practical knowledge; if so, it is not technology.

Now, these distinctions have extraordinarily powerful implications. The most important is probably this: that human persons have no "nature," if by nature we mean that an examination of the physical or biological traits of *Homo sapiens* could reveal—in any generous sense at all—the proper or natural norms for the careers of persons taken individually or collectively. The closest we come to natural norms is marked by the discipline of medicine. But even here, there are instructive limitations. For one thing, medicine is premised on the prudential interests of man—the prolongation of life, reduction of pain and suffering, the effective use of the limbs and organs for characteristic enterprises. But this means that the normative orientation of medicine is essentially borrowed from more comprehensive assumptions about human nature rather than discovered by an exercise of science; the rejection of such prudential objectives, in suicide for instance, is not an impossible or even statistically unlikely condition and is quite compatible with nontendentious views of rationality and the rational use of technological facilities. For another thing, the prudential interests mentioned are merely determinably identified; the particular specification of every such interest, even in medical terms, will be a function of the cultural orientation and institutionalized expectations of each particular society. Distinctions regarding normal and healthy longevity, for instance, are obviously a variable function of the technological achievements of different societies.[6] Hence, in spite of their noticeably conservative nature, the norms of medicine are culturally specified in precisely the same sense in which the "nature" of human persons is specified, that is, in accord with variable doctrines and ideologies, once the minimal but only determinable limits of the prudential conditions for any viable life are conceded. But if human persons have no nature, no discoverable normative nature, then, redundantly, there can be no natural norms for the direction of human culture—*a fortiori,* for the direction of human technology. This, I think, is the source of the pathos of our growing realization that human beings have created an increasingly dense cultural environment or culturally altered "natural" environment and are increasingly focused on the possibility of further and quite deliberate technological alterations both of their environment and of themselves.

The thesis has a direct consequence for ecology. For, no attempt at ecological "corrections" makes any sense if it excludes the distinctive cultural concerns of man, and the cultural concerns of man cannot (in principle) find confirmation in the physical or biological features of the world except in the sense of mere viability. The most dreadful nuclear war, tribute though it may be to our technological genius, cannot exceed the requirements of a stable ecology except in the eye of competing ideologies or, less pertinently, unless it destroys the very conditions under which a human population, however reduced, may reproduce itself from generation to generation. Ecological balance, like homeostasis, is an elastic concept minimally bound by relatively neutral conditions of bare survival and stretched in incompatible but equally plausible directions by culturally generated norms of acceptable human life.

The upshot is this. If human persons are essentially culturally emergent entities and if the inherently rulelike regularities of culture are not reducible to the lawlike regularities of physical nature, then, assuming, conservatively, some minimal but determinable constraints on the viability of human enterprises—that is, congruence both with the laws of nature and with reasonably imputed prudential interests[7]—only a pluralism of values and the relativism that that entails are conceptually defensible doctrines.[8] But if this is true of the cultural genus of human existence, it is true *a fortiori* of technological activity as well.

The nature of the emergence of human persons is admittedly controversial. What has so far been said does not commit us to a particular ontology. It does commit us to a rejection of reductionism and to a form of relativism in values, which are already significant. These are in fact the corollaries of resisting the identity of human bodies and human persons[9] and exposing the indefensibility of the claim to have discovered the function of man as such. The irony is that, though the social roles that human beings fill—the physician, the pilot, the general, the statesman (to remind ourselves of the ancient Platonic thesis)—are properly analyzed in functional terms, man as such, contrary to the eudaemonism of the Greek world, has no discoverable function. Even Plato's favorite specimens of human objectives fail to be uniform in the required way: the objectives of the statesman and the physician are as such determinable only, determinately filled only by reference to the doctrinal convictions of the society involved; whereas the objectives of the pilot and the general (or even of the postman and the fireman) are understood largely within the implicitly contractual terms of quite determinate institutions. The first needs to be informed in order to be determinate at all; the second, in being determinate, is too narrow a model for a theory of human nature. Again, *Homo sapiens* need not be classified in functional terms, just as the baboon and the dog need not; is classified rather in terms of physical resemblance to standard specimens; and human persons have no assignable function at all, unless it be the vacuous one of adopting some viable set of functions within a given culture.[10]

To admit this, however, is utterly to expose the conceptual inadequacy of an army of philosophies of technology which, like Lewis Mumford's thesis about the dialectical contest between monotechnics and biotechnics (however humanely motivated), either assigns human nature a distinct but indemonstrable function or is content to formulate a vacuous and entirely general one. Thus, after years of investigation, Mumford is prepared to claim:

> ... autonomy, self-direction, and self-fulfillment are the proper ends of organisms; and further technical development must aim at reestablishing this vital harmony at every stage of human growth by giving play to every part of the human personality, not merely to those functions that serve the scientific and technical requirements of the Megamachine [roughly, the efficient machine-like organization of human society for the sake of such order itself].[11]

But this is either to pretend a biological confirmation of quite particular ideological convictions or else it is to insist on values that, by their very nature, cannot but be instantiated by every conflicting social program.

It is because human persons are culturally emergent, because culture cannot be reduced to physical nature, and because neither culture nor nature can be shown to disclose norms or rules of conduct sufficient to direct human existence "correctly" that the implicit challenge and threat of a runaway technology is rendered so poignant.[12] We strain to understand the limits of our technological power; our expectations and fears are increasingly given a technologically oriented expression; and we alter ourselves and our environment to accommodate what we perceive to be the relatively autonomous routines of technology itself. All this is true. But the important point to appreciate—that theorists from Plato to Mumford have utterly failed to grasp—remains that this preoccupation with technology or other cultural concerns has not obscured in the least the path to the true vocation of man. Being a creature of culture, the human person has no natural vocation, unless it is the vacuous vocation—the vocation to commit himself to particular vocations.

Human persons are partisans, ideological advocates of competing vocations. There is no exclusively correct vocation for human societies. And there are none that are untenable on any supposed moral grounds, except in failing to meet certain distinctly minimal constraints: for instance (a) those that are incompatible with the laws of nature or of biological viability; (b) those that are incoherent, inconsistent, incompatible with admitted facts of a nontendentious sort, or presuppose the objective discovery of the true norms of human existence or the like; (c) those that are not congruent with the putative prudential interests of the human race, such as the prolongation of life, the reduction of pain and suffering, security, a measure of control over the use of material goods and the like. More argumentatively, social programs are open to dialectical attack, in terms of the ideologically shaped perception and conviction of a people: here (d) conflict with the actually effective values of a society is decisive; or (e) criticism of one society's values in terms of those of another committed to a competing ideology. The logic of the situation is this: the formal or relatively formal constraints on admissible theories of normative social existence can at best provide necessary but not sufficient conditions for validating particular claims; and the dialectical constraints, though persuasive enough to condemn or endorse such claims, cannot be more than the very expression of ideological conviction. It is, so to say, our animal vitality deployed through the world of culture that selects which way of life will be compelling. Thus we become partisans—by adhering to the articulated and variable ways in which we interpret the prudential interests we mean to defend and sustain. The formal and quasi-formal constraints (for instance, requirements of nonarbitrariness and universalizability) permit us only to be *rational partisans*. The optimistic reading of these conditions proclaims, Let a thousand flowers bloom. The realistic reading simply says there is no other way, reminds us that contrary presumptions have been responsible for the greatest crimes and the greatest human suffering. Why then, the argument goes, should we fear admitting that the correct moral direction of our culture—hence, also, our technology—cannot be discovered by any exercise of any known science or any more informal way of discerning the objective features of the real world? The pretense to have made such a discovery is, on the historical record, the principal source of the great calamities of the planet. It makes no difference, here, whether we are Marxists, Thomists, Freudians, phenomenologists, pragmatists, utilitarians, contractualists, or egoists. If the argument sketched is essentially correct, there is in principle no way to *discover* what is required. From the point of view of vitality or *engagement,* it hardly matters; for, as Nietzsche very tellingly observed, the race is markedly prone to life-giving illusions.

What is wanted, then, is an overview of technology consistent with the claims advanced. Needless to say, the very conception of technology is an ideological instrument of its own. But this will be true, on the argument given, even if such a conception were at least minimally plausible or eligible for debate, in the sense that it succeeded in fitting in a fair way the most powerful and least doubtful truths about the human condition. That is all I should claim for the thesis I advance; but it is, it seems to me, rather more than can be claimed, convincingly, for other more influential views.

The single most important thesis of my account holds that the theory of what a person is, is essentially a part, the most fundamental part, of a general theory of culture. The distinctive property of persons, I insist, is the mastery of language and the developed forms of self-reference that that makes possible. Culture, however, is not a substance; hence, no reference to entities that exhibit cultural properties or reference to cultural properties themselves entails any speculation that would threaten the prevailing materialism of science or encourage dualism, idealism, or any other more exotic ontology. The admission of cultural phenomena entails as its most forceful consequence the untenability of reductionism. What one needs to grasp here is simply that materialism need not be reductionistic. It has of course strenuously sought to be such.[13] The minimal requirement of a non-reductive materialism is either the denial of an identity between physical bodies and persons (quite misleadingly but incorrectly construed in terms of bodies and minds) or the denial that discourse about persons and their characteristic properties is a mere *façon de parler* of some sort, however convenient, regarding systems completely describable in purely physical terms.[14] There may well be multiple ways of developing such a nonreductive account. My own preference regarding the analysis of the emergent entities of the cultural domain is to advocate replacing the "is" of identity with the "is" of embodiment, wherever reference to such entities is involved. Here, it is useful to consider that the principal cultural entities include persons, works of art, artifacts, machines, and the parts of language. Other putative entities, for instance culturally significant acts or actions or collective entities like nations and corporations may, I believe, be treated as subaltern entities (that is, grammatical referents that serve as replacements for predications made of more fundamental entities) or as fictional entities (that is, referents treated as if they possessed capacities borrowed from genuine persons).[15] In any case, all other cultural entities of whatever nature they may be are causally dependent on the agency of persons, for it is persons who create works of art, produce artifacts and machines, speak languages, perform the relevant sorts of actions, form nations and corporations and the like. Culture, then, is the context of the characteristic life—the behavior and production—of persons; and persons, embodied as in fact they are (in living bodies), performing linguistically and in whatever other ways presuppose their linguistic competence, exhibit properties, establish relations, and produce objects which form the network of a culture.

Broadly speaking, the cultural domain is embodied in the physical, just as persons are. Sculptures, for instance, are not identical with the marble blocks in which they are embodied: they possess the physical (at least selected physical) properties of their embodying materials, but they also possess, as works of art, such intrinsic properties as intentional design, symbolic or representational features, expressive qualities, interpretive import or the like—properties that are neither analyzable in purely physical terms nor ascribable, except by cultural association, to physical objects as such.[16] Similarly, words and sentences are embodied in sounds and physical marks and are clearly not reducible to them. A fair approximation of the distinctive properties of cultural entities may perhaps be sketched by remarking that they are either rule-following or rule-governed phenomena. We may in fact preempt a familiar term, often used in a somewhat different sense, and call cultural entities Intentional—as being either rule-following or rule-governed.

The cultural domain, then, is a functionally specified domain. It is the world of posited objects functionally associated with physical objects or living systems in virtue of their essential Intentionality. Roughly, relative to some putative order of rules, norms, traditions, institutions, practices, doctrines, ideologies, theories, ideals, and the like, physical or biological systems are marked off as embodying a set of entities that possess linguistic properties, or presuppose such properties, or exhibit properties that are the analogues of them. The fact that only creatures that have mastered language can reflect on such distinctions explains the sense in which the order of cultural things "does not exist" for (that is, cannot be cognitively discriminated by), or exists only incipiently for, the most intelligent mammals that live among humans. The cultural order of things is contextually bound to the mastery of language. But for all that, it cannot be ephemeral, since only the community of persons attends to the permanent and the ephemeral and what might otherwise have been thought to be ephemeral is just what is essential to being a person. Anyone who thinks that reductive materialism is a tenable thesis challenges himself to describe his own distinctive understanding in purely physicalistic terms. Correspondingly, admit the existence of persons: a complex world of cultural objects—including other persons—cannot be resisted.

The full details of this account need not concern us here.[17] Perhaps it is enough to suggest that cultural entities cannot exist except as physically embodied; that embodied entities, though ontologically distinct from embodying entities, possess at least a certain relevant number of the properties of the entities in which they are embodied; and that reference to embodying entities provides for the numerical identity, though not for the qualities, of the entities embodied in them. In short, the cultural domain is cognitively marked only by creatures possessing a language; who, therefore, see themselves as persons capable of conversing and who see their own utterances, behavior, products as significant in terms of the functional rules of language itself or of practices that are suitably similar. This is the sense in which persons are rule-following creatures (even if they improvise and invent new rules), and the sense in which works of art, language, machines are rule-governed phenomena. The "is" of embodiment, then, reflects the Intentional orientation of persons themselves, resistance to reductionism, the functional nature of cultural entities, the need of such entities to be identified by reference to purely physical or biological systems, and their persistence as emergent entities for creatures that cannot ignore their own linguistic competence.

In characterizing the cultural, we have characterized the genus of the technological. The very existence of persons and language must be pre-technological; but of course the irony is that technology knows no proprieties, except what ideology instructs. Consequently, persons and their language—hence, all sectors of all cultures—become the fair materials for technological change.

The question remains, granting all our caveats, What do we mean by the technological? The answer may be put conditionally. *If* basic science may be construed as defining the limits of physical possibility, *if* applied science may be construed as that interpretation of the domain of nature, subject to basic science, that falls within the scope of perception and perceptual testing, *if* practical knowledge is either applied science or informal but effective skills lacking the foundation of science or the union of the two, then technology may be roughly characterized as the intersection of practical knowledge and ideology.[18] In this sense, no human society lacks a technology; contemporary technology is distinguished both by its power and scope and by the incremental importance of applied science. On the other hand, applied science is not technology; it is the interpretation of the perceptual world in terms of the physical laws and limits of basic science. Technology, then, is practical knowledge but not practical knowledge merely; the lion has practical knowledge but lacks a technology. Practical knowledge is knowledge of means and ends, of "know-how," of the effectiveness of particular kinds of action: it is not knowledge merely of effects or effective skills unwittingly exercised; it is, at the very least, knowledge *that* certain skills are effective in determinate ways, though possibly without knowledge of the causal regularities that are involved.[19] Technology, by contrast, is *praxis,* more or less in the Marxist sense, however open to dispute the particulars of Marxist social analysis may be. The key idea is given by the maxim: "Life is not determined by consciousness, but consciousness by life."[20] Technology is practical knowledge dominated by the realities of social existence among humans, that is, of ideological objectives affecting all means-ends sequences of practical knowledge. Such objectives cannot be simply characterized as efficiency or effectiveness, since those notions are bound to be considered in abstraction from social realities. For example, intentional obsolescence is as legitimate or as compelling a technological objective as economy; one cannot generalize here in terms of the putatively rational objectives of practical knowledge.[21]

This means that the criticism of technology is dialectical, in the sense given; that is, that what is required is coherence within the effective ideological commitments of a society. The key consideration—notably denied by Marxists—is that there are no sufficient grounds by appeal to which we may determine the correct ideology: claims of that sort, however effective, are internal to the requirements of given ideologies. The cultural nature of persons disqualifies all such claims and commits us to a conditional relativism, that is, to the flowering of any social system compatible with the minimal constraints already noted. We are not thus committed as partisans; but, as rational partisans, we anticipate the possibility of alternatively compelling social convictions. The issue, at bottom, is a contingent one: the claims of dialectical necessity are themselves instruments of social persuasion.

Nevertheless, there is a point in attempting, within our moment of history, to formulate relatively nontendentious directives for technology in general. These are not moral or political or logical or scientific directives. They are directives for *praxis:* generalizations of prudence projected, within the limits of rational review, for all competing, realistic, or plausible ideologies. They are, as has already been remarked about prudence itself, only minimal constraints and only determinable; hence, like practical knowledge, they are subject to diverse ideological interpretation. In any case, nothing more objective is available; and convergence, in a planetary sense, provides the only grounds for the only political and moral compacts that can be relied on. For example, there is a clear need (a) to establish limits to the total population of the world; (b) to replace our dwindling nonrenewable source of energy; (c) to insure the viability of all plant and animal stocks, whose bearing on human existence is not yet clearly understood; (d) to reduce as far as possible all forms of pollution or other controllable threats to the physical health of the race; (e) to determine the physical conditions for the maximal vigor, intelligence, longevity, social stability, and viability of the race; (f) to formulate, for the world's actual population, a minimally acceptable level of the quality of life and of a mechanism for providing that quality.

In a word, if as we must, we must compete technologically, it would be better to do so as rational agents than otherwise. So that, as particular ideologies rise and fall, promising solutions to just those technological problems that no human society can any longer afford to ignore will be harvested for whatever forms of life human history may eventually yield.

FOOTNOTES

1. Cf. Noam Chomsky, *Language and Mind* (New York: Harcourt Brace Jovanovich, enlarged, 1972).
2. Cf. Joseph Margolis, "Mastering a Natural Language: Rationalists vs. Empiricists," *Diogenes*, No. 84 (1973), 41–57.
3. Cf. David Lewis, *Convention* (Cambridge: Harvard University Press, 1969); and David Schwayder, *The Stratification of Behavior* (New York: Humanities Press, 1965).
4. Karl von Frisch (with Otto von Frisch), *Animal Architecture*, trans. Lisbeth Gombrich (New York: Harcourt Brace Jovanovich, 1974), p. 2.
5. Cf. I. C. Jarvie, "The Social Character of Technological Problems: Comments on Skolimowski's Paper," *Technology and Culture*, VII (1966), 384–390; also, "Technology and the Structure of Knowledge," reprinted (with some changes) in Carl Mitcham and Robert Mackey (eds.), *Philosophy and Technology* (New York: Free Press, 1972).
6. Cf. Joseph Margolis, *Negativities. The Limits of Life* (Columbus: Charles Merrill, 1975), Ch. 7; also, "The Concept of Disease," *The Journal of Medicine and Philosophy*, I (1976), 238–255.
7. Cf. *Negativities*.
8. Cf. Joseph Margolis, "Robust Relativism," *The Journal of Aesthetics and Art Criticism*, XXXV (1976), 37–46.
9. Cf. Joseph Margolis, "On the Ontology of Persons," *New Scholasticism*, L (1976), 75–84.
10. Cf. Stuart Hampshire, *Thought and Action* (London: Chatto and Windus, 1961); and Joseph Margolis, *Values and Conduct* (Oxford: Clarendon, 1971), Ch. 5.
11. Lewis Mumford, "Technics and the Nature of Man," in Paul H. Oehser (ed.), *Knowledge Among Men* (New York: Simon and Schuster, 1966). Cf. also, Lewis Mumford, *The Myth of the Machine*, 2 vols. (New York: Harcourt Brace Jovanovich, 1967–70); and Lewis Mumford, *Technics and Civilization* (New York: Harcourt, Brace, 1934).
12. Cf. Joseph Margolis, "Moral Cognitivism," *Ethics*, LXXV (1975), 136–141.
13. Cf. David Lewis, "An Argument for the Identity Theory," *Journal of Philosophy* LXIII (1966), 17–25.
14. Cf. Joseph Margolis, *Knowledge and Existence* (New York: Oxford University Press, 1973), Ch. 7.
15. Cf. Joseph Margolis, "Works of Art as Physically Embodied and Culturally Emergent Entities," *British Journal of Aesthetics*, XIV (1974), 187–196; revised, in *Art and Philosophy* (Atlantic Highlands: Humanities Press, forthcoming), Ch. 1.
16. I have explored the issue at length in *Persons and Minds*. (Dordrecht: D. Reidel, 1977).
17. Cf. H. P. Grice, "Meaning," *Philosophical Review*, LXVI (1957), 377–388; also, Joseph Margolis, "Meaning, Speakers' Intentions, and Speech Acts," *Review of Metaphysics*, XXVI (1973), 681–695.
18. I think this notion is rather close to the view advanced by Jarvie, *loc. cit.* From other of his views, however, I believe I have construed matters in quite a different way, particularly in favoring a form of relativism; cf. for instance I. C. Jarvie, "Understanding and Explanation in Sociology and Social Anthropology," in Robert Borger and Frank Cioffi (eds.), *Explanation in the Behavioral Sciences* (Cambridge: Cambridge University Press, 1970).
19. This is somewhat opposed to the view of Mario Bunge, "Toward a Philosophy of Technology," reprinted in Micham and Mackey, *loc. cit.*

20. Karl Marx, *The German Ideology*, ed. R. Pascal (New York: International Publishers, 1939).

21. Contrast, here, the views of Bunge, *loc. cit.*, and Henryk Skolimowski, "The Structure of Thinking in Technology," *Technology and Culture*, VII (1966), 371–383.

THE PROBLEM OF SCALE IN HUMAN LIFE: A FRAMEWORK FOR ANALYSIS[1]

Robert E. McGinn STANFORD UNIVERSITY

INTRODUCTION

Understanding technology comprehensively entails grasping the character and structure of its effects on society. One helpful way of thinking about these effects is in terms of the concept of scale. Technology makes possible various transformations in the scale of life in society; these in turn frequently affect the realization of important human values. Effects of scale therefore often mediate technology's interaction with values. Here I am concerned primarily not with assessing the impact of specific transformations of scale on particular values, but with articulating a conceptual framework for understanding one way in which technology affects values. I will elaborate the concept of scale and suggest it might profitably be treated as a variable given serious attention as a matter of course in systematic thinking about technological change.

After spelling out the basic dimensions and components of the notion of scale, I will discuss economies and diseconomies of scale, "scalar analysis," recent value-laden movements to regulate scale transformations and, finally, connections among scale transformations, technology, and human values.

"THE SCALE OF LIFE": A DIMENSIONAL ANALYSIS

In the expression "the scale of life" in, e.g., modern society, the senses suggested by the term "scale" can be subsumed under two categories of dimensions: *magnitude dimensions* and *adverbial presence dimensions* (see Table 1). The basic elements contained in environments in which human life unfolds include people, technics (the material products of technological activity), and social organizations. Magnitude dimensions of scale express the relative greatness of these environmental elements and other elements closely related to them.[2]

The first magnitude dimension is that of *number,* greatness of quantity, e.g., of individuals living in a particular life-environment, or technics in it, or choices open to these inhabitants. The second magnitude dimension, *size,* concerns how large the technics, settlements, social organizations,

Table 1

SCALE: A Conceptual Taxonomy

I. Magnitude Dimensions
   A. Number ("how many" [or: multiplicity])
      i. people
      ii. choices
      iii. technics (material artifacts)
      iv. social organizations
   B. Size* ("how large")
      i. technics
      ii. settlements (e.g., towns, cities)
      iii. gatherings of people (extent of occupied area)
      iv. social organizations (e.g., corporations)
   C. Power ("greatness of capacity" [to achieve a certain effect])
      i. individuals (in various sectors of life-activity)
      ii. organizations (e.g., military, industry, labor, religion)
      iii. technics
      iv. professions
      v. institutions (e.g., family, government)

II. Adverbial Dimensions (adverbial modes of an entity's presence or existence)
  A. Mobility ("how far–how easily")
     i. physical (movement in physical space)
        a. people ⎫
        b. materials ⎬ "the moved"
        c. information ⎭
     ii. social (movement in social space)
        a. vertical
        b. horizontal
     iii. psychological (movement in psychological space)
        a. identity (base[s] of)
        b. self-esteem (base[s] of)
  B. Span ("how long" [or: duration])
     i. lifetime spans (how long something lasts)
        a. individual person
        b. technic
        c. social organization
        d. artistic or cultural style or movement
     ii. activity time spans (how long something takes)
        a. planning time
        b. execution time
           α. construction
           β. performance
  C. Pace ("how rapidly" [or: rate at which an activity or process proceeds])
     i. objective (rate of environmental modification)
        a. rate of interpersonal contact
        b. rate of influx or flow of information
        c. rate of influx of technics
        d. rate of procession of activity in sectors of life (e.g., work)
     ii. subjective (felt rate of environmental modification)

*In the first three categories under I.B., "Size" is tantamount to "extent of physical magnitude." However, with respect to I.B.iv, social organizations, a complication arises. Determinants of size other than physical magnitude come into play, e.g., number of employees, unit and dollar levels of sales, profit level, and amount of assets. However, since size is an important characteristic of social organizations, one that contributes to the overall scale of life in a particular society, it seemed advisable to broaden the notion of "Size" to accommodate talk of the size of social organizations within our conceptual framework. This was done while recognizing that one thereby introduces an element of logical interdependency between magnitude dimensions, viz., criteria for the *size* of one kind of environmental constituent (social organizations) refer to another magnitude dimension: *number*. Finally, note that, although related, social organizational size is logically distinct from the power of such entities. That a corporation or geopolitical unit is large does not in itself make it powerful.

and gatherings in a particular life-environment are. The third magnitude dimension is that of *power,* greatness of capacity to achieve a certain effect, on the part, e.g., of organizations, technics, and individuals.[3]

Several examples of the remarkable transformations of scale that have occurred since the advent of the industrial era follow. Number and size: when William and Mary ascended the throne after the Glorious Revolution, roughly three-quarters of the British people lived in villages whose mean population was between 250 and 450, probably around 300. The median or middle-sized village probably had something over 400 inhabitants. "Much of the Stuart population lived out their lives in settlements so small that in the twentieth century we should regard them as miniatures, curiosities."[4] In contrast, based on the census of 1951, "over half [the English] live in towns of 50,000 inhabitants or more, which," states Peter Laslett, "are so vast that none of our rural ancestors would recognize his surroundings as human."[5,6] Concerning size and power, consider, e.g., the evolution of earth-moving tools. Not long ago, man moved the earth with simple picks and shovels. By 1917, the largest coal-burning steam shovel in use had a capacity of a few cubic yards per bite. Today an electric-powered dragline earth-mover called "Big Muskie" is used. As tall as a 32-story building, its bucket scrapes up the earth as it is hauled toward the machine at the end of a 310-foot boom. The bucket's capacity is 220 cubic yards. One hundred seventy electric motors, ranging from 40 to 2,000 horsepower, give it the power to move 4 million cubic yards of earth per month.[7] The use of this mammoth $25 million technic in strip mining has been much discussed. Equally staggering transformations have occurred in modern military and industrial power and in that of their associated technics (e.g., bombs, computers, engines, presses). Viewed as extensions of human capacities, these transformations have contributed significantly to the increased scale of modern life. They will not, however, be described here.[8]

The second category of scale dimensions, adverbial or character of presence dimensions, refer to relative greatness in *adverbial modes* of the lives or presences of the various environmental constituents. The first of these dimensions is *mobility,* which indicates how these constituents exist in space: with one or another degree of facility of movement, facility being a function of both the extent of movement as well as the temporal and other resources expended in the process of so moving. As here understood, mobility includes movement through social and psychological "spaces" as well as through physical space. The moved may be materials, technics, and information as well as people. The levels of the different components of mobility are interrelated. For example, psychological mobility, as involving, *inter alia,* the degree of variability in the bases of individual identity and self-esteem in a society, is apparently bound up with the level of physical mobility existing in that society.

The second adverbial presence dimension is *span:* either the duration of the (actual or effective) life of an entity or activity (e.g., a person, technic, organization, or artistic or cultural movement), in other words, how long the entity or activity *lasts;* or the duration of the planning or execution of an activity, in other words, how long the planning or execution *takes.* The third adverbial presence scale dimension is *pace,* the relative rapidity or rate of procession of an activity or process involving some environmental constituent. Mobility and span relate to the degree to which the entities in question have overcome (in an appropriate sense) spatio-temporal and other limits associated with the nature, condition, or current state of development of the entity under consideration (e.g., human locomotive capacity and processes of aging) and with the nature of the space in which the entity is mobile or enjoys a certain life-span (e.g., gravity and laws which affect the life-spans of social organizations). Mobility and span measure the extent of the effective or actual presence of an entity or activity; in the case of mobility, the range of an entity's effective or actual presence; in the case of span, the length of an entity's or activity's effective or actual existence or course.

Several illustrations are again in order. Regarding physical mobility, in Colonial days a young Moravian woman wrote the following about her trip from Pennsylvania to North Carolina, "It was not a long journey, only a month."[9] In April 1842, young Richard Wagner, apprehensive about the rehearsals for the premiere of his new opera, left Paris hurriedly by coach, and after a journey of five days, arrived in Dresden.[10] Modern transformations in the scale of mobility give rise to the phenomena of "shuttle diplomacy" and "summitry" as well as to that of mass tourism with its 21-countries-in-14-days syndrome: "Did you enjoy your trip?" asked Jones. "I don't know," replied Smith, "I haven't gotten my pictures back yet." As for the mobility of materials, we seem unimpressed to read on a San Francisco menu: "Maine lobster, fresh daily." Regarding informational mobility, as against the speed of communication in the era of intercontinental "hot lines" and communications satellites, recall that the Battle of New Orleans took place after the signing of the peace treaty in the War of 1812 because the news could not reach the battlefield in time.

With respect to span, life expectancy in the United States and England has increased dramatically from about forty years in the mid-nineteenth century to slightly over seventy years in 1974, an increase of roughly 75 percent.[11] On the other hand, there are the modern phenomena of "disposable" products and planned obsolescence. The phenomenon of nostalgia for musical and sartorial styles of relatively recent vintage is possible only because of the diminishing life-spans of artistic-cultural movements. Taken to extremes, this trend gives rise to Andy Warhol's bizarre notion of the fifteen-minute celebrity. As far as activity-times are concerned, the complex technological systems of contemporary society seem to require extended planning periods. Over twenty years of planning went into the San Francisco Bay Area Rapid Transit System.[12] On the other hand, the execution times of some important human activities have shrunk dramatically in recent years; e.g., abortions of first-trimester pregnancies can ordinarily be carried out in a matter of seconds using a vacuum system, and many food preparation times can be radically slashed with a microwave oven.

Regarding pace, it is helpful to distinguish between the "objective" pace of life and its "subjective" or felt pace. Objective pace is a function of the rates of information flow, interpersonal contact (including the number of old and new contacts per unit of time), the rates of influx and use of technics, and the rates of procession of other life activities (e.g., work, education, eating). Subjective pace depends on the objective pace of life, i.e., the rate at which an individual's environment changes, and the rate of individual assimilation of this modification, hence on habituation and on the capacities of persons and social structures to digest the influx of environmental change. In this connection one may speak of success or failure in processes of individualization or socialization of technics, people, information, procedures, or, most generally, the multiple new; that is to say, of the harmonious integration of a multiplicity of new items into an individual's or society's life way. Significant disparity between objective and subjective pace is apt to bring on psychological or cultural dislocation. Values and social conventions may be strained or forced to adapt in the face of increases or, for that matter, decreases in pace. Pelto's study of the social impact of the introduction of the snowmobile into Finnish Skolt Lapp society during the 1960s, including the resultant rapid increase in the pace of Skolt life, is a salient case in point.[13] On the other hand, American members of a NATO force based in southeastern Italy in the late 1960s were often frustrated and impatient at the leisurely pace with which the locals conducted their affairs (e.g., meals, repairs, etc.).[14]

ECONOMIES AND DISECONOMIES OF SCALE

Economists have devoted much attention to the analysis of "economies of scale," somewhat less to its companion notion "diseconomies of scale."[15] In both cases the foci of their attention have been the pecuniary advantages of increasing or decreasing organizational size. They have inquired whether increasing the size of a productive unit will decrease the per-unit cost of output, how this can be so, and at what point additional increases in scale will effect a reduction in the overall efficiency of the enterprise. It would be useful in decision-making contexts to make this scalar analysis more comprehensive by systematically taking into account possible economies and diseconomies of scale other than financial ones. Moreover, the objects of this extended mode of analysis need not be limited to economic enterprises.

For convenience, let us focus on economies and diseconomies of large scale.[16] Possible nonmonetary economies of large scale associated with the dimensions of scale discussed above include: for *number*, greater variety in one's experience with the psychic benefits such exposure may bring, such as less fear of the different and increased opportunities for affiliation; for *size*, organizational stability and increased security for its members, opportunities for utilizing organizational resources to support innovation and permit diversified activities by members; for *power*, enhanced opportunity for making an impact on one's environment, a possible source of self-esteem; for *mobility*, greater variety of experience of other societies and cultures, hence enhanced prospects for self-understanding and avoidance of conflict by improved communication; for *span*, the pleasures of grandparenthood and the chance of seeing at least some of one's major projects completed and make a difference; and, for *pace*, the oft-remarked invigorating quality of modern metropolitan living.

On the other hand, there are a number of nonmonetary diseconomies associated with large-scale life. These are rarely incorporated into processes of decision-making which bear on the scale of human life. For example, the dimension of number may reach a point in a particular environment where crowding and crowding behavior manifest themselves. Size and pace may do likewise vis-à-vis impersonal, purely instrumental behavior and involuntary anonymity. Increases in organizational size and power can lead to overspecialization and diminished accountability. Increases in physical mobility and pace may engender feelings of deracination and disorientation, while increases in life-span are apt to intensify generational cleavage, particularly when coupled with the decreased life-span of cultural styles. Some hold that the pace of urban living contributes to the development of debilitating psychological stress in certain individuals.[17] Finally Robert Lifton suggests that the increased power wielded by many individuals today, coupled with their pace of life and the amount of information they must digest, engenders the phenomenon of "psychic numbing."[18]

In sum, consideration given economies and diseconomies of scale should attend the full range of benefits and costs likely to result from possible transformations. "Scale" should be construed as including the six dimensions noted above and "economies" and "diseconomies" as embracing non-monetary as well as monetary costs. I share Thoreau's revisioning of the concept of "the cost of a thing" as "the amount of what I will call life which is required to be exchanged for it immediately or in the long run."[19] The view that "increased scale is better" has long been ensconced in the inner chambers of orthodox modern socio-economic thought. This *a priori* bias in favor of ever-increasing scale transformations should yield to sensitive empirical scrutiny of the full range of value-laden consequences of positive or negative scale transformations.

"SCALAR ANALYSIS"

At least two kinds of scalar analysis can be envisioned: retrospective and prospective. Retrospective analysis consists of examining a historical phenomenon and elucidating the scale of life obtaining therein. This analysis may be useful in understanding the special character of contemporary technological society, indeed, in making illuminating synchronic or diachronic comparisons between comparable life-forms. For example, in *The World We Have Lost* Peter Laslett makes a number of comments about various dimensions of the scale of life in pre-industrial England. While not systematically organized, they suggest that one important way of characterizing the process of industrialization is as a change in the scale of life. Here are several statements from Laslett's work which amount to an embryonic retrospective scalar analysis. Besides the figures cited above about the number of persons inhabiting the typical Stuart village, he observes that "Few persons in the old world ever found themselves in groups larger than family groups, and there are few families of more than a dozen members."[20] "Everything physical was on the human scale for the commercial worker in London, and the miner who toiled in Newdigate's [1719–1806] village of Chilvers Coton. No object in England was larger than London Bridge or St. Paul's Cathedral. . . ."[21] "The ordinary person, especially the female, never went to a gathering larger than could assemble in an ordinary house except when going to Church," weekly market days, annual fairs in each locality, or occasions of cooperative plowing or harvest.[22] After naming eight organizations and purposes which brought groups of people together, Laslett concludes: "The fact that it is possible to name most of the large-scale institutions and occasions in a sentence or two makes the contrast with our own world more telling than ever."[23] As regards the mobility of information, it was on Sunday mornings, in the 10,000 parish churches of England, in groups of twenty, fifty, a hundred or two-hundred that the illiterate mass of people were informed in the only way open to them of what went on in England,

Europe, and the world as a whole.[24] Finally, regarding span, given the life expectancy of the day, the death of the master baker or other master craftsman who was also head of the family ordinarily meant the end of the socio-economic family unit.[25]

Besides more detailed and systematic descriptions of the scale of life in pre-industrial times or, e.g., twentieth-century village life, retrospective scalar analysis might also examine how the scale of a particular way of life fostered or frustrated certain values and shaped the forms assumed by its various social institutions (e.g., friendship, family, leisure, rituals). Finally, historical studies could be made of the social and cultural consequences of transformations of scale effected by specific technological innovations. Such inquiries hold promise for enhancing our self-understanding.

Prospectively, scalar analysis may be viewed as a component of a comprehensive version of technology assessment. In this instance an attempt would be made to estimate the nature and magnitude of changes in the scale of life likely to result from adoption or dissemination of a technological innovation. A survey would ascertain public response to these prospective changes in light of individuals' values. This information would then be incorporated into decision-making concerning the technology in question. Examples of this type of scalar analysis will be reserved for another occasion.

EMERGENCE OF LIMITS ON SCALE TRANSFORMATIONS

Aristotle once observed that "Most persons think that a state in order to be happy ought to be large; but even if they are right, they have no idea of what is a large and what a small state. . . . To the size of states there is a limit, as there is to other things, plants, animals, implements; for none of these retain their natural power when they are too large or too small, but they either wholly lose their nature, or are spoiled."[26] Recently there have been signs that Americans and other peoples are beginning to think in terms of limits on further increases in the scale of life. There have been movements in this direction not only in relation to the dimension of size, but to most of the scale dimensions noted above. For example, regarding *number,* in 1976 the Indian government introduced a plan to penalize its employees and New Delhi residents who do not limit their families to two children.[27] The National Park Service has established limits on the number of people admitted to certain parks at any one time.[28] Regarding *size,* in 1971 the community of Petaluma, California, enacted a controversial statute limiting its growth to 500 homes per year. The so-called

"Petaluma Plan" was declared unconstitutional in 1974 by the U.S. District Court for Northern California.[29] However, this judgment was reversed by the Ninth Circuit U.S. Court of Appeals in August, 1975.[30] The Appeals Court held that ". . . the concept of public welfare is sufficiently broad to uphold Petaluma's desire to preserve its small town character, its open spaces and low density population, and to grow at an orderly and deliberate pace."[31] Regarding *power,* the prospect of serious technology assessment and environmental impact assessment, antitrust actions against IBM and ATT, reforms in campaign financing under the Federal Election Campaign Act, and post-Vietnam War legislation placing limits on Presidential war-making powers indicate emerging awareness of the need to place some limits on various forms of growing institutional power.

Regarding *mobility,* American rejection of the SST, the dispute over the granting of landing rights to the Concorde, and the creation in a number of European urban centers of car-free pedestrian zones evidence determination to place limits on physical mobility, at least in the sense of limiting the scope of the domain wherein, and the conditions under which, existing peak levels of mobility may be attained. Regarding *span,* proposals for "death with dignity" laws granting critically ill persons or their surrogates the right to permit their physicians to terminate care extending their lives via sophisticated medical technology constitute a departure from the attitude of (more) life regardless of cost or quality. The dimension of *pace* is the most difficult on which to effect limits directly because it is dependent on, although not reducible to, the other scale dimensions. With one potentially important kind of exception, it is difficult to imagine how the pace of life in modern technological societies might be decreased, or its increases prevented or moderated. For such societies lack any operational category of the sacred which might lend an aura of inviolability to their existing life ways. The exceptions are those limits dictated by apparent resource scarcities, e.g., the 55-mile-per-hour speed limit.

If one becomes convinced that unexamined, unregulated scale transformations may constitute an impediment to the fulfillment of basic human needs and associated values and developmental tasks, then one is more likely to be open to considering proposals for limits on further scale transformations, or at least limits on their rate of introduction; both of these in the name of protecting or enhancing the status of human values and the quality of human experience.

SOME MORALS AND A CONCLUSION

1. Technological progress does not always or necessarily issue in increases in the scale of life in modern society, even if, as seems likely, such

increases are always ultimately rooted in one or another kind of technological change. Indeed, it may well be the case that if, on behalf of certain human values, it is decided to decrease the scale of modern life, then, given the structure and fabric of contemporary industrial society, technology may actually be a *sine qua non* of such diminutions of scale. One thinks, e.g., of contraceptive technologies and of technologies of power, transportation, and communications, which permit decentralization of some industrial production.

2. Instead of vague talk about "rolling back technology" it makes more sense to begin by trying to decide what scale of life is desirable in future society, then proceeding to discussion of the sociotechnical systems best adapted to achieving these goals. This approach may help combat both those who see technology as villain, looking backward or forward to its once and future "absence" as panacea, and those who view all transformations to increased scale as progress. We need a unified rather than fragmented approach to scale phenomena, one doing justice to the fact that scale phenomena themselves form a complex interlocking system. The creation of state or federal agencies charged with formulating a coherent scale policy for the future is worth serious contemplation.

3. One important way scale transformations affect human values and the quality of experience may be illustrated with the notion of "organic wholes." Human beings seem ill-disposed toward the presence of chaos in the flux of their experience. They are disposed to understand the complexes of phenomena they encounter in terms of organic wholes;[32] that is to say, other things being equal, humans wish to relate the particular elements of their experience-complexes to larger wholes of which these elements may be seen to be parts, this in such a way that the parts contribute to the whole and the whole in turn bestows perceived significance on the parts.[33] This disposition underlies the concern of many students with finding a way of relating the knowledge and information absorbed in courses outside their special fields both to their growing specialist's knowledge and to a larger whole that provides a context in which that otherwise undigested material takes on significance. For Nietzsche, "only he who has a clear view of the overall picture of life can avail himself of the individual sciences *(Wissenschaften)* without harm to himself, for without a normative overall picture the sciences are threads which nowhere lead to a goal and make our life's course all the more labyrinthine."[34] The same disposition helps explain favorable Scandinavian worker response to changes in productive processes designed to enable workers to understand how their particular tasks fit into larger productive wholes, or to permit them to switch from assembly-line repetition of meaningless micro-tasks to small-group assembly of a product from start to finish.[35] Finally, the same disposition grounds the deep

attachment to their urban environments felt by many residents of certain cities. They are able to relate their own neighborhood experience to the larger but still intelligible whole of the city which bestows a sense of orientation and place on the smaller part.[36] The connection of all this with scale and human values is that whether one is talking of socio-economic organizations, social units, or the inner space of the mind, the relationship between a human being and the scale of any phenomenon to which he is relating is vitally important to the possibility of meaning in his experience. If the scale of life is too great in terms of any of the dimensions discussed above, it may become difficult for a person to perceive a thread of meaning in his particular experiences within the larger spatial environment, whether physical, social or psychological.

4. The importance of scale phenomena in relation to human values may be seen with the aid of the following formula:

$$B = f(P,E).[37]$$

This is to say, human behavior is a function of the (psychobiological) nature of the person (including his past history and present condition) and of the character of the environment within which that behavior unfolds. Transformations of scale radically alter the character of the environments within which human behavior unfolds, thus affecting the degree to which such environments facilitate or thwart need-fulfilling behavior, as well as the realization of associated human values. Thus one hopes for rapid and widespread achievement of a consciousness of the importance of the overall scale of life in technological society, a consciousness embracing but more comprehensive and systematic than the intuitive awareness of the importance of the man-made environment reflected in Churchill's riposte to a 1943 proposal to construct a new Parliament building: "We shape our buildings and afterwards our buildings shape us."[38,39]

FOOTNOTES

1. An earlier draft of this article was delivered at the Annual Meeting of the Society for the History of Technology, Chicago, Illinois, December 28, 1974.

2. For example, available choice.

3. To be intelligible, discussions of transformations in the scale of power must specify the particular sector(s) of life-activity or experience in reference to which the power claims are being made.

4. Peter Laslett, *The World We have Lost* (New York: Scribners, 1965), p. 54.

5. *Ibid.*, p. 53.

6. In his penetrating essay "The Metropolis and Mental Life," Georg Simmel describes the psychological distancing mechanisms employed by many urban residents in response to such transformations. See *Individuality and Social Forms,* D. Levine, ed. (Chicago: University of Chicago Press, 1971), pp. 324–339.

7. Harry M. Caudill, *My Land Is Dying* (New York: Dutton, 1971), p. 23, and John F. Stacks, *Stripping: The Surface Mining of America* (San Francisco; Sierra Club, 1972) p. 20.

8. An important issue regarding the scale of power, too complicated for discussion here, is identifying those sectors of activity in which individual humans may be said to have experienced significant positive or negative transformations in the scale of their power since industrialization (e.g., in work, education, politics, etc.).

9. *Smithsonian*, Vol. 6, No. 8, November 1975, p. 67.

10. Robert W. Gutman, *Richard Wagner: The Man, His Mind and His Music* (New York: Harcourt Brace Jovanovich, 1968), p. 84.

11. See Joseph Berkson, "Life Expectancy," in *Encyclopaedia Britannica* (Chicago: Benton, 1973), Vol. 13, p. 1091, and *Statistical Yearbook: 1974* (New York: United Nations, 1975), pp. 81 and 83.

12. Stephen Zwerling, "The Politics of Technological Change: Some Lessons from the San Francisco Bay Area Rapid Transit District (BART)," *Public Affairs Report*, Vol. 14, June 1973, No. 3, p. 2.

13. In pre-snowmobile days, round trips by reindeer sled from home to trading post took about three days. They now often take less than five hours. Consequently there have been significant increases in interpersonal contact and in exposure to new items of technology and information among the Skolt. See Pertti J. Pelto, *The Snowmobile Revolution: Technology and Social Change in the Arctic* (Meno Park: Cummings, 1973), p. 73.

14. I am indebted for this example to one of my former students, James L. Johnson, who observed this phenomenon while stationed in San Vito, near Brindisi, Italy.

15. See, e.g., J. S. Bain, "Economics of Scale," in *The International Encyclopedia of the Social Sciences*, David L. Sills, ed. (New York: Macmillan and Free Press, 1968), Vol. 4, pp. 491–495; and *The Dictionary of Economics*, G. Bannock, R. E. Baxter, and R. Rees, ed. (Baltimore: Penguin, 1972), pp. 119 and 135–137.

16. An economy (diseconomy) of large scale is typically a diseconomy (economy) of small scale.

17. See, e.g., Christopher Alexander, "The City as a Mechanism for Sustaining Human Contact," in *Environment For Man*, W. Ewald, ed. (Bloomington: Indiana University Press, 1967), pp. 60–102.

18. "From Analysis to Form: Toward a Shift in Psychological Paradigm," in *Salmagundi*, No. 28, Winter 1975, p. 64.

19. *Walden* (New York: Bantam, 1962), p. 128.

20. Laslett, *op. cit.*, p. 7.

21. *Ibid.*

22. *Ibid.*, pp. 8–9.

23. *Ibid.*, p. 10.

24. *Ibid.*, p. 9.

25. *Ibid.*, pp. 7–8.

26. *Politics*, Benjamin Jowett, trans. (Oxford: Clarendon, 1963), pp. 266–267.

27. *New York Times*, January 8, 1976, p. 3.

28. See, e.g., "Wilderness Restrictions: Sequoia and Kings Canyon National Parks," published by U.S. Department of the Interior, National Park Service, Western Regional Office, San Francisco, Ca., 94102.

29. 375 F. Supp. 574 (U.S. District Court for Northern California).

30. 522 F. 2d 897.

31. *Ibid.*, pp. 15–16.

32. In regard to perception, this claim is tantamount to a thesis of gestalt psychology.

33. Amitai Etzioni speaks of the related "basic human need" of "*context*, variously referred to as the need for orientation, consistency, synthesis, meaning, or 'wholeness'." See *The Active Society* (New York: Free Press, 1968), p. 624.

34. *Schopenhauer as Educator* (Chicago: Gateway, 1965), p. 25.

35. *Wall Street Journal*, October 25, 1974, pp. 1 and 23.

36. British architect Sir Lionel Brett believes that for this attachment to grow a city must provide (a) "centrality," the feeling that one can be at "the hub of the wheel" when one so desires; (b) a sense of comforting enclosure, the paradigm of which might be that of the resident of the medieval walled town as opposed to, e.g., Los Angeles; and (c) "we need to [be able to] stand on some high place and see and hear the city as a whole." See *Architecture in a World* (New York: Schocken, 1971), pp. 146–147.

37. For a recent discussion of Kurt Lewin's original equation, see George G. Stern, "B = f (P,E)," in *Issues in Social Ecology: Human Milieus*, Rudolph H. Moos and Paul M. Insel, eds. (Palo Alto: National Press, 1974), pp. 559–568.

38. "A Sense of Crowd and Urgency," in *Winston S. Churchill: His Complete Speeches, 1897–1963*, Vol. VII, 1943–1949, Robert Rhodes James, ed. (New York: Chelsea House, 1974), p. 6869.

39. I wish to thank my colleague, Edwin M. Good, for his searching critique of an earlier draft of this essay.

TECHNOLOGY, MASS MOVEMENTS, AND RAPID SOCIAL CHANGE: A PROGRAM FOR THE FUTURE OF PHILOSOPHY OF TECHNOLOGY

Joseph Agassi BOSTON UNIVERSITY, TEL AVIV UNIVERSITY

The problems the philosophy of technology encompasses are very broad, starting from the question, Are we better off with technology or without, and with what tool is this decidable? This is an example of a hardly practical question. Consider, however, questions such as, What criteria are used by government agencies to allow the implementation of innovations? How do different agencies and different countries compare? Such questions are of great philosophical-methodological interest, as well as of a great practical value. Is it true, as pilots believe, that runways are improved only after disasters? If so, why? Can this be improved? Questions of this sort are hinged on methodology, on the philosophy and methodology of the social sciences, and on (democratic) social philosophy. It is no surprise that this area is backward, especially in view of the classical opinion that technology is purely physical technology and thus hardly problematic.

The classical philosophy of technology made no provision for the adaptation of society to technology, no provision for social reforms necessitated by technology. Though social changes of this sort were made, they lagged behind. Now, due to population explosion and pollution many ecologists predict certain inevitable calamities, perhaps an irreversible change in the balance of nature that might make mankind extinct. The question I wish to pose here is *a priori* practically hopeless. It is, What changes ought we introduce, and how can we introduce them rapidly so as to avert too much of a calamity? To narrow down the question so as to make even a preliminary discussion of it at all conceivable, I wish to put this question for my present discussion: Can we learn something from the recent mass movements about rapid social change? Can we make the mass movements more effective, more democratic, more instructive? More pointedly, can we focus the mass movements on the solution of what I call the "technological apocalypse"?

I shall, then, divide my time now among the three following topics.
1. What the mass movements were meant to be;
2. The politics of mass movements; and
3. The technological apocalypse.

1. WHAT THE MASS MOVEMENTS MEANT TO BE

I wish to begin by quoting from the third and last volume of the autobiography of Bertrand Russell, who, in a certain sense, was the father of the modern mass movements, or at least a major factor in their evolution. Of course, Russell did not plan things in any manner that resembled the outcome. What he had was an immense sense of urgency, a sense of now-or-never about the choice between abolishing nuclear war and abolishing mankind. What Russell felt was that the choice was in the hands of the fates, whereas it should be made rationally by all concerned. We are prone to forget this because his Ban the Bomb movement ended in a failure of sorts, and because somehow, perhaps miraculously, perhaps not, a precarious balance is kept and we pretend to have learned to live with the bomb. I do not think we can get the proper sense of the events of barely two decades ago, unless we try to empathize with Russell's sense of emergency and his desperate effort to step up his activities.

The stepping up of the Russell activities, in the year 1957, is reported by Russell in detail. It led up to his exciting open letter to the leaders of both the United States and the USSR, which opened up a sort of public debate between them. The debate dried out fast: the public in the United States was not interested enough and the government was too self-righteous. But the effort Russell invested was not in vain. He, and like-minded people who likewise tried to arouse public interest, got together and founded, early in 1958, the Campaign for Nuclear Disarmament, CND for short. The CND organized the celebrated Aldermaston Marches in 1959 and 1960, the mass meetings in Trafalgar Square in the heart of London, and the like. "By the summer of 1960 it seemed to me," comments Russell, "that Pugwash and CND and the other methods that we have tried of informing the public had reached the limit of their effectiveness. It might be possible to so move the general public that it would demand *en masse*, and therefore irresistibly, the remaking of present governmental policies, here in Britain first and then elsewhere in the world." Russell knew what change he wanted, and he knew he needed mass public pressure in order to effect it. But he felt his techniques fell short. "Towards the end of July, 1960, I received my first visit from a young American called Ralph Schoenman. . . . I found him bursting with energy and teeming with ideas, and intelligent, if inexperienced and a little doctrinaire about politics. . . . What I came only gradually to appreciate . . . was his difficulty in putting up with opposition, and his astonishingly complete, untouchable self-confidence." Here Russell describes Schoenman as the archetype of the mass movement organizer, who has since become a familiar part of our scene. "At the particular time of our first meetings," continues Russell, "he acted as a catalyst for my gropings as to what could be done to give our work in the CND (Campaign for Nuclear Disarmament) new life. He was very keen to start a movement of civil disobedience that might grow into a mass movement . . . so strong as to force its opinions upon the government directly. It was to be a *mass* movement, no matter from how small beginnings. In this it was new, differing from the old Direct Action Committee's aspirations in that theirs were too often concerned with individual testimony by way of salving individual consciences."

Before going into a detailed examination of all this, let me conclude the story of Russell's and Schoenman's activities as Russell reports them. Russell himself does not analyze or explain, and his narrative from this point on lays stress on civil disobedience—an indubitably new ingredient in the campaign, though in itself not new, of course. The mass movement began early in 1958; Schoenman entered it early in 1960; the activities in the new style burst into the world in winter 1961 and the momentum picked up and stayed high for almost a whole incredible year—between fall 1961 and summer 1962. In January 1963 the whole affair was over. End of Russell's story and of his direct contribution to it.

What then happened has not yet been sufficiently chronicled, but is still fresh in memory. The movement crossed the ocean and spread in the

United States in diverse directions: student liberation, black liberation, sex liberation, women liberation, gay liberation. But all these movements were, for most of the time, put in the shade by the mass protest against the American involvement in Vietnam—indeed ever since the day Martin Luther King, Jr., declared he could not go on in good conscience leading the black liberation movement without joining the anti-Vietnam War movement as well and until the end of the war. The movements, especially the student movement and the anti-Vietnam War movement spread all over the world. Their techniques included, as had the black liberation movement before, both civil disobedience and violence. What the students introduced first were the teach-ins. These were immensely popular and successful, I think, but some viewed them with suspicion as possible means of slowing down the movements and thus dampening their impetus and robbing them of their mass character. I shall return to this soon.

Soon after the Vietnam War was over, much of the impetus dissipated. Some of it went into a new mass movement—the ecology movement. This was aborted by its own mass hysteria and by the energy crisis. The other movements had some measure of success and some consolidation, whether in the women's organizations, or elsewhere. Large and powerful and beneficial as their results may be, they can hardly be viewed as mass movements *á la* Russell and Schoenman, since they either consolidated and became proper organizations or merged into the general background.

So much, I think, for the most pertinent facts. There are successes and failures to be assessed in many respects. What I shall discuss is the possibility of enlightened, even sophisticated, mass movements that can deal with much harder problems than how to force governments to stop a war or nuclear armament; problems such as how to cope with the world's most urgent problems. For this, clearly, democracy and rational exchange are of the essence. Is there no conflict between rational dialogue and mass movement?

There is one further incongruity here, but one which bothers me less and which I shall touch upon only tangentially: democracy, like any other political form, presupposes some organization, some stability, whereas the mass movements of all types, being movements, are not so well defined for the question of democracy to hold. To this one can answer, first, that even the most amorphous body has some rules of procedure, however loose, and rules and procedures may be more democratic or less democratic. Second, all movements, being purposive, may have ends, more democratic or less. And with this I shall leave my discussion for a while and outline in brief the recent history of mass movements.

2. THE POLITICS OF MASS MOVEMENTS

It was simplistic of me to say that the Ban the Bomb issue and the Vietnam issue were simplistic. I confess I found myself, personally, in a painful state of inactivity during the great events of the sixties, since I greatly sympathized with the movements but could not possibly endorse their simplistic positions. Nor was this a mere marginal affair: this is obviously inherent in the very conception of mass movements.

It is a historical fact that the leaders of the mass movements, from Bertrand Russell to Noam Chomsky and Howard Zinn, declared their cases to be clear and unarguable. Of all of them only Martin Luther King was right. His case, while he advocated it, was clear-cut and unarguable. I do not mean that racial equality is unarguable, since world peace is unarguable too; I mean that while all King demanded was the imposition of the law, the defense of certain admittedly basic rights of people who were certainly admittedly deprived of them, one could join King unhesitatingly. Not so with the other issues, since one could hear arguments on both sides that made sense.

It seems clear to me that all his life Bertrand Russell conceded that there was room for debate, that mass movement suppresses debate, and that hence his support for a mass movement is a matter of expediency. In his autobiography he says nothing about the failure of his movement, partly no doubt in order to avoid squabble but partly also due to this difficulty. Chomsky and Zinn were in much less trouble. Chomsky declared, "I belong to the party that says that the grass is green." During the days of peak activities Boston University students stood on top of cars on the edge of Boston Common under the glare of the television cameras and yelled through bull horns to thousands of their peers who filled the common. We tried all possible peaceful means, they declared earnestly; all means failed: it is now time for the revolution.

This simplistic view of things stems from the line of thought of Marx and Lenin, and was conveyed to these youths in the successful classes of Professor Howard Zinn. But that is not how things work. The charming, earnest, sincere youths had not tried all the peaceful means.

The idea of a mass movement then was really too simple. The issue must be clear, the awareness of it wide enough, governments have to be as stubborn as the Pharaoh, and all the organizer has to do is call all the children of Israel to gather together and force their way on one concrete issue. They can force governments to pass certain laws, or populations to give up certain bigoted attitudes. And the change has to be such that reversal is practically out of the question.

All this is much to my dislike, yet I think mass movements can be terrific and their services are urgently needed right now in terms of our technological problems.

That mass movements can do better is, I think, as unarguable as that they can do worse. There were evil hysterical mass movements, most famous amongst these were the Flagellists after the Black Death in the late Middle Ages, at the very dawn of the Renaissance. There were attempts to emulate them in the United States as soon as the peak of the mass movements here passed—Jesus freaks and their like. Fortunately it did not work. This may indicate that through mass movements we can achieve only what is adequate for the intellectual level of the population. But there are different components to the education and sophistication of any given population, and, moreover, such things can be developed, and in mass movements they can be developed rapidly. On this point, on the possibility of rapid mass education, I think Marx was right. It is possible, and, as he said, it can be achieved only through combining the struggle with the education of the struggling masses.

I am thinking of the teach-ins, which were immensely successful but were stopped, first by organized heckling of the Students for a Democratic Society or the extremist branch of that movement known as Weathermen, and then by the organizers of the movements.

Who organizes mass movements? Who has the right to do so? On what issues?

Perhaps the most wonderful thing about mass movements is that anyone who thinks he is able enough and has the right conditions can try. It is cheap to dismiss mass movement organizers as charismatic. Perhaps this is false: Paul Newman and Marlon Brando are extremely popular and they joined mass movements actively, but they were not the leaders that Chomsky was. Of course, if the very success of a mass leader proves him charismatic then that is that. Yet democracy is best served by allowing any one to act as a mass leader. Even the fact that the major tool of some mass movement is civil disobedience does not change this. And of violence and inciting to violence I shall not say anything except that the bitter experience of the Black Panthers proved by tragic means the obvious truth that violence can serve a mass movement only in a popular overthrow of a tyrannical government.

The movement that has the greatest promise for technological problems and that should undertake the greatest and most important and urgent roles is the ecological movement. That movement developed rapidly—as rapidly as other movements—partly because a vacuum was there to be filled in the space of mass movements (the vacuum is still there), partly because of the new and intolerable level of pollution (the situation is rapidly deteriorating). The movement was defeated—as a mass movement, I mean—by its inadequacy. It claimed that the Alaska pipeline issue was clear enough and won; but the energy crisis immediately reversed the decision. And it declared the supersonic commercial flights clearly unacceptable and won there too, but the battle on this is not finished, unless technology will improve the level of performance of supersonic flight and thus take the issue off the agenda. There were no other clear-cut issues, not even in cases where no one questions the urgency of the situation. Mass hysteria about Lake Erie does not help decide how to save it. But, no doubt, the movement did much to bring public awareness and make ecology a profitable political plank everywhere.

But this is not the end of mass movements. As I have described, the modern-style mass movements simply sprang into being under stress. I suppose it is of their nature that they come and go as they do, with intense stress and while heightening the excitements. Hence they must remain to a large extent spontaneous. Of demonstrations the last word still is Lenin's. Demonstrations, he said, are hard to organize; spontaneous demonstrations, he added, are particularly hard. I should go further and say, not only are mass movements spontaneous demonstrations and so hard to organize: democratic mass movements are particularly hard to organize.

I suppose interest in mass movements will not die easy, and lessons from them will be seriously considered by the able organizers of future ones. It is therefore easy to notice that as long as the West, where mass movements take place, is democratic, only those mass movements succeed which manage to present a clear-cut issue and effect a change, in public morality, in administrative or court procedure, or in the law, which takes root at once in the society in which it is effected.

It therefore seems to me that the future of mass movements is ripe for two major changes. First, that its organizers will have to present not only a clear-cut case, but also a clear-cut argument to explain why the expected change will take place and not be reversed as soon as spirits quiet down. Second, that the major tool for forging clear-cut cases can be the teach-ins plus the mass media. There is everything to say against masses acting either administratively—government by the people is impossible, or legally—mass tribunals are barbaric. But there is everything to say for the teach-in mass demonstration to emulate the practices of congressional committees which can invite expert witnesses and interrogate volunteers. Here, I think, lies the future of this medium of social change. For this the organizers will have to start their procedures a bit ahead of time so as to be in phase with the turns of events that prescribe rapid action. For, the main rationale for this medium is that time is short.

Here I come to a philosophical aspect of the matter. The problem of induction as a problem of empirical justification of action, social or private, is insoluble. We never know whether we are too slow or too fast in implementing an innovation. Different societies have standards regulating all this, and the standards are regularly tested and altered. But some innovations are not subject to standards, some standards vary greatly depending on the urgency of the situations. Military establishments take greater risks in testing and implementing innovations since they fear the greater risk of unpreparedness; market mechanisms push corporations to similar considerations. Pilots say runways only improve after blood is spilled on them; because, I presume, runways conform to standards but standards are inadequate and improve too slowly.

That population control and pollution controls are matters of emergency is commonly admitted. That standards to deal with them are either grossly inadequate or nonexistent is likewise admitted. The mass movement can come in here, and of course it will make mistakes like any other movement, and more. This should be no discouragement if it is *a priori* admitted beforehand, especially since the mass movement, being so spontaneous and almost entirely amorphous, can be more flexible than any organized body.

3. THE TECHNOLOGICAL APOCALYPSE

The wedding of mass movement new style and apocalypse new style into the ecological movement was as obviously propitious as ill-fated. As the first phase is complete we may try to consider or plan the next one.

Apocalypse, meaning revelation, has traditionally meant a prophecy of doom, especially war, famine, and pestilence, perhaps also the end of the civilized world or of humanity or of earth as a whole. The ecological apocalypse is not new, and its modern prophet was Aldous Huxley, who wrote about it extensively in his *Point Counter Point,* in his *Ape and Essence,* and elsewhere; and also Julian Huxley, one of the most ardent campaigners against population explosion. But the discussion on whether technological progress as a whole is really progress is old. That is to say, admitting that every innovation is implemented because someone finds it worthwhile; and assuming the questionable thesis that my progress is not your regress; even then we can ask, is it on the whole worthwhile to introduce technology or not?

We do not have the intellectual tools to ask such a question, since we study questions within intellectual frameworks, and frameworks take for granted answers even to some global questions. Indeed, intellectual frameworks constitute sets of answers to some given questions such that they generate some research programs, as I have explained elsewhere.

Also, the question is of no practical importance. We simply cannot stop the march of technological progress. We can, at most, impede it.

Moreover, as we cannot stop the march of technological progress globally; it is mere folly, an ostrich policy, to try to impede it or ignore it locally. One who eats natural foods but breathes polluted air and drinks polluted water is but a fool. And soon all air on earth may be polluted.

This, however, is not to say it is never wise to impede progress. Quite possibly the success of the American ecological movement to impede the implementation of supersonic civil aviation will lead to the evolution of better techniques that will not risk the environment more than subsonic flights do. No doubt, the rapid implementation of Western technology in underdeveloped countries with little or no planning causes severe cultural lags there, creates new tensions there, and so on. But I cannot enter all this now. Rather, let me say some general things about the growth of technology and its social implications.

The ideology of social planning for technological innovation was, and still is, that of Adam Smith: planning is best done by entrepreneurs. Marx, incidentally, with his extreme hostility to laissez-faire, only added force to Smith's ideas. Under capitalism, said Marx, no government intervention has any force, since the government is an instrument of oppression in the hand of the ruling class. And, dialectically, he felt that this was as it should be: free trade is inadequate but we should not protect it; rather we should let it run its course as fast as possible and then give way to socialism. This was a view which Marx held from the beginning to the end of his career. His celebrated speech on the question of free trade of January 1848 ends up defending free trade against protectionism because it sharpens social problems and hastens the revolution.

I need not register my disagreement with this view. I have already said that the first thing to learn from the experience of the mass movements is that in the West improvement can be implemented only into and within democratic systems; and I shall not elaborate. Rather, let me add another obvious point, and it is that if laissez faire ever existed, it was over by Marx's times. Not only labor legislation proved effective, as George Bernard Shaw mentions when arguing that Marx was unempirical and doctrinaire. In the mid-century an ecological crisis took place, especially in England, which was in part averted by legislation that gave local authorities sanitation and control functions—akin to the legislation that enabled the local government of London to legislate against coal fires only two decades ago and thus clean the London air for a while.

But, of course, what is characteristic of today's ecological crisis is, first of all, that on the national scene of every advanced country where it is a problem, it sharpens the conflict between production and preservation:

while production is run by a well-organized capitalist market, preservation has no spokesmen of any force; and second, the crisis has become international or global, with no spokesmen for global interests to speak of.

Let me mention one very obvious point that has barely reached the mass platform though everyone who has paid attention to such matters must have noticed it. Population control has different levels and problems on each level. On the nuclear family level it is best soluble in advanced families in advanced societies, though there are variants between, say, Catholics and Protestants. Moreover, the nuclear family that prefers quantity to quality does not complain of excess population. Hence, on the nuclear family scale the problem only occurs when more pregnancies occur than desired. For them the solution is largely technological, and the pill and the IUD are almost ideal. Of course, for this they need some information, education, enlightenment. But the unspoken thesis of the traditional working philosophy of technology is that enlightenment and technology grow together or atrophy together. In addition this calls for the relaxation of certain taboos, religious or legal, against birth control or against abortion. This, too, is agreeable to the traditional philosophy of technology on its laissez-faire creed.

It is, as usual, when we move from the individual to the social level that traditional views prove inadequate. Somehow, following along the laissez-faire theory, all through the nineteenth century, common sense prescribed national policies of minimum planning with maximum effect.

When we come to the global level we are stymied. The founders of the ecology movement felt this very keenly. Some ecologists said explicitly that it is a scandal that Western governments allowed themselves to offer Ceylon large-scale means of overcoming epidemics, especially malaria as it happens, without coupling the offer with some means of population control. I find such comments both unintelligent and immoral. But I mention this to illustrate the low level of our present ability to cope with the problem of population control on the global level.

Some ecologists said, zero population growth begins at home, on the family and the national level. But suppose the West keeps its level constant, or suppose Protestants keep their level constant, while the others grow. This will cause a rapid demographic trend that not all will welcome. For my part, I suppose there is much consolation in those enlightened people who would rather teach than breed. But I cannot simply see here a solution to the global problem—at least not without an extensive debate leading to a radical change in attitude, i.e., at least not without a mass movement. But sooner or later the agenda will be, How can one country influence progress in another and how can global planning develop soon and effectively to avert the coming apocalypse?

Obviously, the laissez-faire theory allowed first nineteenth-century imperialism and later the tendency of governments of advanced countries to help governments of backward countries consolidate, no matter how backward these were, so as to be able to trade with them, to invest there, etc. The paradigm is the oil-producing countries, and it is really of no import at all whether the official organization in charge of the process is a Western company, a joint Western and local concern or a local concern. The local elite is backward and prevents progress at home; it sells oil for some luxury items and for arms and for almost no goods and services to distribute to the large masses which are still mostly illiterate.

The ameliorating move of the West, the programs of foreign aid, failed since they were purely economic: they took no notice of local impediments to economic progress and so failed even economically; moreover, they were based on the hope that in the long run economic progress will bring all sorts of progress. Perhaps; but the long run is too long. I shall leave this topic referring the interested and concerned to D. V. Segre's excellent *The High Road and the Low,* London, 1974.

One still better move was the Peace Corps and Care and their like. They failed; the unenlightened leaderships of backward countries found ample reasons, good, bad and indifferent, to put an end to such programs. But there are countries that might still welcome the Peace Corps, perhaps if and when jointly organized; there are countries that can be made to accept the Peace Corps; and there is the Bourguiba plan of shipping masses of students from backward countries to be trained in highly skilled jobs in the advanced countries. These things need more thinking out and strong pressures on governments—and since time is short, teach-ins and mass movements may be called for.

There is, however, no substitute for proper world coordination of world population growth, of world economic planning, and of world-wide arms control. The failure of the United Nations organization, even in the attempt to control nuclear proliferation, is a fact. The absence of an alternative is likewise a fact. That these harbor a disaster is also a fact. If time were available, we could begin by analyzing what prevented the development of that body into a platform of rational debate and how to force the parties participating in it to start where Bertrand Russell—to return to my earlier example—left off. Of course, the United States former U.N. representative, Daniel P. Moynihan, tried to impose an alteration on those standards of diplomacy which preclude this: he saw a conflict between diplomacy and dialogue and I have to say he was not the first. But there was no mass movement demanding to reinstate him; nor can I say that he would be successful were he to stay in the U.N. longer.

But I must leave it here: I have already entered deeper into politics than some might deem in good taste when in a symposium on the philosophy of technology. So let me just say, no program for a philosophy of technology can be viable unless it is highly political in orientation: the result of two centuries of effort in the direction of physical technology without attention to social and political technology have caused a lag, and the lag must be filled as rapidly as possible, since time is short and the catastrophe may be around the corner. All I can pose today is the questions on tomorrow's agenda for philosophy of technology. A major one is, I say, Can there be democratic mass movement for world planning and peace?

PART II: THE UNIVERSITY OF DELAWARE CONFERENCE, 1975

TOWARD A SOCIAL PHILOSOPHY OF TECHNOLOGY

Paul T. Durbin UNIVERSITY OF DELAWARE

One happy result of the recent torrent of criticism that has come down on scientists, technologists, or the "technological society," may be the spawning of a new and needed academic discipline—the philosophy of technology. A number of serious, albeit beginning, attempts have been made to address the problem of the lack of a respectable philosophy of technology in English-speaking philosophical circles.[1] By "serious" here I mean books or articles with sufficient philosophical sophistication to be taken seriously by academic philosophers.

What I propose here is not another beginning; I have already added my voice to the chorus of those lamenting the lack of an adequate philosophy of technology.[2] I want, instead, to follow up my earlier call to action by proposing the broad outlines of a substantive contribution to what I hope and expect to be a growing literature on philosophy of technology. Since I argued in that earlier paper that what a particular philosophy of technology amounts to is a statement as to what one feels a good technological society *ought* to be like, plus some persuasive arguments aimed at getting influential others to agree, what a "substantive contribution" means is no more than a reasonably well-articulated—but admittedly partial, even partisan—*approach* to a philosophy of technology.

INTRODUCTION

What I attempt to do here is quite bold. I begin by stipulating that for me "technology" will mean (a) the communication of (b) meaning in (c) the technical community (something I will further define later on). Since the small number of academic philosophers who have so far commented on technology have used the term in senses different from this, my problem is ready made: to play off my philosophy of technology, based on this definition, against theirs. Where the boldness comes in is with the facile sliding back and forth that I will do between *communication* and *meaning* and the criticism of *social science* approaches to science and technology; I will sometimes focus, in discussing the alternatives, on communication, at other times on meaning or the adequacy of certain social science approaches, without paying much attention to the niceties of philosophical distinctions or the appropriateness of mixing philosophy of social science with the theory of meaning. I feel justified in taking such liberties with contemporary philosophical analysis for two reasons. First, I make a simple-minded assumption—that working in a technical community has meaning for technologists, and that that meaning is shared in by them—which makes such sliding back and forth justifiable if one adopts a particular philosophical approach (see below). Second, if someone adopts one or another different philosophical approach with differing epistemological assumptions, presumably he could if he chose (though it might be an extensive project) establish linkages between his theories of meaning and communication and questions of the philosophy of social science.

I will next sketch out my philosophical approach; then—as the final step in this long introduction—I will discuss the theoretical model I employ in elaborating my view as to what a good technological society should be like.

APPROACH

My approach I would say is in line with the general outlook of American pragmatism, but especially with the approach of G. H. Mead. What I will sketch out here is an outline of a Mead-like *social* philosophy of technology.

Two comments:

(1) What I mean by "social philosophy," in speaking of a social philosophy of technology, involves three things: (a) a theoretical perspective broad enough to allow criticism of, in the form of suggesting alternatives to, standard social science theories about social structures; (b) a properly social, i.e., nonreductionist, social theory grounded in a nonreductionist, social epistemology or theory of knowledge (this is where Mead comes in); and (c) a concrete statement, taking a stand on urgent social problems. What the first of these means will be seen more clearly when the *problem* of the paper is presented in terms of philosophical alternatives. The second point will be taken up in a moment in elaborating on the references to Mead. The final point needs clarification now.

Most commonly "social philosophy" as a phrase is taken to be a close relative of "political philosophy"; some authors even prefer the composite "social and political philosophy." The object in these subspecialties within philosophy is to give reasonable answers to questions as to why men live in society at all, who should rule in society, and within what limits, what are the rights of rulers and subjects, leaders and led, and what are the goals of a good society. Where the focus is more political than social, the questions are likely to focus on institutionalized government; where it is otherwise, as in the case here, the focus is likelier to be on hierarchical arrangements and social structures of all kinds. In either case, however, the questions seem to be unalterably *normative;* that is, they demand of the answerer that he or she take a stand on who *should* rule, what *ought* to be the limits of authority or the rights of citizens or workers or members of any group, what *should be* the goals of a community.[3] A social philosophy of technology, accordingly, should take a stand on the rightful place of technology in modern society.

(2) Now for a few comments on my claim to be offering a "Mead-like social philosophy of technology." In one sense, this may be a misleading expression: Mead never really wrote a philosophy of technology, social or otherwise.[4] He did make a number of incidental remarks about technology, about what we would today call technological society, and especially about applications of the scientific method in the area of social problems. Some of these remarks are worth dwelling on for a few minutes, as is the context in which they were made. However, as I will also note, there is a serious flaw in Mead's overoptimism about applications of the experimental method; correcting this flaw—while still, hopefully, remaining faithful to the implications of Mead's social theory of knowledge—is what will allow me to elaborate a Mead-*like* (rather than a straight "Meadian") social philosophy.

A goodly number of Mead's remarks on technology and the promise of a technological society are to be found in Part IV, Value and the Act, of

Mead's posthumously published papers, *The Philosophy of the Act*. I will focus in particular on three essays, "Science and Religion," "Back of Our Minds," and "Experimentalism as a Philosophy of History."

"Science and Religion": In this rather slight contribution, in the context of comments on the place of science in the post-Renaissance modern mentality, Mead has some interesting things to say about the technological aspects of the early twentieth century.

> Men have set upon [the] task of refashioning the world in which they live because of the analytical method of modern physical science. Unquestionably the most important results have been the enlarged societies and the increased interplay of social forces which these inventions of means of transportation and intercommunication have brought with them (*Philosophy of the Act*, p. 471).

He then gets more specific:

> Certain . . . changes can be recognized while they are taking place, notably in education and in public hygiene. . . . Perhaps as yet no one could give a satisfactory statement of the social values which the newspaper has created and may create in the future. It has developed as a huge institution out of inventive genius and narrow-range curiosities and advertising possibilities. . . . The changes that science has initiated have reconstructed our universities and created many other institutions of higher learning. . . .
> . . . In industry, scientific control of the means of production and distribution has produced its most outstanding result (pp. 471–72).

Clearly Mead was a man of his age, enthralled with the rise of the great metropolitan newspapers and universities, and with the rapid expansion of industry and industrial research.

"Back of Our Minds": This is an essay on what happens when change outstrips our social ability to adapt because of the influence of traditional ideas and values. Mead gives three illustrations, but the most interesting here is the third:

> The development, on the one hand, of the biological and social sciences and, on the other, of social organization, has brought human society to the point at which it may ultimately control the conditions that presumably determine human misery, disease, surplus population, and the propagation of undesirable types. We cannot eliminate earthquakes, or hurricanes or cyclones, but we can conceivably learn to live with them or avoid them (p. 490).

Here surely is technological optimism virtually unbounded. But Mead does add: "Now I am not so silly as to suppose that, if we were simply willing to be intelligent, we could in the immediate future solve any of these fundamental social problems''—though he does feel it may at least help to point out that it is traditional ideas and values that often block progress.

Before turning to the third of Mead's essays to be noted here, I want to say a little about the intellectual context in which Mead wrote these social philosophy essays. Andrew Reck has caught it well in his introduction to *Selected Writings: George Herbert Mead,* a collection of papers published by Mead during his lifetime. After locating Mead, along with John Dewey, in the progressive tradition of the first quarter of the twentieth century, Reck notes:

> Education did not hold a monopoly on Mead's readiness to contribute to social philosophy and practical affairs. Social justice and governmental reform were also close to his heart. He wrote articles on such topics as settlement houses [''The Social Settlement: Its Basis and Function'' (1908)], criminal justice [''The Psychology of Punitive Justice'' (1917–1918)], philanthropy, and what later became recognized as ''social security'' [''Philanthropy from the Point of View of Ethics'' (1930)]. Sensitive to the sorry condition of aesthetic values in a mechanized, industrial civilization, he even ventured to comment on the motion picture as a form of artistic expression and aesthetic satisfaction [''The Nature of Aesthetic Experience'' (1925–1926)] (Reck, ed., *Selected Writings,* p. xxxv).

In these essays Mead's contribution is not particularly distinctive, as contrasted for instance with Dewey or William James. As with all the pragmatists, what he recommends is the application of the experimental method to social problems; cf. ''Scientific Method and the Moral Sciences'' (1923). If there is anything particularly Meadian about Mead's papers, it lies in his single-minded application of his thoroughgoing social view of mind, self, and society.

''Experimentalism as a Philosophy of History'': Mead argues that we need philosophy of history to deal with the *general* conflict of values: ''Logic . . . aesthetics . . . ethics. . . . Each within its field undertakes to determine the process by which conflicts of its values should be resolved, but no one of them has control over . . . all the strife between essential values'' (*Philosophy of the Act,* p. 499). Mead then relates his theme to an example which, though it may be unfair to take it out of context, is instructive:

> But it is more convincing as well as more grateful to present this in an example, and let the example be that of the problem of eugenics. Shall society deliberately breed men in accordance with the laws of heredity, as it breeds sweet peas . . . and beef cattle? (p. 499).

Apparently this problem was one heatedly discussed in Mead's intellectual circles. He was neither a eugenics extremist—"The most radical eugenicists do not profess to know the breed of men that should multiply" (p. 502)—nor what he would call an "Augustinian" absolutist: "There are two approaches to this problem. One of them starts from present unquestioned values and . . . proposes social regulations that tend to restrain the reproduction of certain strains of human stocks that by these standards are undesirable" (p. 503). If anything, Mead can be called a moderate on the issue.

Nevertheless, Mead's technological optimism led him to defend a position that many today would find humorous at best, pernicious at worst:

> Consider now the import of the technique of eugenics in approaching the problem of race prejudice. Race prejudice is an unthinking emotional attitude based upon an equally unthinking sense of group superiority. It has had its function in the past of heightening group solidarity, but any valuable function that it may have had is lost, and it has become in our composite communities one of the most serious evils because of the extreme difficulty of bringing it under the control of the individual. Such control will be reached, however, when accepted judgments of superiority and inferiority are based upon the same secure analysis as that which selects strains of cattle or wheat. The most fundamental attack that can be made upon race prejudice is through the careful and scientific application of the intelligence test (p. 510; note that this was written before 1931).

No comment is needed except to note the ironies of history!

H. D. Duncan, a Mead follower and maverick social theorist we will be looking at later on, has commented on Mead's near-utopian optimism:

> For a generation living with the terrible memories of the torture and murder in Hitler's death camps and the torture and murder of Negroes in our own country, Mead's world of benign others who, if they can only be reached, will help the individual to realize himself to his fullest capacity seems a utopian dream. . . .
>
> Mead's assumption that an increase of knowledge would bring an increase of virtue does credit to his heart if not to his head. His extraordinary optimism and deep belief in human perfectibility were evidently a matter of choice. It was by no means the prevailing attitude among the artists and intellectuals of Chicago during Mead's life in the city. Few cities in the world were rocked by more violence. . . . Yet in the midst of all this exposure of the dangers of a democratic society, Mead was consistently optimistic.[5]

Duncan attributes Mead's failure to see the darker side of human nature to his lack of a sense of the tragic and proposes as an antidote a theory of society based on the symbols of art. Without discounting entirely this reference to artistic criticism, I would propose another antidote for the lack of a sense of the tragic. Mead and the other pragmatists were so committed to liberal progressivism that they may have failed to take into account the good points of other political persuasions. Mead's comments on Marxism are nearly always negative. But more to the point, Mead may have been so turned off to traditionalist conservatism that he failed to consider at all the dark side of human nature so often emphasized by that political persuasion.

In any case, an adequate philosophy of technology for the latter part of the twentieth century—however faithful to Mead's social approach in other respects—must temper its optimism about the potential contributions of science and technology, taking into account the darker side as well as the triumphs of technology since Mead's death in 1931 in the light of the whole range of political alternatives. This is what I mean by a Mead-*like* philosophy of technology.

THEORETICAL MODEL

I need to spell out the *leading metaphor* for my social philosophy of technology, the main outlines of my *model* of a good technological society.[6]

There are two main features of the theoretical model I intend to propose here as a social philosophy of technology. It is *antireductionist* (a point that will not be made until later, in the main body of the paper), and it is based on the concept of *communication*.

Because the term "communication" has so many meanings, it is essential at the outset to define what I mean by it. Human communication is an essential constituent of group activity as such—i.e., as irreducible to the individual acts of members of groups—because group activity presupposes common meanings. In fact, and this is a crucial point, effective communication creates a cohesive community of shared meanings, however shortlived. This leads directly to my definition: By "communication," I will mean *the self-conscious establishment and/or maintenance of shared meanings, which very effort sustains a "meaning-community," a group in which members understand one another's signals or symbols and can be expected to respond to them in relatively predictable ways.* (The predictability is only relative in the sense that it allows behavioral-science-type predictions without eliminating the unpredictable novelty that is required for groups to be flexible, dynamic, and progressive.) In short, what I take to be the leading concept or metaphor of an adequate social philosophy is *deliberate meaning-community establishment and maintenance*.

It should be noted at once that this definition is, and is intended to be, normative. A deliberate countereffort, an effort not to share meanings or to destroy the meaning-community would be socially reprehensible—a "sin" if you will—according to the standards of the community in question. This is not to say that all deception is taboo in all communities; the definition/prescription only proscribes deception which in principle violates the precept/concept of "shared meanings."

It may be useful at this point to spell out some implications of a normative social theory of technology based on the concept or metaphor of "meaning-community." This can be done most easily by drawing a number of what appear to be natural inferences that seem to follow from the theory relative to the *science* portion of what I will later call the "technical community." (I could just as easily do this for the engineering or industrial-chemical applied science communities—I am currently working on a survey of research and development in Delaware—but there are two advantages in doing it this way. It allows tie-ins with (a) philosophy of science discussions via the reference to Thomas Kuhn, and (b) with sociology of science literature; there is not much "sociology of technology" as such to tie in with.)

What the theory implies, with respect to the scientific part of the technical community, is that each specialty within science, including very small subspecialties, is a small scale "meaning-community." Such scientific "schools" differ in status, with some ranking higher than others in the informal hierarchy of science. The hierarchy is somewhat different in the whole of what I will refer to as the "technical community," but there is a hierarchy there too. It should be noted that the sample inferences drawn here—while all of them refer only to the science portions of the larger technical community—could, with modifications, be extended to any science-based technical community.

Within the set of subsystems of science, one could say, using terminology borrowed from small-group research in sociology and social psychology, that the norms of some scientific groups are not as well established as others. It has been aptly said that, "What is proper scientific method or sufficient scientific evidence depends on what the leading practitioners of the science believe them to be."[7] What my theory would imply is that this does not apply equally or in the same manner to all scientific groups.

Furthermore, the degree of communicational success, the degree of success in sharing common meanings or solutions of problems, is an index of the rank of a particular subdiscipline within the scientific hierarchy.

Adding the two preceding hypotheses together leads to the further conclusion that the criteria or standards for the acceptance of new theories in science are conditioned by what happens in the emergence of a new "meaning-community." This might seem mysterious, or even perverse, but the hypothesis is not without evidential support, in the observable ability of an emergent scientific group to solve a particularly vexing problem which had been insoluble before. Those aware of recent controversies in philosophy of science will recognize this as the thesis of Thomas Kuhn, in *The Structure of Scientific Revolutions* (1962),[8] now given as a consequence of broader social theories.

A final inference that can be drawn is that there should be a relationship between the set of scientific subcultures and other intellectual subcultures in the larger society. If communicative acts of a particular kind—those that bring about the resolution of problematic conflicts over the interpretation of nature—bind scientific communities together, it is likely that there are other types which will tie together other intellectual communities. This is simple common sense. And the most obvious example is applied science or technology, where the problem to be solved is not usually one of interpreting nature but of controlling it or eliciting from it new technological possibilities. (This often leads, by the way, to new "social inventions" as well, to take care of often unforeseen consequences, but that is another issue.) In other communities, the highest ranking will not be given to groups that excel in problem solving, but in some other communicational activity that binds together the members of these communities—such as the expression of feeling or the satisfaction of religious needs. The informal hierarchy in the overall culture with respect to its "thinkers" or "knowers" will depend on the relative value that each culture assigns to different communicational activities. Such status is, of course, a temporary thing, with groups constantly struggling, in most cultures, to legitimate their respective values. In our complex technological culture this final inference is an important one to recall in attempting to understand and evaluate the so-called "two-cultures" controversy (cf. C. P. Snow) of the last decade.

After this exhausting list of preliminaries, it is time to turn to the problem at hand.

STATEMENT OF THE PROBLEM

There are difficulties with the view to be presented here, and an excellent way to present a philosophical problem is to frankly lay out the alternatives. Although philosophy of technology is only at a beginning stage in English-language philosophy, I see four basic types of alternatives to my view that either have already been proposed—in preliminary form in some cases, or implicitly in nontechnical, social-commentary philosophizing in other cases—or are likely to be proposed in the near furture. I. Meaning, including the meaning of technology is *absolute*. II. Behaviorism: behaviorist-reductionist accounts have been given of both meaning and society (though not specifically of technological society unless we consider B. F. Skinner's *Beyond Freedom and Dignity* [1971] such). III. The current academic division of labor, assigning the theory of meaning to philosophers, and treatments of technological society—e.g., in sociology of science or of mass society—to sociologists. IV. Contemporary Marxist-oriented critiques of technology especially in capitalist society.

The views will be presented as alternatives to the metaphor/definition of communication as "meaning-community establishment and maintenance," as outlined earlier. I begin with *the* philosopher *par excellence,* Plato, and a modern-day Platonic philosophy of technology.

Problematic Position I:

"*Meaning*" *is absolute,* something objective to be participated in, or it is nothing; "socially relativized meaning" is no meaning at all.

Plato, in the dialogue *Parmenides,* argues that, whatever difficulties arise with respect to the existence of the Forms, they remain essential to meaningful discourse (by which I take him to mean social communication). "If, in view of all these difficulties and others like them, a man refuses to admit that forms of things exist . . . he will completely destroy the significance of all discourse" (Cornford translation, 135b–c). Plato attributes the statement to the Parmenides of the dialogue, but there is no reason to believe that the view is not Platonic at least in the sense of being a position Plato thought worthy of serious discussion.

The chief difficulties of the *Parmenides* have to do with the concept of "participating" or "sharing in" (the knowledge of) the Forms. How can "parts" (particulars) be said to share in "wholes" (universals) when the two have a totally different epistemological—and perhaps also ontological—status? If one falls back on such concepts or metaphors as "imitation" or "image" or "reflection" as in a mirror (as Socrates does, 132d), how does one know that the (particular) image is like the original Form (universal) except in terms of a further Form of "likeness"? And so on indefinitely. A. E. Taylor takes Socrates' immediately preceding suggestion, that particulars are alike "in thought" only, as a foreshadowing of nominalism or conceptualism[9]; whatever the merits of this view, nominalists and conceptualists right down to the present have utilized the "form-of-likeness-*ad-infinitum*" argument as a telling one against Platonic Forms.

Yet the fact remains, contrary to the nominalists, as the enigmatic remainder of the *Parmenides* shows, "If there is no one [universal], there is nothing at all"—no particulars, no nominalistic "fictitious" likenesses, etc. (166b). We are back to the earlier point: "If a man refuses to admit that forms of things exist . . . he will completely destroy the significance of all discourse."

Therefore, following in what has been taken to be a Platonic tradition down through the centuries, there *must* be some "sharing" of particulars in the Forms, a spiritual or intellectual "incorporation" of the knower in the known and vice versa. Plato's classic metaphor is that of intuition, of "looking directly at the sun" (*Republic,* allegory of the cave, bk. VII).

There is a simple, straightforward appeal to this Platonic theory of communication, and it, or (especially Oriental) variations of it, has always found supporters. Given an otherwise meaningless jumble of shadows, or competing claims to "the truth," it is appealing to assume that the darkness will be clarified or the conflicts resolved when all see "the one same truth."

It might be possible to elaborate a *Timaeus*-like twentieth-century Platonic philosophy of technology on the basis of this theory. Fortunately, I do not have to attempt this. Carl Mitcham and Robert Mackey have now made available in English some portions of the partially Kantian, fully Platonic philosophy of technology of Friedrich Dessauer.[10] There is no point in giving an elaborate account of the view here; it is enough to note that a Platonic/idealist philosophy of technology would maintain that there is some absolute essence of technology, or else (as in Dessauer) that technology gives us entry to Forms or absolute essences of some sort. It is as likely as not that some such view would have been proposed even if Dessauer had not done so.

Problematic Position II:

Behaviorism: There is no "group behavior irreducible to the behavior of individuals"—or at least as a program of research it makes no sense to assume *a priori* that there is.

At the opposite end of the spectrum from Plato is a strict stimulus-response behavioristic account of communication. The best known contemporary version is that of B. F. Skinner, most explicitly stated and persuasively argued in *Beyond Freedom and Dignity* (1971).[11] Skinner's view, though it is built around what Skinner calls a "technology of behavior" and though he calls for the wholesale adaptation of his laboratory techniques in what he feels can be a humane and democratic social engineering, cannot properly be called a full-fledged philosophy of technology. Skinner's attempts at philosophizing are too crude for that. Still, enough strict behaviorist accounts of communication have by now been proposed in the literature for someone to develop a serious behavioristic philosophy of technology. (Not all the authors of these accounts would agree with Skinner's "democratic" but elitist, behavioral-scientist-technocrat assumptions; however, enough would to make it legitimate to keep these assumptions in the back of our minds as we review a behaviorist account of communication.)

Behaviorally inclined communicologists or communication theorists tend to use some variant of the following quasi-mechanical model of the communication process: sender-encoder-channel-decoder-receiver, all relative to a message or information.

Rigid behaviorists would strictly insist on a mechanical interpretation of the model: there is no distinction whatsoever, they would say, between human, subhuman animal, and mechanical (e.g., electronic information transfer) communications, in terms of information processed and quantification of the process. To say anything else would be to open the door to claims of "consciousness" and all sorts of mysticism and nonsense. Skinner, while sympathetic, is not all that rigid. He goes out of his way to explain, on behaviorist principles, how such notions as "consciousness" and "freedom" have come to be widely accepted.

What Skinner is rigid on is his insistence that all aspects of social behavior are in principle explainable in terms of differential reinforcement of external, environmental (social or physical) stimuli; he is equally strong in his condemnation of the "literature of freedom" for its resistance to unlimited expansion of behavioral technology. The outstanding social science theoretician to extrapolate this sort of strict behaviorism into a social science theory—he adopts the phrase "methodological individualism" as descriptive of the approach—is George Homans, in *The Nature of Social Science* (1967).[12] For Homans, who rejects even such standard social categories as "role" and "culture" because he thinks they are nonexplanatory "nonoperating definitions" or at best (involved in) "orienting statements," the only real explanations in social science are "propositions about the behavior of individual human beings [based on differential-reinforcement-type experiments], rather than propositions about groups or societies as such" (p. 40).

Where all this stands in opposition to the definition of communication as "meaning-community-building" is in its rejection of the concept of "meaning" as social, as anything but behavioristically explicable. Behaviorism is programmatically reductionistic.

Problematic Position III:

Current academic approach: A division of labor between philosophical (primarily empiricist) accounts of meaning, and empirical-sociology theories of science or mass society.

I do not intend, here, to summarize either the voluminous analytical philosophy literature on the theory of meaning or the abundant and growing material by sociologists on the sociology of science or of mass society. What I do give is just enough of each to contrast it with my own view. The first (A) portion of this section deals with some analytical treatments of meaning; the second (B) with some criticisms of recent work in sociology of science.

A. Most behaviorists, if they refer to such terms at all, are likely to be nominalists. Conversely, one of the leading philosophical nominalists of the day, W. V. Quine, has said that he is "as much of a behaviorist as any sane man could be." The qualification is significant, and the difference deserves to be explored in some depth. The threads I shall try to tie together here are somewhat nebulous.

The first two threads are brief quotations from Quine's *Philosophy of Logic* (1970).[13] At one point, in arguing against the notion of "propositions" as bearers of information, he outlines two possibly clear meanings of "objective information":

> Ideally, a particle physics does offer . . . an absolute concept of objective information. Two sentences agree in objective information, and so express the same proposition, when every cosmic distribution of particles that would make either sentence true would make the other true as well. . . .
>
> A different way of reckoning objective information is suggested by the empiricist tradition in epistemology. Say what difference the truth or falsity of a sentence would make to possible experience, and you have said all there is to say about the meaning [information] of the sentence. . . . Some epistemologists would catalog [the] alternative [here] by introspection of sense data. Others more naturalistically inclined, would look to neural stimulation (pp. 4–5).

Quine does not agree with either of these views—"Either way, a doctrine of propositions as empirical meanings runs into trouble" (p. 5)—but that is not the thread to be pulled from the quoted material. Later, Quine refers back to these "two august notions of information, one of them cosmological and the other epistemological" that might give a "clear notion of information"; there he notes that if neither of the accounts can be made good, we have not passed "beyond the stage of metaphor." A few sentences later he comments, "Where the myth lies . . ." (pp. 98–99). It is this idea that if a complete, literal, empirical account of the meaning of a concept cannot be given, we are in the area of metaphor or myth. This would say, if we pressed the view very hard, that, for instance, if behaviorism cannot be shown to be literally true then it is at present at best a metaphor (about man as a mechanical or electronic information-exchange system), at worst a myth.

The second thread I would like to pluck from Quine is this "[Rudolf] Carnap, in his linguistic theory of logical truth, has represented language as *analogous* to a formal deductive system. . . . We do better to abandon this *analogy* and think in terms rather of how a child actually acquires his language" (p. 101, italics added). Not metaphor this time but analogy. Still, the point is the same: in Carnap's approach to philosophy of language what we are dealing with is something other than literal truth. (I would add that the same is the case in philosophy of communication; I would also add that there are strong similarities between this theory of communication and that of Ludwig Wittgenstein's early *Tractatus Logico-Philosophicus*.)

Is the later Wittgenstein of the *Philosophical Investigations* still dealing in analogy or metaphor in his "therapeutic" view of philosophy of language? That point can be set aside for the moment, while we note how a leading Wittgensteinian, Max Black, criticizes G. H. Mead—and thus would also presumably criticize my definition of communication—for failing "to attach a literal and nonmetaphorical meaning to the key expression . . . 'taking the role of the other'."[14]

Wittgenstein's approach is again contrasted with what Black calls the "Mead-Dewey theory of communication" (p. 232) in a concluding summary: "Behind [any] accords there is a profound disagreement. Wittgenstein preached the end of all metaphysics, and had less sympathy for a metaphysics that aspired to be empirical than for one that was avowedly transcendental or mystical. . . . Now Dewey [or Mead] . . . must be viewed as a metaphysician in the grand manner" (p. 245). Black is guilty of exaggeration in the final phrase he uses here, but he seems to be basically correct in contrasting the later Wittgenstein (at least insofar as I understand him) with the Mead-Dewey theory of communication. For in the latter theory there is an assumption of an *object* referred to or "meant" in the use of a communicational symbol, whereas for Wittgenstein a word's (or other symbol's) *use* is its total meaning. Black is probably correct in castigating John Dewey for certain careless expressions in which the latter refers to meanings as "ideas" or "essences"—or even (pragmatic) "uses" where these are poorly defined. However, in the much more careful Mead, it is clear that "objects" referred to in meaningful communication are *necessary social constructs:* communication would be pointless if there were nothing, no object, to communicate about, yet all such "objects" communicated about have their reality only as social constructs.[15] There is no evidence that Wittgenstein ever considered this careful version of pragmatism, but presumably he would have rejected it too. Even the most minimal of realisms would, probably, have struck him as "metaphysical"—indefensible, if not necessarily rejectable as "metaphysics in the grand manner."

Summarizing, we have a chorus of significant philosophers—Quine and Carnap, Wittgenstein and Black—agreeing, in spite of subtle disagreements, that a literal, nonmetaphorical account of language (and communication?) is desirable and that such an account needs no (objective) "meanings"—which seems to mean that, if anything, "meaning" is subjective. Quine perhaps strays farthest from the orthodoxy of this now-standard analytical theory of language-communication in questioning whether any literal account now exists. He seems to imply, in a passage cited earlier, that experimental psychology may ultimately tell us "how a child actually acquires language"; but this does not commit him to saying we know this, literally, now.

In fact, much of what passes for literal accounts of language or communication, in the literature of analytical philosophy, is clearly metaphorical. "Language games" and (philosophical) "therapy," two of the most common expressions are simply the most obvious examples. Where Mead differs—and one should recall here that the pragmatism of Mead or even Dewey is every bit as "scientific" in intent as the approaches of Carnap or Quine or Black—is in recognizing the explicitly metaphorical character of his analysis. For him the primal experience of "sharing meaning" comes in the (social) experience of manipulating physical objects; physical as well as social objects are "socially constructed realities," yet in experience they do not appear as such—they appear as given. Assumed goals of social acts, and physical objects "out there," however much socially constructed, are *given,* and communication is meaningless, pointless, without them, notwithstanding any claims to the contrary of "philosophical therapy."

B. In the current academic division of labor there is no theory of technology that ties in *directly* with the analytical philosopher's parsimonious theory of language-communication. Black is, as it happens, one of the directors of Cornell's Science, Technology, and Society Program, but, at least publicly, that program is as neutral to the truth claims of various philosophies of technology as a good Wittgensteinian could hope for. What is the case, something that suggests a connection to me, is that academic philosophers of an analytical bent seem to take seriously only certain types of social theorists. These academic sociologists of science occasionally stray into the area of what I will call the "technical community," and it is possible that someone could in the near future elaborate a general theory of technological society. The outstanding academic social theorist today is probably Talcott Parsons, and the Parsons-influenced sociologist of science *par excellence* is Robert K. Merton.

What I view as a likely dominant attitude among academic analytical philosophers, if philosophy of technology (as I expect) begins to catch on, is one of continuing detached criticism. The facts about technological society will be left to the social scientists, with philosophers continuing to criticize social science theory from the sidelines. The most common sort of criticism, now, in the literature on the philosophy of the social sciences, is to lament the failures, either in generality or in rigor, of social science theorizing—usually by contrast with theories in the physical sciences. To the extent that this has, or will have, any influence on social theories, it is probable that it will tend to push in a reductionist direction. We have seen how reductionist George Homans's "methodological individualism" is, and that work bears the strong impress of the Carnapian philosopher of science, Carl Hempel. Other work may not be as narrowly based, or as behavioristically inclined, but almost any social science theorizing influenced by analytical philosophy (especially philosophy of science) will in some degree lean toward reductionism and/or a limited behaviorism.

Certainly this is the case with Norman Storer, the editor of a recent collection of Merton's essays, *The Sociology of Science* (1973).[16] In his own earlier book, *The Social System of Science* (1966)[17] Storer refers to his approach as a "sort of reductionism, at least in assuming that it is important for us to pay attention to *why* people want to do what we sociologists think their groups *need* them to do if these groups are to continue their corporate existence" (p. 167; italics added). Applied to science, this comes out as: "Why do most scientists want to play exactly those roles that turn out to be functional for science as an institution?" And presumably something similar could be asked about technologists, even though Storer's theoretical model does not fit them.

It may seem unexceptionable to ask why scientists or technologists want to do what their professional communities expect them to do. Where exception could be taken is to the claim, if indeed Storer is making it, that the motivations of individual scientists *explain* the social values of science, rather than the other way around or in terms of mutual influences in both directions.

The view I am suggesting here, and with which even a minimal reductionist should disagree, can be approached by taking cognizance of a commonplace, that the technical community has its own values. Since technical people relate to these values in substantially the same way that members of a religious subculture, or a group of nonscientific intellectuals, relate to their values, it is legitimate to suggest studying the subcultural experience of technical communities using something like the techniques of literary criticism. That is, by recognizing a difference between treating all documents or public statements in science as on a par, and introducing an "element of discrimination" that would recognize some technical literature as distinctly superior to run-of-the-mill productions. Scientists and other technical people, it seems clear, do make such distinctions; they have, I feel, what can only be called a "(sub) cultural feel" for the really superior scientific product.

Whether Storer's—or Merton's or Parsons'—model of the scientific social system does justice to this element in the scientific subculture is a crucial question for a social analysis of the technological community. How scientists "appreciate" (note the term) the "competent response" that Storer makes central in his "exchange model" of science is at least as important as the response itself. The danger in Storer's admitted reductionism, however minimal, would be for it either to ignore this factor of "appreciation," or, more perniciously, to explain it reductively, reducing appreciative responses to pure "commodity-exchange" responses.

The nonreductionist understanding of "meaning-community establishment and maintenance" with which I would agree is succinctly expressed by H. D. Duncan:

> Now either we deal with meaning through sociological methodology which allows us to interpret symbols in their social categories, or we disregard symbols and assume that they are but "reflections" of some kind of extra-symbolic reality. We "know" (like Pareto) that behind the mask of the symbol there lie interests—economic, political, sexual, as the case may be—which "really" determine human relationships. This, of course, reduces symbols to epiphenomena.

Again:

> If we cannot explain *how* symbols function to achieve order in social relations, we cannot invoke "symbolic action" as a constituent element of the "social." Instead we must reduce symbols to "referents" in nonsymbolic elements, such as sex, work or religion.[18]

If we apply this to the technical community, what Duncan would be saying is that scientists, etc., have social purposes that cannot be adequately explained by appeals to the economic interests or psychological makeup or sexual hangups of individual scientists; there is a distinct mode of social interaction in science or technology, which is not captured in terms of some other mode. (This is not to say there is something mysterious about social interaction as such; Duncan's point would be that the right kind of analysis of the literature of science or technology would yield a *genuine* sociology of the technological community.)

At this point I do not want to argue for this view, but only to note its contrast with reductionism even of a minimal sort. Presumably, analytical philosophers would find positively abhorrent such terms as "constituent element" of "the social as such" or "a distinct mode of social interaction." Certainly Duncan's manner of social theorizing, however much praised on book jackets and in reviews, has never caught on as a major force in American sociology—and least of all in sociology of science. Since the point here is to suggest difficulties with the position being maintained, perhaps the point has been adequately made simply by presenting Duncan's peculiar-sounding approach.

Problematic Position IV:

Marxist accounts of technology and technological society. I do not intend to discuss Marxist accounts of technology in any detail in this paper, but it seems appropriate to mention them as another potential alternative. In my earlier paper on technology and values,[19] I supplied brief accounts, with quotations, from Herbert Marcuse, and John McDermott, as well as brief criticisms. Neither of these authors, in my opinion, has yet proposed what I am here calling "serious" philosophy of technology. Marcuse's many efforts come closest, and I suspect that a serious Marcusian philosophy of technology will be forthcoming in the near future—either from Marcuse himself, or from a disciple tying together elements from many of Marcuse's works, or from some other Marxist-oriented social theorist deeply in debt to Marcuse. My reason for omitting Marxist philosophies of technology from consideration here is that, in terms of my concept/metaphor of "meaning-community," and in terms of my antireductionism, Marxists would be likely to respond along essentially the same lines as I would to the alternatives listed above: They are "idealist" or "reductionist" in pejorative senses, or else they are value-neutral in a way that leaves the technical community open to the worst kind of reactionary, militaristic manipulation. Beyond any such agreement, on the other hand, Marxists would likely abominate my view as "idealist" on other grounds, at the same time they would lump it, and other pragmatist approaches, with empiricism generally, repudiating them as "liberal" in the bad sense, or "pseudo-socialist."

AN ARGUMENT FOR A NORMATIVE VIEW OF THE TECHNICAL COMMUNITY

Step One: "Technology" and the "Technical Community":

Before I now, at long last, give some details of the social philosophy of technology that I want to propose here, a clarification is needed—namely, of the word "technology" as I will use it. Jacques Ellul, far and away the best known philosopher of technology translated into English, prefers the term "technique." He defines it so broadly that it includes any means-to-end rational organization of behavior, whether or not it utilizes or depends upon machines, computers, or scientific knowledge of any kind. "Technique" in this broad sense Ellul takes to be the spirit of the age of contemporary Western culture. And he takes this "spirit of technique" to be an enslaving force whose shackles contemporary man is not likely to break.[20]

The difficulty with sweeping assertions about ours as an "age of technique," or an "age of anxiety," etc., is that such claims are almost impossible to deal with; their acceptance or rejection depends on the persuasiveness of an image rather than on evidence. Consequently, I do not feel that I need to claim I have a perfectly objective definition of "technology" in order to reject Ellul's in favor of a more restricted definition.

What seems to me to be one common factor in most broad treatments of technology is its association with scientific knowledge. While some purists argue for a clear distinction between the two, I will propose a limited counterargument against them in a moment. A second common denominator in most treatments broad enough to be called philosophies of technology is a recognition that there is a definite social group which is the carrier of technology—or at least of technological knowledge. (The qualification is added in deference to the popular view—with which I am in profound disagreement—that technology is a morally neutral tool, to be used for good or evil by corporate managers or political leaders or the public.) This carrier is generally referred to in the literature as the "technical community." It includes a large number of scientists, nearly all engineers and technicians, and research managers in government, industry, and research institutes.

I focus, then, in my approach, not on technology in the abstract but on this scientific-technical community. My social philosophy of technology is concerned with its inner subdivisions, structures, and functions, with its relationships to other subcultures in society, its products, its values, and its (sometimes implicit) view of human nature and society. For me, the phrase "social philosophy of technology" thus means a set of generalizations, a systematic treatment in philosophical terms, of at least some of these aspects of the technical community.

Step Two: Technical Hierarchies as Normative:

Earlier in the paper, while first outlining my model, I argued that the technical community is hierarchical with relative status determined by success in achieving shared meanings. This is not, so far as I know, as yet an established empirical finding, but it seems a fair hypothesis to say that the highest rank is usually reserved for the solution of purely theoretical problems—e.g., in theoretical physics or pure mathematics—with the lowest ranking being given to those whose task it is to solve what are often disparagingly referred to as "purely technical problems."

Hierarchies and stratification systems in science are the object of two recent major studies in the sociology of science—Diana Crane, *Invisible Colleges* (1972),[21] and Jonathan and Stephen Cole, *Social Stratification in Science* (1973)[22]—with the latter focusing on the internal hierarchy within the American physics community. The flavor of the work can be seen in this quotation from chapter 9: "At some point, usually early in his career, the average physicist realizes that he is not going to win the Nobel Prize and that he even has very little chance of getting a tenured position at a major department. He will then turn his attention to achieving success in what might be called the 'minor leagues'." (The parallel with other sciences and with nonscientific academic specialties seems obvious.) Both the Coles and Crane conclude, on the basis of their empirical studies, that most scientists do not contribute much to scientific progress. Yet surely what they mean is that they do not succeed, to any great degree, in terms of what is publicly defined as "scientific success"—not that they do not succeed in achieving the local recognition of the "minor leagues." Most of them succeed in sharing the (low-rank) "meaning" of their particular community.

To suggest a social philosophy of technology is not to suggest substituting speculation for such empirical work in the sociology of science or technology. The point is rather to suggest questions that might be asked with respect to the theoretical frameworks applied in these studies. Most contemporary philosophers who concern themselves with social theory end up criticizing it as, for example, "lacking in explanatory power" relative to physical science or even the biological sciences. I would say that we ought also to criticize social science questions or problems as well as social science theories or answers. What are the "right" questions to ask, and according to what standards of correctness? Are the questions being asked the ones that ought to be asked if we want really informative, or "relevant," answers or some such thing? While social science theories may indeed be lacking in explanatory power relative to the "harder" sciences, they often also seem to lack boldness, imagination, relevance to real-life social situations, or commitment. Applied to sociology of science or the sociology or psychology of work (there is not much available in the way of direct sociology of technology), this would suggest that some current models of the sociology of science would not appeal—revealing as they often do an emulation of science or of certain exemplar sciences—to those who view it as their task prophetically to move scientists and other members of the technical community toward greater responsiveness to human needs.

Again, the point is not to criticize sociologists simply for the sake of criticism. The point is to emphasize by questioning, that there may be a variety of theoretical models available which could lead to different interpretations of the "facts." In other words, the point of the social philosophy approach recommended here is not to disparage accomplishments in empirical work on the social systems of science and technology, but to improve them by suggesting possibly more fruitful theoretical frameworks.

This brings us back to the issue of a social philosophy approach as *normative;* to the idea that what a social philosopher of technology ought to do is take a stand on who *should* rule, on what *ought* to be the rights of citizens, on what should be the goals of a *good* technological society. Restated now, this leads to asking what is, and *ought* to be, (a) the *proper* hierarchy within the technical community, and (b) what ought to be the proper relationships between the technical community and either other subcultures in society or the larger society as a whole, in terms of political power, socio-economic status, and ranking with respect to what the society defines as "culture."

Mine is not the only kind of normative social philosophy of technology that could be proposed. Reference has already been made to Herbert Marcuse's neo-Marxist view of technology based on the Marx-like concept of "advanced alienation," i.e., the "one-dimensional" character of advanced industrial society in which all are alienated—largely through technological means of control—from their true humanity. John McDermott, the "new left" critic of capitalist technology mentioned along with Marcuse, focuses on the "internal contradictions" of contemporary capitalism. Neither of the two categorically repudiates Marx's own vision of a future classless society in which all men will be selflessly devoted to the common good. All this is normative in the sense that it calls, not for intellectual assent, but for *action,* to—in Marx's words—"overthrow the State." The oppressed *must* do this, for "they find themselves directly opposed to the form in which individuals have always given themselves collective expression, that is, the State."[23]

I argued against this view, in the earlier paper mentioned above,[24] on political grounds. Contemporary radicals, for good Marxist reasons that I do not find completely convincing, seem to me to despair of the political process; too often, they do not seem to trust that the oppressed classes for whom they claim to speak will develop the kind of consciousness needed to fight against technology-wielding elites. Marcuse in particular is pessimistic almost to the point that even fellow radicals get concerned. However, as I said earlier, I do not intend here to argue against the Marxist view; if it is to be proved wrong, it will be by political action, not argument.

Step Three: The Continuum of Science and Technology:

One effective way of elaborating a new view is to set it against an established view, to see how the two stack up against one another in argumentation. The closest anyone has so far come in American academic-philosophy circles to the establishment of a well defined philosophical approach to technology is in a series of articles going back to the Summer 1966 issue of *Technology and Culture.*[25] Henryk Skolimowski there set the tone for the proceedings. After noting that a comprehensive philosophy of technology might have to discuss the moral implications and underlying assumptions of technology, he says: "The weight of these problems cannot be underestimated. However, they are outside the scope of my considerations. In this paper I shall be concerned with what I call the philosophy of technology proper, that is, with the analysis of the epistemological status of technology. Technology is a form of human knowledge."[26] In the ongoing epistemological discussions of technology, one of the *Technology and Culture* papers has provoked a fair amount of discussion—Joseph Agassi's "The Confusion Between Science and Technology in the Standard Philosophies of Science."[27] While Agassi has, in the course of later arguments, continually modified his position, he still remains faithful to an overall Popperian-falsificationist epistemology,[28] on the basis of which he concludes that science and technology must be kept categorically separate. It is this conceptual stand that I wish to argue against.[29]

The point I would like to make and defend is that *however different conceptually science and technology may be, at least in the United States today they share important institutional similarities* that simply *cannot* be overlooked in an adequate philosophy of technology. Furthermore, such an "adequate" philosophy must be normative (a point Agassi would not necessarily dispute).

I will not try to document these points here; instead I will argue on the basis of some personal reactions I have had while collecting materials for a proposed book of readings on philosophy of technology. I think this experience can be generalized into an argument for a normative social philosophy of technology. One chapter of the book will now lump readings relevant to recent federal science-policy decisions together with readings on proposals for a no-growth economy. Originally I had planned these as separate chapters, one focusing on federally supported science, the other on industrial research and development. Without explicitly intending to reflect Agassi's conceptual distinction, I reasoned in a more or less commonsense way that there was a natural split here, that the oft-voiced warning not to confuse science and technology at least suggested a practical if somewhat arbitrary distinction between often federally funded "basic" research and "mission" or "product-oriented" applied research (mostly in industry). However, as I proceeded through various revisions of introductions to and commentaries on the readings, I became more and more impressed with the fact that the similarities were at least as important as the differences.

Originally the discussion of federal science was focused around a thesis that current science education has tended to make leaders of the federal science establishment excessively narrow and defensive when science and technology are criticized. The discussion of industrial R & D and the no-growth issue, on the other hand, centered around the narrowness of definitions of professional objectivity that leave questions of the social responsibility of technology in the hands of corporate managers, politicians, or consumers. The first sign of a problem with the scheme was that some of the readings could just as well fit in either chapter. This suggested rethinking the matter of separating the material and questioning whether or not the artificiality of the separation might be more serious than previously thought. This led to a consideration of affinities as well as disparities between the two kinds of technical community as originally proposed. What stood out in bold relief, on reflection, was a common note summed up in the single adjective "narrow" that showed up prominently in the two analyses—whether in terms of the narrowness of scientific (but also professional) education, or the narrowness of definitions of "professional objectivity" whether in science, or engineering, or elsewhere.

In recent years when top science administrators have spoken out on policy issues,[30] they have seemed to many non-scientists to be extraordinarily sensitive, touchy, defensive. Almost without knowing it, they seem to have become ardent defenders of the *status quo,* of "the good old days" of post-Sputnik scientific affluence. Most serious is the tendency, when deciding on criteria for policy decisions with respect to science, to act as though scientific and technical decisions are so complex, so much a matter for experts, that the rich tradition of studies of policy-making standards and criteria, going back over thousands of years in the West at least, were somehow irrelevant. (I will not substantiate this here; one must look up the documents and decide for himself.) This might seem to suggest a worrisome degree of blindness toward any help that the humanities, political science, etc., might provide in the solution of technical problems if there could be a fruitful interaction between scientists, other technical people, and non-scientists on policy issues.

On the other hand, a recent spate of proposals for a "no-growth" or "steady-state" economy—for instance, the dramatically presented *Limits to Growth*[31] of Dennis Meadows, *et al.*—seemed to be in interesting contrast with the typical justification of industrial R & D as "good for the economy" or "a profitable investment for the corporation." The social-philosophy issue that arises in this connection has to do with a range of views on professionalism extending from absolute objectivity and policy-spurning in favor of solving strictly technical problems, at one extreme, to the view that corporations ought to encourage their professional employees to do work on social issues at company expense even if it involves crusading against supposedly irresponsible corporations. Here the surface issue is conflicting definitions of professionalism; however, at a deeper level the issue is very close to our other questions of narrowness and scientific/professional education.

In short, the scientific and technical communities—however different Agassi and others may argue they are in terms of logic or methodology—share fundamental institutional characteristics. Both emphasize objectivity as a group norm; they demand rigorous quantitative training; they confer the highest status and awards on persons with extraordinary technical expertise; and finally, both groups are paradigmatically "professional" in the current sense of the word. On this last point: in the universities nowadays, while so-called "pure science" departments are still located (largely for historical reasons) in colleges of arts and sciences rather than in the professional schools, in the current hierarchy of "professional experts" scientists often rank above all others and are the envy of such long-established professions as medicine (in its practice or non-research dimension) and the law, as well as of the rapidly "professionalizing" social sciences and humanities.

Step Four (Conclusion): Hidden Value Assumptions in the Technical Community:

What this suggests is that the technical community might well be in danger of becoming identified with "the establishment" in the same dangerous way that, many people allege, medicine and the law are. This would, of course, be ironic, considering the fact that if it happened it would be largely a function of the claim to "expert professional objectivity." A recent study of academic voting patterns[32] points up what most observers probably suspected anyway—a tendency for business-oriented applied scientists and professionals to display a voting pattern in line with business interests and at odds with the unions and reform Democrats. In the current alignment of political groups in the United States, does it make sense for the technical community to side with the rich and powerful rather than the weak? The question does not assume an answer one way or the other, or even that there is an answer to the question as posed. Nonetheless, if it does turn out that there is some "logic" in sizable portions of the technical community leaning in one direction or the other, it will be because there is some affinity between certain social values and science or technology—at least in the United States today. This would make science and technology not value-neutral. And even if it were proved that science at least is neutral, this would not preclude social philosophers from arguing that a neutral science nonetheless ought to be used, for instance, for all segments of society equally and not just for special interests.

REPLIES TO ANTICIPATED CRITICISMS FROM OPPOSING VIEWS

It is time now to tie all this together, and one effective way to summarize the view and move on toward a conclusion is to reply to possible objections that might be forthcoming from defenders of opposing points of view.

1. Marxists: Marxist criticisms have already been dealt with by—glibly, I am sure, in their eyes—turning back upon them Marx's claim that problems are solved not by intellectual debate but by action. Whether or not technology can be tamed and put to work in the "social humanism" of a future classless society is a matter for history to determine. In the meantime, Marxists are surely right that any other view must turn out to be (by Marxist standards) "reactionary," committed to political resistance of one sort or another to Marxist goals or at least Marxist methods.

2. An Idealist Philosophy of Technology: The Platonic/neo-Kantian philosophy of technology of Friedrich Dessauer[33]—along with any ultraconservative views that would see technological problems as no more than passing aberrations, mere temporal blocks to the working out of the "tried and true ways" toward the ultimate realization of a Divine Design—can be equally briefly dismissed. What I find wrong with Dessauer's view, that technology gives man entry to a world of ideal forms when he discovers the "perfect" solution to a technical problem—and thus confers on the technologist aware of the meaning of his work an unlimited right to pursue his Godlike creative goals—is that it is subjective in the bad sense (the variety of views as to what is "ideal" is almost limitless) and consequently leaves the technologist who claims to know the meaning of his work unchecked by democratic claims of the rest of us poor mortals.

The remaining two alternatives listed above are (a) a Skinner/behaviorist-type philosophy of technology, and (b) an analytical philosophy of technology which would, along with Henryk Skolimowski as quoted earlier, "be concerned with . . . the philosophy of technology proper, that is, with the analysis of the epistemological status of technology," while leaving "the moral implications of technological progress . . . the weight of [which] cannot be underestimated" to others to solve, presumably with the help of the best sociology of science and technology available at the time

3. Behaviorism: The primary objection I have to a behaviorist program for technology is that it is elitist and technocratic. However much Skinner may feel that his "behavioral technology" can be managed democratically, it seems difficult to escape the counterclaim that—if we are to "take conscious *(sic)* control of our destiny"—*someone* is going to have to be the technical psychological expert teaching (?) others how to implement the behavioral technology appropriately.

This final word, "appropriately," introduces a second, more narrowly philosophical (in the academic sense) objection to any behaviorist account of anything. What I would like to dwell on is an aspect of group behavior that behaviorism—on its own terms and whatever the version—may not be able to handle at all or at least not adequately. Individuals cannot be presumed to "just know" how to behave in groups, how to respond to symbols or to play roles. "Appropriateness" and "inappropriateness" are fundamental features of group behavior: if we did not know how to behave we would not be able to behave at all. This is obvious, but it is one of those obvious facts that are often ignored, with disastrous consequences. Much of our "how to" knowledge is of course so habitual that we are unaware of it; some people even maintain that a large part of it must be unconscious if we are to act or behave "smoothly." Self-conscious individuals, for example, unaware of how to behave in social situations, often cannot function at all. Unconscious or not, the point is that "how to's," norms for appropriate and inappropriate behavior (a broader category than moral norms or values, since we must know how to apply them appropriately), seem to be essential to the operation of groups as such. Without forms or patterns, group behavior would be nothing more than something like stylized rituals of inexplicably coordinated atomic individual behaviors. Why? Behaviorists (or "methodological individualists" like George Homans, mentioned earlier) think they can explain "appropriate" behavior, along with good and bad behavior, (supposedly) free choices, obedience to norms presumably even of appropriateness, etc. At least Skinner is explicit in saying that he can. To do so, they must rely on nothing more than differential reinforcement especially in terms of introjecting parental or superego norms. But this explanation runs smack-up against a pair of difficulties: (a) Does not the child need to know how to introject parental norms appropriately? and (b) Can undifferentiated responses—a behaviorist allows himself no others—explain appropriately or inappropriately differentiated responses? A behaviorist would usually say Yes, falling back on the environment, physical or social, to provide the differentiation. However, *this response will not work: adaptive behavior, however much adjustment there is to external environmental stimuli, is still just bare response in behaviorist terms*. There must, therefore, be an internal adaptive mechanism to provide the differentiation dubbed "appropriate." In groups what this means is that *without individual violations of role expectations, group behavior must endlessly repeat itself even in its mode of adaptation to external stimuli*. In short, for behaviorism there can be no "appropriate" responses to stimuli not reducible to bare responses.

These counterarguments are not going to prevent Skinner and others from continuing to propose their "behavioral technology" model as a program for the future. But at least they explain why I choose another model. I believe there is appropriate and inappropriate behavior, and I am fearful of technocracy—perhaps most of all when it masquerades under the label of democracy.

4. Analytical Philosophies of Technology Combined with Sociology of Technology: The objection I have to (anticipated) analytical philosophies of technology must be somewhat more circumspect. For one thing, I am trying to anticipate rather than respond to actual arguments. For another thing, the term "analytical philosophy" covers a very broad range. Nevertheless, two general arguments can be proposed after an initial note.

The note has to do with W. V. Quine's statement quoted earlier, "I'm as much of a behaviorist as any sane man could be." Some analytical philosophers are openly sympathetic toward behaviorism—if not necessarily also toward "behavioral technology" politics—even when they are careful to note that it is only a "programmatic" (i.e., not a dogmatic) behaviorism. Against them the same argument as above would hold. Against anti- or nonbehaviorist analytical philosophers of technology, however, new arguments are needed.

First argument: If we look at the Max Black version of an analytical theory of language/communication (above), its predominant feature, in opposition to the view proposed here, is its hostility toward metaphysical, ontological or speculative assertions of any sort—even the limited socially constructed "realities" of pragmatism. The argument against this view is a long way from our present concern with philosophy of technology, but it is interesting on its own merits (at least to philosophers); there may be a connection with technology (as spelled out earlier and as will appear in the second argument); and in any case examples for illustration can be chosen from technological discourse.

The first part of the argument depends on the fact that language *seems* to involve references to extralinguistic entities. Certain Wittgensteinians like Black would explain away these references, avoiding any suggestion of an extralinguistic "meaning" or "intention" by saying that meaning equals use, the only "meaning" is the ability to use a term appropriately. To examine this claim we can look at some examples: "State the matter quantitatively and I can deal with it"; "That is an inefficient use of materials"; "The spacecraft (plane, bridge, etc.) failed because it was poorly designed"; "Your job is to find an alloy with a, b, c qualities so it can stand up under x, y, z conditions."

What people in the technical community do in response to expressions such as these is not simply to respond but to respond *selectively,* as if they had *something further in mind:* a technological product, a good design, efficiency, quantitative precision. Now if we are to explain these selective responses adequately, we must assume either (a) that the meaning-use includes a reference to products, designs, efficiency, precision or (b) that it does not. If we assume (b), we seem either to be involved in an infinite regress—the reference is explained by its use, but this use is normally self-conscious or at least can be reflected on, which requires explanation by another use, etc.—or else we are back in some version of behaviorism where "selective response" is reduced to "bare response" and self-conscious reflection is not allowed. If we assume (a), we are *not* committed to some sort of Platonic "objective essences" or phenomenological "intentions," though we are committed *at the very least* to something like the minimal socially-constructed "objects" of Pragmatism.

Second argument: A more pertinent argument, however, may be constructed in proper social philosophy terms as outlined earlier. (The above argument depends on a social epistemology, but that is another matter.) As I already mentioned once in passing, there is often an affinity between analytical philosophers and such academic sociologists as Robert Merton—for instance in the Science, Technology, and Society Program at Cornell. In fact, I predicted earlier that if philosophy of technology catches on among analytical philosophers, it is likely that they will continue to concentrate exclusively on conceptual analysis, leaving the fact-gathering with respect to the technical community to the social scientists. What I see this as doing is institutionally reinforcing *academic specialization;* the trend today is often even justified as a *professionalization* of philosophy. I do not want to argue here against such a trend. (I have argued elsewhere against its excesses.)[34] But what I would remark upon is that there is the same danger here of ending up pro-establishment that I noted earlier with respect to scientists and other expert professionals. Being pro-establishment is, of course, no vice, but it does represent an anomaly if it is a result of "professional objectivity," and it *could* lead to considering anti-establishment or other establishment-critical philosophers as "un-professional."

CONCLUSION

There is a place today for the development of an academically respectable philosophy of technology in American and other English-speaking circles. The movement seems inevitable and welcome. Not inevitably, but in all likelihood, a whole variety of philosophies of technology will spring up: Hegelian or other idealistic, phenomenological, Marxist or other radical or leftist or socialist, behaviorist-technocratic, analytical.

What I have argued for here is a social philosophy of technology that is explicitly *normative*. What I would like to see is a continuation—even an expansion notwithstanding the current job market and pressures toward vocational education—of efforts begun a generation ago but never brought to fruition to broaden the training of scientists, engineers, biomedical researchers, etc., in the humanities and social sciences. I would like to see this as part of a redefinition of "professionalism" in the technical disciplines in two senses. John W. Gardner, the eminent critic of contemporary institutions who began the citizens' lobby Common Cause, has said, "I would lay it down as a basic principle of human organization that the individuals who hold the reins of power in any enterprise cannot trust themselves to be adequately self-cricitical";[35] because of the ever-present threat of developing vested interests, organizations need to build in institutional self-criticism. Though science and the other technical professions are built on the principle of peer review, there is no doubt in my mind that vested interests do infect the scientific and technical establishment. Again, Abraham Edel has argued brilliantly that in an era of "big science" the old dichotomy between scientist-as-scientist and scientist-as-citizen no longer has its former validity; he strongly urges the scientific community to get involved in collective political action—for instance, in strikes for specific demands, in exercising what he calls a "ferment-function" to raise consciousness about problems, in institution-making and even in marketing certain discoveries for the public welfare.[36] Finally, I would hope that the scientific and technical professions, if they do recognize their political function, would get more involved in such liberal causes as civil rights, better education and housing for the poor, environmental crusades, etc., and less involved with such things as the military-industrial complex.

I would hope this view can be debated in open public forum against alternative views. My original criterion for a "serious" philosophy of technology was that it be taken seriously by academic philosophers—even if only seriously enough to be attacked. Whether my approach qualifies, then, remains to be seen.

FOOTNOTES

1. See Carl Mitcham and Robert Mackey, "Bibliography: The Philosophy of Technology," *Technology and Culture*, Vol. 14, no. 2 (April 1973); also, Mitcham and Mackey, eds., *Philosophy and Technology: Readings in the Philosophic Problems of Technology* (New York: Free Press, 1972); John Burke, ed., *The New Technology and Human Values*, 2 ed. (1972); Willis Truitt & T. W. G. Solomons, eds., *Science, Technology, and Freedom* (1974); Alex Michalos, ed., *Philosophical Problems of Science and Technology* (1974).
2. Paul T. Durbin, "Technology and Values: A Philosopher's Perspective," *Technology and Culture*, Vol. 13, no. 4 (October 1972).
3. Alan Gewirth, *Political Philosophy* (New York: Macmillan, 1965).

4. For Mead's works, see G. H. Mead, *The Philosophy of the Present* (Chicago: Open Court, 1932); *Mind, Self and Society* (Chicago: University of Chicago Press, 1934); *The Philosophy of the Act* (U. of Chicago, 1938); Andrew Reck, ed., *Selected Writings: George Herbert Mead* (Indianapolis: Bobbs-Merrill, 1964); Anselm Strauss, ed., *George Herbert Mead on Social Psychology: Selected Papers* (Chicago: University of Chicago Press, 1964); also David L. Miller, *George Herbert Mead: Self, Language, and the World* (Austin: University of Texas Press, 1973).

5. H. D. Duncan, *Communication and Social Order* (New York: Bedminster, 1962; Oxford and New York: Oxford University Press, 1968), pp. 102–103.

6. On models and leading metaphors in different philosophical approaches, cf. Stephen C. Pepper, *World Hypotheses* (Berkeley: University of California Press, 1942, 1966).

7. Bernard Berelson and George Steiner, *Human Behavior: Shorter Edition* (New York: Harcourt, Brace and World, 1967), p. 57.

8. Thomas Kuhn, *The Structure of Scientific Revolutions* (Chicago: University of Chicago Press, 1962; 2d ed., 1970).

9. A. E. Taylor, *Plato: The Man and His Work* (Cleveland: World; Meridian edition, 1956), p. 357.

10. Mitcham and Mackey, eds., *Philosophy and Technology* (see footnote 1), pp. 22–25, 317–334.

11. B. F. Skinner, *Beyond Freedom and Dignity* (New York: Knopf, 1971); also Skinner, *About Behaviorism* (New York: Knopf, 1974).

12. G. C. Homans, *The Nature of Social Science* (New York: Harcourt, Brace and World, 1967).

13. W. V. Quine, *Philosophy of Logic* (Englewood Cliffs, N.J.: Prentice-Hall, 1970).

14. Max Black, "Dewey's Philosophy of Language," *Journal of Philosophy*, Vol. 59 (1962), pp. 505–523; reprinted in Black, *Margins of Precision* (Ithaca, N.Y.: Cornell University Press, 1970); reference there, p. 234.

15. See Peter Berger and Thomas Luckmann, *The Social Construction of Reality* (New York: Doubleday, 1966), for both an insightful title and an apt expansion of a basically Meadian view. In his own terms, Mead defines "reality" as a (social) "confluence of (individual) perspectives." He is perhaps clearest in defining what he means relative to science—his paradigm of human knowledge. The "real world out there" is simply the given (socially constructed) world taken for granted which gives rise to problems: without it, nothing could ever be called into question as problematic, yet if its givenness is exaggerated, as in naive realism, there is no possibility that a given world could, at a later time, itself be called into question. Cf. Mead, *Philosophy of the Act* (see footnote 4), pp. 42–45.

16. Norman W. Storer, "Introduction" to Robert K. Merton, *The Sociology of Science: Theoretical and Empirical Investigations*, ed. Storer (Chicago: University of Chicago Press, 1973).

17. N. W. Storer, *The Social System of Science* (New York: Holt, Rinehart and Winston, 1966).

18. H. D. Duncan, *Symbols in Society* (New York: Oxford University Press, 1968), pp. 6 and 19–20.

19. See footnote 2, above.

20. Jacques Ellul, *The Technological Society* (New York: Knopf, 1964).

21. Diana Crane, *Invisible Colleges: Diffusion of Knowledge in Scientific Communities* (Chicago: University of Chicago Press, 1972).

22. Jonathan R. Cole and Stephen Cole, *Social Stratification in Science* (Chicago: University of Chicago Press, 1973).

23. Karl Marx and Friedrich Engels, *The German Ideology* (New York: International Publishers, 1960), p. 78.

24. See footnote 2, above.

25. Mitcham and Mackey, in their book of readings (footnote 1), include two of those articles, under the aptly titled section, "Conceptual Issues."

26. Henryk Skolimowski, "The Structure of Thinking in Technology," in Mitcham and Mackey, *Philosophy and Technology* (see footnote 1, above), p. 43.

27. Joseph Agassi, "The Confusion Between Science and Technology in the Standard Philosophies of Science," *Technology and Culture*, Vol. 7, no. 3 (Summer 1966); cf. J. O. Wisdom, "The Need for Corroboration: Comments on Agassi's Paper," *ibid.;* Agassi's reply to Wisdom, *Technology and Culture*, Vol. 8, no. 1 (January 1967); and the references in footnote 29, below.

28. See Karl R. Popper, *The Logic of Scientific Discovery* (New York: Basic Books, 1959), *Conjectures and Refutations* (New York: Basic Books, 1962), and *Objective Knowledge: An Evolutionary Approach* (Oxford and New York: Oxford University Press, 1972).

29. The reader interested in following up on Agassi's constantly refined views will have to keep up with a (probably to be continued) series of articles by Agassi: "The Logic of Technological Development," *Proceedings of the XIVth International Congress of Philosophy* (Vienna, 1968); "Positive Evidence in Science and Technology," *Philosophy of Science*, Vol. 37, no. 2 (June 1970); and a paper delivered at the Philosophy of Science Association biennial conference 1972, to be printed in *Boston Studies in the Philosophy of Science*, Vol. 11.

30. E.g., in "Toward a Science Policy for the United States" (1970) and "Technology: Processes of Assessment and Choice" (1969), both reports put out by the Committee on Science and Astronautics, U.S. House of Representatives, the latter by the whole committee, the former by the Subcommittee on Science, Research and Development.

31. Dennis L. Meadows, *et al.*, *The Limits to Growth* (New York: Universe Books, 1972).

32. E. C. Ladd, Jr., and S. M. Lipset, *Academics, Politics and the 1972 Election* (Washington, D.C.: American Enterprise Institute for Public Policy Research, 1973); cf. also Ladd and Lipset, "The Divided Professoriate," *Change* (May–June 1971), especially p. 55.

33. See footnote 10, above.

34. Paul T. Durbin, "The 'Professionalization' of Philosophy: An Essay in the Sociology of Philosophy," *Proceedings of the American Catholic Philosophical Association*, Vol. 47 (1973), pp. 98–109.

35. John W. Gardner, *No Easy Victories* (New York: Harper and Row, 1968), p. 42; cf. also pp. 127–129, and *Self-Renewal* (New York: Harper and Row, 1963), esp. chap. 8.

36. Abraham Edel, "Scientists, Partisans, and Social Conscience," *Transaction*, January 1972, pp. 33–39 and 52; original, unabridged article in the *Proceedings* of the Center for Philosophic Exchange, Summer 1970.

THE EXPLANATION OF TECHNOLOGY

Albert Borgmann DEPARTMENT OF PHILOSOPHY

UNIVERSITY OF MONTANA

There is at present no highly developed philosophy of technology. One reason is the lack of agreement on the sense in which technology calls for an explanation. There is little fruitful controversy because it is not clear what precisely one should disagree on. There are greatly varying views of technology, but they do not reflect a variety of explanations of a certain phenomenon; more often we have various phenomena which are called by the same name.[1] To be sure, once the homonymy is removed, substantive differences remain. But they are then located in different established fields, social and political philosophy, philosophy of science, or philosophy of psychology and there is still no topic proper to the philosophy of technology. It seems that there is nothing in particular to explain for the philosophy of technology. And yet the word technology is often used to characterize what is distinctive of our era. We have at least intimations that the distinction of our age is both taking on a definite shape and is poorly understood. There *is* a problem we would like to have explained.

But how can we address the problem when it is so elusive or vague? Let us consider this possibility: Rather than starting with an attempt at delimiting what needs to be explained, we might reverse the approach and examine our methods of explanation. They may be such that they do not let the nature of the present world come clearly into view. Perhaps the philosophy of technology can be established as a distinctive field only if a distinctive manner of explaining and understanding is developed at the same time. Moreover such an understanding of the field may not just be distinctive of our world but decisive for its welfare.

Our understanding of explanation is shaped by the natural sciences. This is apparent in formal and informal ways. When someone asks us for an explanation of this or that, we feel the more secure and responsible the more nearly our account approximates the scientific model. When a ten-year-old asks us about the phases of the moon, we will not resort to mythology. When we owe an account of our irritability, we will not give it in the language of religion. Rather we will resort as best we can to astronomy and psychology. This inclination agrees well with empirical studies which indicate high prestige and firm acceptance of the sciences.[2] But the inclination not only points to the prevalence of the sciences and their mode of explanation but also to the structure of scientific explanation. We claim and grant success in explaining the phases of the moon when we have succeeded in deriving them from the general features of the solar system, and when asked to explain those, we aim to subsume them under the general laws of astronomy and physics. Similarly in psychological explanations. To explain is to subsume the particular phenomenon under general scientific statements or laws.

We cannot do justice to the manifoldness and gradations in which scientific explanations have penetrated and helped to shape the everyday world. Let us proceed from the rough indications just given to the formal structure of scientific explanation as it has been laid bare by philosophers of science. Carl Hempel and Paul Oppenheim have distinguished between the problem that is to be explained, the so-called *explanandum,* and the set of propositions which provides the explanation, the *explanans*.[3] The latter has two subsets, one consisting of scientific laws, the other of the initial conditions of the explanandum phenomenon. The explanandum then follows from the explanans; i.e., the laws and conditions are the premises from which the explanandum proposition is deduced as the conclusion. Hence the name "deductive-nomological explanation."[4] Depending on the accidents of time, what is deduced may be a prediction (the explanation of a future event) or a retrodiction (the explanation of a past event).

Hempel and others have not only explicated the structure of scientific explanation which is approximated in the everyday world. They have also labored to bring out the dominance of scientific explanation by showing that whatever advances a claim to explanatory force must more or less meet the standard structure of scientific explanation. In particular they showed that explanations of a supposedly different structure or kind, e.g., functional, teleological, or historical explanations, either fail to explain or fail to be genuine alternatives.

The case for the dominance of scientific explanation that emerged from these arguments is strong. It is of special interest in regard to historiography. Though history is an ancient and venerable discipline and has had the benefit of much philosophical reflection, no sort of historical explanation has come to the fore which is widely regarded as a clear and viable alternative to deductive-nomological explanation.[5]

What must arouse dissatisfaction with the dominance of this kind of explanation is its tendency to obscure definite gaps in our understanding of what it means to explain something. These gaps open up around scientific explanation itself. To begin, we must remember that a scientific explanation normally gets underway only when the scientific laws are given. Even in the case where the laws are discovered in an attempt to solve a problem, the discovery itself, though it is part of an explanation, is not thereby explained. We have no general explanation of how scientific laws are discovered. This is not for lack of attention. Historians and philosophers of science have devoted much ingenuity and diligence to the study of how new scientific laws and theories arise. But what understanding we have of these matters is not derived from deductive-nomological explanations. More particularly, rationalist or inductivist explanations of the emergence of scientific laws which at least emulate the rigor of deductive-nomological explanations have proven quite inadequate. It is worth remarking that the philosophers of science who remind the general historians that their successful explanations are of the deductive-nomological cast do not attempt to cast their own explanations of the history of science in that mold.

Another gap is found on the opposite side of scientific explanations. An explanation gets underway only when it is clear what problem is worthy and in need of explanation. But again we have no general explanation of how problems get stated.[6] A closer look at the first gap will lead us to a better understanding of the second.

The problem of the rise and succession of laws and theories in the history of science has many aspects. An important thread in this history and its discussions emerges when we consider that there is apparently no rule whose application leads to progress. This appears from a study of the details and circumstances of any scientific breakthrough. The lack of a rule is equally well demonstrated by the failure of those allegedly possessing such a rule to achieve genuine and consistent progress. Scientific progress seems to be unpredictable in any strict sense, and that is to say, unexplainable.

Yet if we speak of progress in the succession of scientific discoveries, there must be a pattern in this development, and it must be one from weaker to stronger stages. These two findings are not really incompatible. Scientific advances may well be inexplicable and incomprehensible when we look at the future. The hitherto unthought is as of now unthinkable. But this is a psychological limit. Looking at the past, after a great thinker has thought through what seemed unthinkable, the advance will exhibit theoretical ties to the past which are clear to all experts.[7] Progress up to the present is theoretically and hence psychologically perspicuous. Progress in the future is psychologically veiled and hence theoretically opaque.

Thomas Kuhn, whose thought on these matters has become well known and influential, has never denied that scientific research makes progress. He has certainly denied that progress is steady and by accumulation. However, there is an issue on which Kuhn seems divided, and it is the truly controversial and interesting one. It regards the question how radical the discontinuities in scientific progress are.[8] We seem to face a dilemma. On the one hand the periods ushered in by new scientific theories or paradigms cannot be so radically different from their predecessors that they are incomparable or incommensurable with them. We would then have change but not progress.[9] It is the immersion in the details of history which makes us hesitate to belittle and reduce the differences among eras to degrees of crudity and ignorance. One is tempted to posit not just psychological or epistemological differences, but differences in the nature of reality, i.e., ontological differences.[10] On the other hand, we cannot deny that in the development of science, theories supersede one another in attaining ever greater explanatory power.[11] But as noted before, the power has not expanded so as to cover its own history and character.

The result of the history of scientific progress does not explain itself in the deductive-nomological sense in which we have taken "explanation" so far. But the history does exhibit a pattern which can be pointed out, and such pointing out is a kind of explanation in its own right, one that I will call *deictic explanation*. I will concentrate on one feature of this pattern and point it out very tentatively and briefly. The early scientific theories of the Western world had both world-articulating and a world-explaining significance. Thus Aristotle's physics and astronomy contained laws which permitted deductive-nomological or subsumptive explanations.[12] But these laws were moored in the singular structure of the cosmos which was articulated and pointed out in Aristotle's theory or vision of the world. The articulated world order of Aristotelian physics and astronomy is more or less of one piece with the world of Aristotelian metaphysics, ethics, and all his other disciplines. In this world order, everything had its place and rank. The movements or changes of things could be predicted on the basis of laws which reflected the privileged dimensions of the world and the rank of things. The Aristotelian laws were of limited explanatory power in that each held only for a small class of phenomena (e.g., for sublunar horizontal motion), in that they yielded only rough or relative predictions, and in that they were inconsistent with one another.

The progress of science is marked by improvements in the scope, precision, and consistency of the laws. In thus gaining greater explanatory power in the deductive-nomological sense, the laws lost their power of deictic explanation. Einstein's theories of relativity no longer reflect or point up a singular world. They do have deictic power in the weak sense of delimiting a set of possible worlds and ruling out certain impossible worlds.[13] We can observe a similar pattern in the development from alchemy by way of chemistry to nuclear physics. Alchemy reflected in its laws a definite life world of a limited number of stuffs and transformative forces and processes. Nuclear physics, being a microtheory, allows for an indefinite number of molar worlds.

This pattern in the progress of science has no *a priori* character. It is an empirical fact that the world can be explained in the powerful scientific theories that we now have. The pace of the discoveries of these theories is a matter of historical fact. But given these two facts, it was inevitable that the deictic power of the sciences waned and all but vanished. This is not a failure of science. Nor is it the case that the *deictic* achievement of the earlier sciences was unquestionable or unique. *Art* has always been the supreme deictic discipline. Art in turn has sometimes been one with philosophy, religion, and statesmanship; at other times these disciplines have complemented or competed with one another as disciplines of deictic explanation.

We now have a framework and a hypothesis for the explanation of modern technology. We may think of modern science as having rendered the world perspicuous by setting it within the matrix of the scientific laws. In this matrix it appears as one possible world. It might within the same matrix be differently arranged. Or in other words, any definite state or event in the world can be subsumed by way of its initial conditions under scientific laws. And any such state or event might have been prevented or modified if the conditions had been different. Thus the change of conditions in accordance with scientific laws yields great transformative power. Modern science lets the world appear as actual in a realm of possible worlds. Modern technology reflects a determination to act transformatively on these possibilities.[14]

But neither science nor technology has a theory of what is worthy and in need of explanation or transformation. Given an explanandum or transformandum, they will explain and transform the problematic phenomenon. But neither has a principled way of problem *stating*. To be sure, science has authentic access to the problems that arise within a research program.[15] But these are not the problems whose solutions constitute the technological transformation of the world. Technology in its turn, merely as the determination to transform, faces an indefinite number of transformative possibilities and cannot provide principled guidance to problems.

But guidance may be expected from the artist. In the periods of great art, the poets, the architects, and the sculptors revealed and articulated the world in its crucial dimensions. They pointed out what was worthy of attention, what had a claim on our reverence, what had to be feared and warded off, what ultimately constituted disaster or grace. The artist thus establishes substantive as opposed to formal truth. He determines not how something is generally and indifferently true, but what is importantly and crucially true.[16]

According to the hypothesis we can explain technology as set between two modes of explanation, the deductive-nomological explanation of the sciences which yields transformative, but not guiding power, and the deictic explanation of the arts which sets standards of action and conduct, but provides no effective means of compliance. Modern technology would then be the intermediary, the determination to employ the power, made possible by the sciences, on behalf of the standards, set by the arts.

We must now test the hypothesis against the major objections to which it is subject and against the facts of the matter themselves.

The most important and general objection contends that the scientific enterprise guides not only itself but can also guide us in understanding and changing the world at large and the technological world in particular. The general objection divides in two of which one holds specifically that the scientific enterprise *discovers* guiding forces and the other claims that the scientific enterprise *constitutes* a guiding force.

The first of these objections is best discussed under the heading of functionalism.[17] Functionalists maintain that the study of organs, organisms, populations, and communities reveals them to be systems, consisting of parts which are so interrelated as to achieve or maintain a certain state for the system. To explain a system is to show how the functions of its parts contribute to its standard state. Such an understanding is based on scientific demonstration and so has the cogency of deductive-nomological explanation. But it has in addition deictic or guiding power because it enables us to identify and correct dysfunctional items or tendencies of a system. These are the claims of functionalism. Clearly, they can be extended to an understanding and critique of technology, and this is largely done for instance in the arguments of the ecological and environmental movement.[18] Though the movement's endeavors are admirable, they rest on uncertain foundations; for so does functionalism generally.

At least certain species of functionalism employ cogent scientific demonstrations, i.e. deductive-nomological explanations, to elucidate the interrelations of parts of a system, to predict deviations from the preferred state of the system, and so forth. But it is fallacious to infer from the cogency of the *explication* of the functional system the cogency of the *delimitation* of the system, its parts and states. If the system and the parts are granted, the standard state may be cogently derivable. If a concise characteristic of the functional system is granted, we may be able to show compellingly that it has a unique physical realization. But in every case cogency rests on a grant.[19] It can only be provided by a deictic explanation. There can be no objection of inconsistency to a functionalism which is put forward as a deictic or substantive theory. But then a claim to anything like deductive-nomological cogency can no longer be made. When environmentalist and ecologist critics of technology are accused of selfishness and bad faith, aim is taken, however unfairly and vaguely, at the inconsistency of their basis. These critics are vulnerable not because they failed to secure absolute foundations (everyone must "fail" in this sense), but because they often pretend to such a foundation, and their claims are then shown to be unfounded. So technology essentially escapes the guidance and strictures of the functionalist critique.

Apart from the well-known functionalisms to be found in the biological, social, and ecological sciences, there are other theories of similar promises and deficiencies. One is the theory of evolution. It is certainly not tautologous. There are empirically ascertainable features and testable laws of evolutionary history. But what this history teaches about technology is not incontrovertibly apparent from those features and laws but needs to be pointed out through an explanation in its own right. Another is the theory of values. If such a theory is to go beyond the preliminary sharpening of tools and sighting of available evidence in order to shed light on technology, it is likely to secure or presuppose its basis in an evolutionary theory or in psychology.[20] The psychology in turn may revert to a mental functionalism or to the kind of physicalism that is proposed in the Identity Theory. But if a physicalist explanation of the mind is guided by the physicochemical theories alone, it is incapable of delimiting (though not of explicating) in a principled manner what needs to be explained. Hence it follows that the psychological veil which hides the future of scientific progress from our view cannot be lifted by a psychological theory, but always and only by the great scientist who reveals a new and better scientific theory.[21]

According to the other major objection, the scientific enterprise embodies a substantive way of taking up with the world, positively or negatively. The positive case can be made historically by showing that science was a liberating event, a breaking of the fetters of superstition, ignorance, and dogmatism.[22] These forces science replaced with rationality, honesty, and a public and inquisitive attitude.[23] A more straightforward argument holds that an inspection of the scientific enterprise reveals that the practitioners of science are held to singularly stringent and august criteria of achievement.[24] Finally, the application to the problem of technology is made when it is held that the deplorable chaos of the contemporary world results from our failure to carry the scientific enterprise to its conclusion by explaining and shaping human behavior according to the best available scientific knowledge.[25]

The first of these three arguments is the strongest because it can point to the very real clashes of scientists with traditional world views. In light of our earlier remarks on the progress of science, it must be admitted that as the scientific theories advance, they more and more withdraw their endorsement of established world views. If a totalitarian power demands such an endorsement, withdrawal is often both undertaken and acknowledged as a resolution. But it is one thing no longer to fulfill a task and quite another to fulfill a task in a new way. Scientific progress can at most be *liberation from;* it can never constitute or provide the thing that it is a *liberation for.*

More specifically, when scientific endorsement is withdrawn from a world view, the latter is required to abandon in light of that withdrawal those of its elements which hitherto provided or implied deductive-nomological explanations, those elements, that is, from which together with particular conditions, empirically testable predictions could be derived. For the withdrawal of endorsement in scientific progress is not a wanton shift of allegiance, but the reflection of the discovery of new and more powerful laws and new explanations which are thereby possible. But it is a mistake to think that a world view must shrink to nothing after it has given up its scientific elements. The Aristotelian hierarchy of being need not be given up with Aristotelian mechanics and dynamics. Accordingly Einstein's relativity theory has no counterpart or counterargument to the Aristotelian hierarchy. To be sure, withdrawal of scientific endorsement forces a world view back to its deictic resources. If these were slim or unwholesome to begin with, this will now become apparent, and the world view may collapse. Conversely, the more purely and fully a world view is by its nature articulated in a deictic manner, the less it is affected by scientific progress. That is true of poetry and art in general. The more complex world views of politics and religion, however, are required not only to expel their scientific elements, but also to rearticulate themselves in light of the new scientific laws. Both enterprises are laborious and encourage conservatism. But it is pointless to call for a substantive controversy between science and theology as Paul Feyerabend does.[26] That call will be frustrated not necessarily by the meekness of theology, but inevitably by the fact that modern science cannot embody a substantive world view of a scientifically authenticated sort.

We can leave it undecided whether the scientific enterprise as a sociological or psychological phenomenon is singularly edifying or pernicious, whether it would lead us to happiness or ruin.[27] In neither case would the guidance originate from the center of the enterprise, that is from the best current theories of science.

On the other hand, the guidance that can be had from the arts may not be heeded. In that case the rise of science as a power without guidance for the world may have substantive consequences in its own right, and technology may be foremost among them. A world whose articulation disintegrates may as such display definite and consequential traits. This is roughly the thesis of Hans Jonas.[28] More particularly he holds that modern science has not only withdrawn its support of established world views, but promoted their dissolution and the establishment of an alternative vision. The world's cosmic architecture is denied and replaced by the infinite manifold of one homogeneous substrate. Manipulation and novelty are integral parts of this promotion, and it has technology as an inevitable if not immediate consequence. Technology ceaselessly transforms the world along abstract and artificial lines.

This is a considerable argument, and it can be complemented by pointing up the close sociological and disciplinary ties between science and technology.[29] It is certainly important to consider carefully, as Paul Durbin does, the empirical facts and consequences of this association.[30] Yet a theory may well point up distinctions beneath the present facts, and this is both possible and desirable in the present case. Joseph Agassi and Mario Bunge have shown that it is possible to distinguish in a principled manner the scientific from the technological procedure.[31] In particular, the scientific methodology is shown to be detached from the common criteria of success as one would expect it from a discipline which is not committed to the establishment of a particular world view.

The distinction between science and technology is also eminently desirable for a critique of technology. Jonas's thesis is so strong because he does not derive technology primarily from science as a sociological phenomenon, i.e., from the habits and characteristics of the scientific community. Rather he derives it from the core of science, from the nature of the scientific theories and of the explanations they make possible. But this strength is also a weakness by its consequences. Current science at its core is true; true in the sense that its theories give us the best representation of the general structure of reality. The truth which a realist claims for science can be denied from an instrumentalist position; but the latter is plausible only as long as it avoids precision. The instrumentalist cannot draw a definite line between everyday knowledge which has access to real states of affairs and scientific knowledge which merely deals with convenient and useful formalisms. Yet if we accept the realist view of science and admit that what our current scientific theories and explanations say of the world is true, then we must also admit that technology, if it is the necessary consequence or companion of science, is equally true. Putting it more discursively, technology on that view is a mode of taking up with the world which is entirely and necessarily in accord with the true nature of the world. One can then deplore the truth of science and technology. But one can criticize technology only in violating the truth.

As we have seen, however, a proper appraisal of the core of science and of the methodology which immediately surrounds and serves that core does not warrant an inference from science to technology. That inference does not fail *a priori*, but it certainly does in fact. A concise and consistent formulation of Jonas's principal thesis fails to agree with crucial features of the technological world. The thesis holds that modern science renders the world homogeneous, infinite, devoid of an encompassing structure and goal. If these processes and their results are not just necessary for technology but sufficient, then technology is nothing but the reduction of the world to unbounded, unstructured homogeneity. Any thesis can be saved by modification, and the present thesis holds if the presence and effects of technology are restricted appropriately. But it is clear from Jonas's discussion of the industrial revolution, of mechanics, chemistry, electricity, and electronics that he does not accept the drastic restriction of the significance of technology which the consistency of the thesis would require.[32]

But if technology harbors formative forces which cannot be delineated through recourse to modern science, how can they be delineated? Jonas never mentions the arts as a formative power in the technological world. Apparently that thesis is not worth attacking, and he is certainly right. Here is a crucial flaw in our hypothesis. Art may have been the supreme deictic discipline of the past. But it is doubtful that modern art can advance such a claim, and if there is such a claim, it is certainly not heeded. But then there corresponds to the crucial flaw in our hypothesis a crucial gap in our understanding of the modern world. We found that no definite world view emerges from modern science, neither from physics nor from functionalism, from evolutionary theory, value theory, or psychology. It is doubtful whether modern art articulates a complete and structured world, and few seem to care at any rate. But between modern science and art, there is not nothing. The modern world does not in any plain and indisputable sense tend toward greater homogeneity and loss of structure. On the contrary, where technology is most advanced, the world is most radically and tightly restructured.

But is this account of our ignorance fair? Is it not rather true that the various problems of the modern world are quite well understood and so also the social and political forces which promote or impede solutions to these problems? A survey of these matters as given for instance by Victor Ferkiss is certainly of some value.[33] But as an analysis of technology it fails because it departs and must depart from a superficial pluralism. After all, it is not the case that some factions of the technological community build dams and others take them down, that some raise the standard of living and others are interested and successful in lowering it, and so forth. All the factions worth reckoning with vie with one another within definite conventions, and the pluralist analyst leaves these unexamined.

It is true that there is much knowledge of this and that in the technological world. But there seems to be little knowledge of the basic forces and patterns that need to be explained in the area between the sciences and arts. What we are lacking is the comprehensive and penetrating vision which was first articulated under the philosophical term *theoria*.

A theory of technology, however, cannot be set between science and art as a foreign body. It must be distinguished from either, but fruitfully so. Our discussion so far is a delimitation of technology and its theory from without. It is now a matter of reviewing these limits and of going past them to a positive characterization. On the scientific side it is clear that no deductive-nomological explanation of technology is possible. But the explanation of technology must take account of the cogency and transformative power that is derivable from such explanations.

Looking to the arts, it is clear that their mode of world articulation is not the same as the technological.[34] Unlike the technician, the artist does not articulate the world in physically shaping the cosmic material and in rearranging all the items in the world. Nor does he, of course, explain the world by proposing or subsuming under laws. Rather he reveals the world in the concrete and singular work of art, in the epic, the temple, or the symphony. He so articulates the coherence and significance of the cosmos. And thus he engenders insight and uncovers substantive truth. That is what we mean by deictic explanation.

When we say of art, or at any rate of the great art of the past, that it has explanatory force and makes truth claims, this is not to be taken as a metaphor or a generous concession to the artist. It is rather to insist that there must be some deictic explanation by a discipline which articulates the singular shape of the world and that the artist is the foremost deictic explicator of the past in concert or competition with the early scientist, the thinker, the prophet, and the statesman.[35] There must then exist a deictic discipline even today, but it is no longer either science or art. Technology now articulates the world. But its articulation, unlike the traditional deictic explanations, has no focal point. Art in creating its work reveals a world. Technology articulates a world, but it is in comparison a silent and implicit articulation because it is not concentrated and eloquent in a concrete work of art, treatise, proclamation, or constitution. Therefore the articulating power of technology must itself be articulated. This will be a quasi-deictic explanation, not one which reveals or is deictic of a concrete work, but one which uncovers *a pattern by which* we can uncover the concrete workings of technology. Let us call this paradeictic or paradigmatic explanation.

Thomas Kuhn has recently called attention to the paradigm as an instrument of elucidation and explanation. His turn to the paradigm and the attention which his endeavors have received reflect an explanatory need which art and science have not met and which the paradigm seems to fulfill. But Kuhn's employment of "paradigm" is beset by ambiguities which often border on misunderstandings of the distinctive force of this notion. Especially in his later elaborations, Kuhn seems inclined to subsume the stronger under the weaker, i.e., to reduce the paradigm to a matter of psychology, sociology, evolutionary theory, or value theory.[36] These are the disciplines, as noted before, whose promise of guiding power must remain empty. Another limitation of Kuhn's investigation is the straightforward one of explicitly excluding the everyday world from the influence of paradigms and their changes.[37] Hence one must go beyond Kuhn in pointing up the paradigm as an explanatory device in its own right and in extending its application to the changes in the everyday world.[38]

This is how far we can fruitfully go in delimiting the technological paradigm from without. Positively speaking, the paradigm of the contemporary world is the technological device.[39] It can be briefly sketched and illustrated as follows. The technological device is the radical and increasingly sharp separation of means from ends. A heating plant has the sole purpose of providing heat. This is its function, and it remains relatively stable. The machinery of a device is indefinitely variable within the boundaries of the function. A totally new and different machine will constitute a better but not a different device. Thus it makes no difference whether the heat comes from coal, gas, or oil, whether it is pumped, blown, or radiated. Given that variability there cannot be such a thing as the essence of the machine which would characterize technology as Le Corbusier and

Mumford had thought.[40] But there is something like the essence of the device.

In the pretechnological world, a means is always more than merely a means. In the wood which burns in the stove there is the work of felling, sawing, and splitting, there is the age of the trees and the species which the land and the climate favor. The stove will bespeak an origin and a history of ownership. Correspondingly, there is no mere end. The burning of the wood indicates the weather when the draft is poor or the stovepipe is red in response to the cold. The stove constitutes a focus, i.e., hearth, of warmth and comfort and thus concentrates the house. A heating plant provides warmth and nothing else.

Although the function of a device remains stable in comparison with the machinery of a device, its stability is not absolute. But whereas the changes in the machinery are as unpredictable as the inventions on which they rest, the changes of the function exhibit a consistent pattern. In the progress of technology, the function increases in prominence and purity whereas the machinery shrinks and recedes. This is a fairly concrete phenomenon that can be seen in the development of the home environment.[41] The improvements of the television set and of the electronic computer are especially striking examples.

But the device comes fully into view only when we consider the human needs which the functions serve. At first the needs seem as firm as the functions. But closer inspection shows that as a function increases in purity, it isolates and transforms a need more and more radically. There is a mere and unmediated need for warmth only once there is a device which singles out and satisfies that need without addressing any other sensitivity or interest. The end of the device is the full and exclusive termination of function and need without disturbance or presentation of further relations. In their isolation, all needs appear as equally important, and hence their number can be increased indefinitely.

The complete fulfillment of a need without further distractions or demands requires that the function of the device is ubiquitous and comes into play instantaneously; it must be easily manipulated and safe. These four traits of a function's presence can collectively be called availability. We can say conversely that whatever is truly available is a device. This suggests that things other than instruments or implements can be procured as devices. The machine device exhibits the technological paradigm most apparently. But social institutions and works of art can also be procured as devices. Technology is more than a matter of machines, as Ellul has emphasized.[42]

These problems belong to the general question of how widely and incisively the paradigm is applicable. It requires more detailed considerations to show, for instance, how it pertains to matters of politics, labor, and leisure.⁴³ Assuming such applicability, important questions regarding the explanation of technology remain open. There are chiefly three. If technology is the procurement of devices, then: (1) Are there any limits to the progress of procurement? (2) Is there any guidance in the expansion of procurement? (3) Can there be a critique of the device paradigm? The answers to these questions will conclude our investigations.

In replying to the first question, we must remember the crucial tie between technology and science. Modern science is a necessary condition for modern technology because it is the transformative power derived from scientific explanations which allows us to effect mere ends, something that was previously conceivable only in magic. In the matrix of the laws of science, reality becomes perspicuous, and hence we are able to disentangle from the complex web of causality those causes which procure a precise and intended set of effects either for the first time at all or in a manner which is unprecedented in its safety, ease, ubiquity, and instantaneity. But that transformative power has at least at the moment a limit which even the boldest and best informed students of technology and its future have overlooked until very recently. The transformative power has so far failed to unlock resources of energy which are both safe and inexhaustible. This puts a definite check on the availability of machine devices. If that barrier turns out to be permanent, organizational devices will increase and improve the more rapidly.⁴⁴ These too are based on science, physical, social, and formal, and they will benefit from scientific progress.⁴⁵

The energy limit is both physical and systematic. It is physical in that there is as a matter of fact a limited amount of available fuels. It is systematic in that the system of science and technology has not produced discoveries or inventions to assure the availability of energy. A systematic advance is always possible. But the systematic boundary, though elastic, remains. Just as there is no scientific law of scientific progress, there is no technique of technical invention.⁴⁶ It is clear that the promotion of the device paradigm is aided by technical inventions. But it has enough guiding and propelling forces of its own to make headway without inventions properly so called. We now turn to these forces.

The device paradigm taken as a prescription to transform the world seems powerful because nothing apparently can resist it. Anything and everything can be made available. But such power is also an impotence because to say that everything can be done comes close to saying that nothing in particular should be done. This deficiency infects the paradigm's explanatory adequacy because obviously things are not randomly procured. By and large, devices do not cancel or conflict with one another. A certain extent of wantonness is properly brought out by the paradigm, but there seems to be no account of the boundaries of that extent. It is indeed impossible to delimit them definitively, but some restraining and guiding factors can be pointed out. One is the use of tradition as a resource. Many functions simulate the presence of a traditional thing or event where that presence has the prominence and isolation of the technological function, and that is to say that the historical depth of these things and events has been removed and replaced by the machine or organization of the device. Thus technology presents the grain of wood in Formica, the taste and whiteness of whipped cream in Dream Whip, the sound of an orchestra in the stereo set. Tradition provides an inexhaustible resource of tastes, shapes, and experiences and thus provides challenges and fuel for procurement.

A second guiding and propelling force is the challenge of procurement itself. To annihilate the risks, the demands of skill and attention, and the spatial and temporal recalcitrance of things is the paradigmatic achievement. Think of the invention of artificial light, warmth, and transportation, not to mention the procurement of health. To the signal procurements correspond privileged fulfillments of needs. The first experience of a clear step from a thing to a device is felt with elation and triumph. But unlike tradition, the world of present things is shrinking. A few recalcitrant cases aside, technology in the advanced countries now must seek out the last reservations for its conquests and turn to such slight tasks as can openers, pencil sharpeners, and snow shovels. But not only the quantity of such conquests is diminishing. Each such conquest has, unlike things and works of art, a sustaining force which vanishes rapidly and irretrievably.

Both these factors find their final significance in the endeavor to procure the entire world as a device. The attempt to render the world's commodities and appurtenances ubiquitous and instantaneous by means of transportation and communication has early been a discernible technological tendency. But comprehensive safety and ease of manipulation have been neglected, and the neglect shows itself in pollution and depletion of resources. This failure of procurement has the undivided attention of the world community. The principal debate is on *how* this encompassing procurement is best achieved. That man might be something other than a procurer is seldom considered. Further investigation is required to determine how devices are combined and readied for final assembly into the one device that is sometimes called spaceship earth.[47] But the general tendency seems to be clear.

The final question is how the device paradigm can be appraised and criticized. That question can be taken theoretically and practically. We can first ask whether the paradigm is a helpful explanatory instrument and an appropriate pivot of a theory of technology. Assuming a positive answer, i.e., granting that the paradigm brings out the pattern of the technological transformation of the world, we can then ask on what grounds the employment of that pattern in the practice of technology should be supported or opposed.

One can give a negative answer to the first question at different levels. There is still the possibility that paradeictic explanations turn out to be fanciful in light of better *kinds* of explanations, deictic, deductive-nomological, or some other sort *sui generis*. If one agrees to the need for paradigms, one may disagree with the *device* paradigm and favor some other. But it seems impossible to decide between competing paradigms because of the circular character of paradigmatic explanations which Kuhn has noted.[48] One cannot test the paradigm against what one sees because the paradigm first tells one what to look for. A similar circularity obtains in scientific theories also. Perhaps it is inevitable and not vicious as Kuhn contends. But the question remains why one should step into or out of a particular circle.

We can see here the tie between the theoretical and practical sense of the guiding question. A critique of technological practice must address the question whether man indeed moves in the circle of the device paradigm, whether he is caught in it, whether he has any possibility or desire to escape from it, and if so what recourse he has. All these questions are pointless if theoretically circularity is the final word. If the questions have a point, then we have come to a limitation of paradeictic explaining and its theory.

Before I turn to those fundamental questions, I want to note that the questions themselves have a clarifying force. It seems to me that most critics of technology, John Galbraith the most notable among them, are divided as to whether they disagree with the present system because it is *essentially* technological or because it is *incompletely* technological, whether they should advocate that we curb or fully adopt the device paradigm.[49] To probe the roots of their and our discontent is to ask how we fundamentally justify our actions.

It seems to me that in a thoughtful conversation with most anyone it is possible to find in that person's past focal experiences, experiences that stand out from among the others as revealing and sustaining. They may be experiences of nature, or art, or of others. Such experiences require openness on our part, but openness cannot produce and guarantee them. They are essentially unforeplanned and amazing. Even where they are preceded by calculation and preparation, when they truly come to pass, we acknowledge them as surpassing our shrewdness and merit. What is so experienced is the strict counterpart to the device. It is in principle unavailable and it is so procured on pain of destroying it as a truly focal thing or event.[50] It is the focal things and events which sustain us and which we see threatened by the expansion of the device paradigm.

The appeal to such forces is clearly an inchoate deictic explanation. Deictic explanations have no cogency. We can see now that this is not a regrettable deficiency but an indispensable trait. To give a deictic explanation of some thing is not to command it and produce it to everyone's satisfaction but to reveal it or to testify to it as something unavailable, as a power other and greater than ourselves. Since there is no cogency, the subjective turn can always be taken. Every realist's thing can be converted into the subjectivist's value; testimony is turned into mere assertion. There is still the possibility of thoughtful appeal.[51] But the subjectivist can subvert all testimony by his countertestimony that *he* has had no focal experience. No one can be forced out of a paradigmatic circle.[52]

In the poetical realist's view, however, paradeictic explanation is moored in deictic explanation. And so a critique of technology is possible. The critique will employ the device paradigm theoretically in order to comprehend and contain the practical employment of the paradigm. But has it not been said before that what the deictic explanation used to reveal is now entirely eclipsed? That eclipse must be accepted. But an eclipse is something less than annihilation. And there is reason to hope that an understanding of the pattern of darkness will help us in overcoming it.

FOOTNOTES

1. For surveys and critiques of theoretical approaches to technology see Paul T. Durbin, "Technology and Values: A Philosopher's Perspective," *Technology and Culture* XIII (1972), 556–576; William Leiss, "The Social Consequences of Technological Progress: Critical Comments on Recent Theories," *Canadian Public Administration* XIII (1970), 246–62; Carl Mitcham and Robert Mackey, "Introduction: Technology as a Philosophical Problem," in *Philosophy and Technology*, ed. Mitcham and Mackey (New York, 1972), pp. 1–30; Donald W. Shriver, Jr., "Man and His Machines: Four Angles of Vision," *Technology and Culture* XIII (1972), 531–555.

2. See the ranking of occupations according to the National Opinion Research Corporation to be found, e.g., in Richard Sennett and Jonathan Cobb, *The Hidden Injuries of Class* (New York, 1973), pp. 221–25. See also Todd R. La Porte and Daniel Metlay, "Technology Observed: Attitudes of a Wary Public," *Science* CLXXXVIII (April 11, 1975), 121–127, and G. Ray Funkhouser, "Public Understanding of Science: The Data We Have," in *Workshop on Goals and Methods of Assessing the Public's Understanding of Science*, ed. Funkhouser (University Park, Pennsylvania, 1973), pp. 18–21.

3. See Carl G. Hempel, "Studies in the Logic of Explanation," in *Aspects of Scientific Explanation* (New York, 1965), pp. 245–95, and "Aspects of Scientific Explanation," *ibid.*, pp. 331–496.

4. Here and in what follows I disregard deductive-statistical explanations since they are structurally similar to the deductive-nomological ones.

5. For recent surveys of this controversy see Rudolph H. Weingartner, "The Quarrel About Historical Explanation [1961]," in Alex C. Michalos, ed., *Philosophical Problems of Science and Technology* (Boston, 1974), pp. 165–180; and Howard Cohen, "*Das Verstehen* and Historical Knowledge," *American Philosophical Quarterly* X (1973), 299–306.

6. Cf. Marx W. Wartofsky, "Is Science Rational?" in Willis H. Truitt and T. W. Graham Solomons, eds., *Science, Technology, and Freedom* (Boston, 1974), pp. 204–206.

7. After the breakthrough, it may even be difficult to see the original problem. Cf. Wartofsky, "All Fall Down: The Development of the Concept of Motion from Aristotle to Galileo," in *Conceptual Foundations of Scientific Thought* (New York, 1968), pp. 419–473, pp. 449 and 456 in particular.

8. See Thomas S. Kuhn, *The Structure of Scientific Revolutions,* 2d ed. (Chicago, 1970); also "Logic of Discovery or Psychology of Research?" in Imre Lakatos and Alan Musgrave, eds., *Criticism and the Growth of Knowledge* (Cambridge, 1970), pp. 1–23 and "Reflections on my Critics," *ibid.*, pp. 231–278.

9. Paul Feyerabend, "Consolations for the Specialist," in Lakatos and Musgrave, p. 202.

10. Kuhn sees the ontological question, but does not believe that a clear answer can be given. See his "Reflections," p. 265, and *Scientific Revolutions,* pp. 184 and 206–207.

11. With Feyerabend's exception, this is stressed in one way or another by all contributors to the Lakatos and Musgrave anthology.

12. For details see Wartofsky's "All Fall Down." Wartofsky does not share the thesis presently to be developed. But much of his account is compatible with it and illustrates it.

13. Strictly speaking, the revolutionary scientist's work has deictic significance also. We may call it *global* deictic significance to distinguish it from the poet's *singular* deictic explanations. The latter's explanations will be discussed in more detail below.

14. For a more detailed and qualified account of the relation of modern science and technology, see Mario Bunge, "Toward a Philosophy of Technology," in Mitcham and Mackey, pp. 62–76.

15. For details see Kuhn, *Scientific Revolutions* and Lakatos, "Falsification and the Methodology of Scientific Research Programmes," in Lakatos and Musgrave, pp. 91–196.

16. The classic statement of this position is Martin Heidegger's "The Origin of the Work of Art," in *Poetry, Language, Thought,* tr. Albert Hofstadter (New York, 1971), pp. 15–87. This position has been elaborated and qualified by Hans-Georg Gadamer in *Wahrheit und Methode,* 2d ed. (Tübingen, 1965). His hermeneutic method has many similarities to the deictic and paradeictic explanations here proposed. But the hermeneutic method remains confused as long as its relation to science and technology is not clarified. For discussion of the explanatory status of hermeneutics, see Hans Albert, "Theorie, Verstehen und Geschichte," in *Konstruktion und Kritik* (Hamburg, 1972), pp. 195–220; and the contributions by Apel, Bormann, Bubner, Gadamer, Giegel, and Habermas in *Hermeneutik und Ideologiekritik,* ed. Jürgen Habermas (Frankfurt am Main, 1971).

17. I have dealt with these problems in more detail in "Functionalism in Science and Technology," in *The Proceedings of the XVth World Congress of Philosophy* (Sofia, 1973–75), vol. 6, pp. 31–36.

18. A more formal and explicit critique of technology, based on functionalism, has been given by Ervin Laszlo, "Human Dignity and the Promise of Technology." *The Philosophy Forum* IX (1971), pp. 165–201.

19. There are many varieties of functionalism, and they have all found their critics who more or less agree on the uncertainty of the foundation of functionalism. Among those critics are Ernest Nagel, Carl G. Hempel, I. C. Jarvie, William Kalke, and Richard Rorty. See my "Functionalism."

20. Examples of the preliminary sort of work on technology and values are provided by Kurt Baier, "What Is Value? An Analysis of the Concept," in Baier and Nicholas Rescher, ed., *Values and the Future* (New York, 1969), pp. 33–67; and by Rescher, "What Is Value Change? A Framework for Research," *ibid.*, pp. 68–98; and "A Questionnaire Study of American Values by 2000 A.D.," *ibid.*, pp. 133–147.

21. This is a deictic achievement in its own right. Cf. note 13 above.

22. Sometimes liberation harbors new enslavement. See Paul K. Feyerabend, "On the Improvement of the Sciences and the Arts, and the Possible Identity of the Two" in Robert S. Cohen and Marx W. Wartofsky, eds. *Boston Studies in the Philosophy of Science,* vol. 3 (Dordrecht, Holland, 1967), pp. 387-415.

23. See Paolo Rossi, *Philosophy, Technology and the Arts in the Early Modern Era,* tr. Salvator Attanasio (New York, 1970); and Wartofsky, "Is Science Rational?" in Truitt and Solomons, pp. 202-210.

24. For arguments pro and con see the selections by Leo Tolstoy, Jacob Bronowski, Karl Deutsch, Joseph Wood Krutch, and Herbert J. Muller in John G. Burke, ed., *The New Technology and Values* (Belmont, California, 1966), pp. 24-49. For discussion see Wartofsky, *Conceptual Foundations,* pp. 403-415.

25. See José M. R. Delgado, *Physical Control of the Mind: Toward a Psychocivilized Society.* (New York, 1969); B. F. Skinner, *Beyond Freedom and Dignity* (New York, 1971); and R. Buckminster Fuller, *Operating Manual for Spaceship Earth* (New York, 1970).

26. In his "On the Improvement of the Sciences and the Arts," p. 404.

27. For illustrations, see note 24 above.

28. "The Scientific and Technological Revolutions: Their History and Meaning," *Philosophy Today* XV (1971), 76-101. (A revised version is given in Jonas's *Philosophical Essays* ([Englewood Cliffs, N.J., 1974], pp. 45-80).

29. See Boris Hessen, "The Social and Economic Roots of Newton's Principia," in Truitt and Solomons, pp. 89-99.

30. Paul T. Durbin, "Toward a Social Philosophy of Technology" (this volume, p. 79).

31. Joseph Agassi, "The Confusion between Science and Technology in the Standard Philosophies of Science," *Technology and Culture* VII (1966), 348-366. For Bunge see note 14 above.

32. Science must be something more than an instrument if it has the kind of power which Jonas ascribes to it. He speaks in fact of "the ontological breakthrough" at the beginning of the modern age and modern science (p. 77). But Jonas speaks like an instrumentalist when he compares the scientific view with the view of the life world (pp. 87, 88-89, 90). Jonas is similarly divided on the question whether the new scientific world view has dissolved human spontaneity or rendered it omnipotent (pp. 92-94).

33. Victor C. Ferkiss, *Technological Man* (New York, 1969).

34. "Articulating" as used here is ambiguous between *revealing* and *producing.* The ambiguity is resolved in what follows.

35. Cf. Heidegger, "The Origin of the Work of Art," pp. 61-62.

36. Cf. Karl Popper's criticism in "Normal Science and its Dangers," in Lakatos and Musgrave, pp. 57-58. No one will fault Kuhn for drawing on these disciplines. But I think the final significance of his *Scientific Revolutions* will depend on whether Kuhn takes the paradigm as the pivot of an explanation *sui generis* or whether the paradigm is an abbreviation for a sort of explanation which is fully spelled out and justified in one of the disciplines named above. For references see note 8 above.

37. See his *Scientific Revolutions,* pp. 90, 93, 207-08. The presumed solidity of the everyday world allows Kuhn to be so unconcerned ontologically (cf. note 10 above), and it also provides a basis for the commensuration of paradigms.

38. Despite these shortcomings there is much to be learned from Kuhn about paradigmatic explanation.

39. I have given more detailed but less circumspective accounts of the device and its mode of presence in "Technology and Reality," *Man and World* IV (1971), 59-69, and in "Orientation in Technology," *Philosophy Today* XVI (1972), 135-142. Those articles also indicate my debt to Martin Heidegger.

40. See Le Corbusier (Charles Édouard Jeanneret), *Towards a New Architecture,* tr.

Frederick Etchells (London, 1931) and Lewis Mumford, *Technics and Civilization* (New York, 1963) pp. 333–337.

41. See Melvin M. Rotsch, "The Home Environment," in Melvin Kranzberg and Carroll W. Pursell, Jr., eds., *Technology in Western Civilization*, vol. 2 (New York, 1967), pp. 217–233.

42. Jaques Ellul, *The Technological Society*, tr. John Wilkinson (New York, 1964).

43. That the paradigmatic approach is generally the one that is appropriate to the explanation of social and political phenomena has been urged by John G. Gunnell, "Social Science and Political Reality," *Social Research* XXXV (1968), 159–201; and by Sheldon S. Wolin, "Paradigms and Political Theories" in *Politics and Experience*, ed. Preston King and B. C. Parekh (Cambridge, 1968), pp. 125–152.

44. Skinner's *Walden Two* (1948) is an example of a highly technological society with limited machine technology.

45. Bunge has emphasized that technology is based both on substantive and on formal sciences. See his "Toward a Philosophy of Technology," pp. 62–64.

46. Cf. Agassi, "The Confusion," pp. 348 and 361.

47. Cf. Fuller, *Operating Manual for Spaceship Earth*.

48. See his *Scientific Revolutions*, pp. 90, 93, 207–208 and "Reflections," pp. 236–237. There are really two circularities at issue, one pertaining to the explanatory force of a particular paradigm, the other attaching to the explanation (metatheoretical relative to the former) provided by the general search for and exhibition of paradigms. The remarks above apply to both types of circularity.

49. See John Kenneth Galbraith, *The New Industrial State*, 2nd ed. (Boston, 1971) and *Economics and the Public Purpose* (Boston, 1973). Indicative cases of Galbraith's ambivalence are his treatments of the technostructure and of culture and the arts.

50. See Henry G. Bugbee, *The Inward Morning* (State College, Pennsylvania, 1958).

51. The necessity and difficulty of giving the discourse of testimony and appeal legal standing is explored by Laurence H. Tribe, "Ways Not to Think About Plastic Trees: New Foundations for Environmental Law," *The Yale Law Journal* LXXXIII (1974), 1315–1348.

52. Cf. Kuhn, *Scientific Revolutions*, pp. 151–152.

VALUES IN SCIENCE

Willis H. Truitt UNIVERSITY OF SOUTH FLORIDA

Science changes. It changes because the guiding ideas, paradigms as Kuhn calls them, that lie behind it change. These ideas are both theoretical and experimental. And the ideas are at least partly unsupported value assumptions, or determined by unsupported value assumptions. There are values at the base of science that were, or are, arrived at unscientifically. There are values in the scientific enterprise itself that are arrived at in the same way. This is true of all phases in the history of Western science, beginning with the Ionian physicists and extending right down to the present. There is a model by which this can be understood and explained. It is a model based on modern social organization, but it has its counterpart in all historical societies that possessed a science. If one examines the model, it will be evident that the distinction between science and technology is unclear, i.e., science and technology are parts of the same continuum. Therefore, a philosophy of technology must also include at least part of the philosophy of science and in particular that part of the philosophy of science with which Agassi concerns himself, namely, discovery and concept formation, although I think for reasons different from those that Agassi is inclined to isolate. Thus, a philosophy of technology will need to examine a range of issues: problem shifts in conceptualization, reorientation in research, and so forth. And these issues ultimately spring from needs and decisions of an extrascientific character. The model I have in mind has three parts: (1) industry and government, (2) technology, and (3) scientific research and development.

In the first place, industry and government, through a series of decisions that are primarily productional or political in nature, set the tasks for technological innovation and growth. Secondly, industry, concerned mainly with the production and sale of goods and services, and government, concerned mainly with the protection of industry, defense and expansion, and political control, exercise extensive control of the condition and use of technology. This control is instrumented through the allocation of social resources only for specific technical tasks which are seen to be consistent with pre-established requirements. Finally, the requirements placed on the technological enterprise are transmitted to the scientific enterprise. Scientific research most conducive to solution of the technological problems imposed by productional and political demand is supported with vast resources.[1] Scientific work which seems immediately unrelated to the dual level requirements of technology and government-industry is abandoned or is undertaken in an ad hoc fashion under undesirable circumstances.

The transmission of political and productional needs from the top to the bottom, so to speak, from extrascientific institutions down to "pure" research, has a propagative effect. For the chain of demand is laden with values which originate from outside the scientific enterprise. These values are the products of societies and are ultimately determined by social organization, distribution of wealth, privileged status and access, and political hegemony.

I do not want to discuss the origination of these values, although this certainly is an important problem. Rather, I will try to describe the values after they are introjected, almost unnoticed, and become implicit in the conduct of science. The tracing of them back to their origin I will leave for a later series of papers, or for someone else to do. Durbin's paper partially undertakes this task in a contemporary context. These values will be sketchily formulated here because of time, for a thorough exposition of them would, indeed, involve following them through the whole social process. But, perhaps, their most basic elements can be grasped.

In order to develop this discussion I will treat four historical phases of science separately. They are not really that separate or isolated, but they show difference in emphases both in procedure and in value orientation. They are: (1) Ancient Science, (2) Medieval Science, (3) Modern Science, and (4) Contemporary Science. Following the focus of this conference, I will expand my remarks on the fourth, contemporary phase.

1. ANCIENT SCIENCE (SCIENCE AS PRAXIS OR PRACTICAL HUMANISM)

Surely, the best understanding of pre-Socratic science is that of an attempt by human beings to acquire mastery over their environment. The beginning of science in Ionia was based on techniques, the arts and crafts. The source of this science was experience and its objectives were practical. Although there was considerable theorizing about the constitution of things, this theorizing was always brought back into contact with those things. The usefulness of this early science to humanity has ample evidence ranging from meteorological forecasting, maps, water clocks, and medicines to primitive machinery. Indeed, even the arts were affected. As Farrington has shown,

> In the Athens of the middle of the fifth century the great ideals of *Philanthropia* and *Philotechnia*, love of mankind and love of science in its application to society, were made the theme of a major work of art. In a setting worthy of such high debate, the Theater of Dionysus itself, the problem of accommodating them to the contemporary structure of society was discussed. The great drama of Aeschylus . . . unfolds how the supreme god Zeus, the symbol of authority in the universe and in society, has declared war upon the titan Prometheus for his love of mankind.[2]

Pre-Socratic science was humanistic and democratic, corresponding, in fact, with the rise of Greek democracy. These value dispositions (humanism and democracy) were implicit in its methods and assumptions and in the idea of praxis itself. Praxis is that very notion of practical mastery over the environment, the capacity of human beings to control and direct the forces of nature, and to utilize nature in the satisfaction of collective human needs. This science was methodologically democratic in its insistence that knowledge is public and arrived at through observation. All people were thought to be capable of acquiring knowledge and attaining truth. This conception of knowledge must be contrasted with the later Platonic conception of knowledge and truth which teaches that only the exceptional few can achieve ultimate understanding. It should be noted, however, that there is continuity in one sense with the Platonic doctrine. This is the indistinguishability of value and fact. In Plato the object of knowledge is itself an intrinsic good. For the pre-Socratics the good lay more in the activity of knowing. There is a marked difference in emphasis announced in the shift from observation and praxis to contemplation and stasis. For Plato, values were still inherent in the scheme of things. They no longer served as prescriptions for human action. With Plato they had become reified; of independent existence; ultimate objects of thought; not directives for inquiry. It has been said that Plato's defense of a specific social class requires a new kind of science in which the focus is on the mastery of some men and women over others. And, of course, this is what the *Republic* is all about right down to the detail of how to deceive and control the masses through propaganda.

Up till Plato, it seems that democratic and humanistic assumptions pervaded the scientific enterprise. Hippocratic medicine clearly evinces these value characteristics. The physician was obligated to secure the health of all patients regardless of their status. These doctors were aware that every treatment they applied to a patient had its experimental as well as its humanitarian side. For example, the tendencies of those in authority to obscure the nature of both physical disease or social injustice is described in a Hippocratic treatise thus:

> My own view is that those who first attributed a sacred character to this malady (epilepsy) were like the magicians, purifiers, charlatans, and quacks of our own day, men who claim great piety and superior knowledge. Being at a loss, and having no treatment that would help, they concealed and sheltered themselves behind superstition, and called this malady sacred, in order that their utter ignorance might not be exposed. . . .
>
> But perhaps what they profess is not true, the fact being that men in need of a livelihood will contrive and devise many fictions of all sorts. . . .[3]

2. MEDIEVAL SCIENCE (ANGELIC SCIENCE)

In the sixth century A.D., Cosmas Indicopleustes disclosed that the motive power of celestial bodies was supplied by angels.[4] And there is general agreement among historians of philosophy and culture that nearly all scientific and philosophical treatises from the beginning of the Christian era on into the Renaissance were essentially theology.[5] This church philosophy and science was not new in its methods. In its early stages it simply adopted the doctrines of Plato and somewhat later it incorporated the Aristotelian science and philosophy.[6] Following the tradition of Plato in which only the person of superior intellect, a philosopher skilled in the logical analysis of ideas, could hope to attain truth, medieval science was constituted as a kind of privileged knowledge. The values implicit in this approach were authoritarian and antidemocratic, esoteric and idealized. One characteristic of the Platonic philosophy that pervaded medieval thought was the attribution of ontological (substantial) status to logical ideas. When logical ideas were given "content" in this way, a number of things followed. First, there was assured a belief in the fixed order of nature. The ontological interpretation of Aristotle's laws of "contradiction" and "excluded middle" were also to reinforce this belief. Second, it was supposed that truth was to be ultimately derived from the logical analysis of ideas. This sort of emphasis suppressed empirical procedures and praxis. Third, following both Plato and Aristotle, great stress was placed on (a) the fixed nature and immutability of objects of knowledge in Plato, and (b) the fixed nature of natural species in Aristotle (which were the theoretical bases of his defense of slavery and inequalities of women). The idea that change is a basic characteristic of the material universe was ruled out. The ethical and political consequences of such a science are all too obvious: the belief in a fixed human nature (involving original sin and atonement), fixed places and classes in political theory (divine right and serfdom), fixed goals in the theory of morality (involving means of salvation). The classical doctrines of fixity and logical realism lent themselves easily to the feudal order of the Christian Middle Ages and were the primary influences on the whole philosophic and scientific development (or better, absence of development) of the epoch from Augustine to Abelard, Ockham and Aquinas. A theory of knowledge as privilege which began as a defense of oligarchic hegemony in the Platonic dialogues was absorbed and redeployed in defense of feudalistic institutions and values. The principal basis of cognition and action (inaction) became: (a) *revelation* (corresponding to the Platonic doctrines of recollection and contemplation) and (b) *authority* (based on the Aristotelian conception of a fixed hierarchical order of nature).

The Aristotelian notion of a science of fixed species was absorbed by Aquinas and only questioned later under the impact of modern European science. The theory of knowledge that accompanies it is an abstract theory according to which a truly scientific understanding of the various kinds of objects in the world of nature is achieved by abstracting the forms of these objects from experienced cases, just as (supposedly) a truly scientific understanding of geometrical shapes is achieved by abstracting geometrical natures from experienced cases. Scientific method is that which conduces to the grasping of an ordered manifold of truths about a kind of object, by bringing about the grasping of the substantial form of an object of that kind; therefore, substantial form becomes the ontological basis of truth: metaphysics precedes epistemology, and empirical inquiry and praxis are diminished. The roles of active observation and interaction were minimized. This was an "angelic science" which cognized the intelligible forms of things as opposed to an active, inductive and predictive science with practical capacities for changing the world.

3. MODERN SCIENCE (SCIENCE AS DOMINATION)

"Knowledge is Power," the phrase best known from Francis Bacon, is further elucidated in the subtitle to his *Novum Organum*. It reads, "Aphorisms Concerning the Understanding of Nature and Man's Dominion Over It." Bacon's main contention was that human success in dominating nature constituted the test and proof of correct understanding of nature. Thus, for him, truth and practice, knowledge and power, were united. In this claim lies the originality of his theory of science; that element which separates it from the traditions of Plato and Aristotle. Academic philosophers are usually prone to treating Bacon's writings on science as a contribution to the logic of induction. This it may have been. But his theory is far more general. It is a theory of knowledge which expresses fundamental changes in the conception of science. It articulates a shift away from the contemplative doctrines of aristocratic and angelic science to a more active, exploitative and experimental view. In behalf of this "new science," "Bacon asserted the identity of knowledge and domination, and with it he uttered his famous condemnation of the works of Plato and Aristotle, those light planks, as he called them, while the weightier matter of the pre-Socratics had sunk to the bottom."[7]

It was not his appreciation of the Greek physicists alone that led Bacon to a conception of "science as domination." Even more important were emerging tendencies in the practice of science and the power and mastery exhibited in its application. This new science took root in the practice of the mechanical arts as early as the late fourteenth century. Between 1300 and 1600 there were three levels of intellectual activity in the urban centers of southern Europe. These activities were conducted by university scholars, humanists, and artisans. The first two of these groups were trained in logical disputation and their methods tended to be abstract, rhetorical and exegetical. The university scholars and humanist literati marked a sharp distinction between the liberal and mechanical arts. They despised manual labor, experimentation, and dissection. The very craftsmen and artisans who practiced these depreciated activities were, however, the pioneers of causal and technical thought. Artists, engineers, makers of nautical instruments, surveyors, cartographers, navigators, gunners employed experimental and quantitative methods and thus expanded human domination over the environment. The primitive measuring devices of the navigators, gunners and other highly skilled craftsmen were the forerunners of the later instruments of the physical sciences.

The latent powers of craft practice were constricted, however, by a division of theory and practice that rested on a social barrier. Whereas the literati disdained the mechanical arts, the craftsmen lacked the methodical, intellectual training of the schoolmen. These two components of scientific method were eventually brought together when the progress of technology, its great successes in subduing and reshaping nature, became so significant as to overcome the social prejudice against manual labor. Galileo is the chief example of the combining of these two components. It was the power of this synthesis and the new world created in its wake that impressed upon Bacon the view that knowledge is power.[8]

The values inherent in the new scientific enterprise struck down the esoteric and hierarchical pretentions of the pre-industrial culture. In its place was set an ethic that coupled human progress with domination and exploitation of the natural world. As a higher synthesis of ancient, practical humanism, the new science differed in that in place of the emphasis on the observational power of human sensory faculties was set the vastly more powerful factor of technical instrumentation.[9] But even as the progressivist doctrine of human well being through domination of nature (Du Pont's "better things for better living through chemistry"?) took on the characteristic of a theological dictum, there were countervailing tendencies at work.

Up to the twentieth century the epochal values assumed in the conduct of science are more or less evident. They are, of course, attached to specific philosophic preoccupations which are, in turn, not detachable from technical and societal processes (which I have not gone into in this short paper). But as hope is a product of history, so is its sundering. And the loss of uniform allegiance to the possibility of a scientific and technical utopia finds its beginnings in the working out and extension of that science itself. Marcuse has put it thus:

> The new science (did) not elucidate the conditions and limits of its evidence, validity, and method; it (did) not elucidate its inherent historical denominator. It (remained) unaware of its own foundation; and it (was) therefore unable to recognize its servitude; unable to free itself from the ends set and given to science by the pre-given empirical reality. . . . (Hence) . . . Reason loses its philosophical power and its scientific right to define and project ideas and modes of Being beyond and against those established by prevailing reality. I say: "beyond" the empirical reality, not in any *metaphysical* but in a historical sense, namely, in the sense of projecting essentially different, historical alternatives. . . . The ideational realm of Galilean science (idealized, mathematical

science) no longer includes the moral, esthetic, political Forms, the *Ideas* of Plato. And separated from this realm, science develops now as an "absolute" in the literal sense no matter how relative within its own realm it may be, absolved from its own, prescientific and nonscientific conditions and foundations.[10]

4. CONTEMPORARY SCIENCE (SCIENCE WITHOUT VALUES)

The historical "dehumanization" of science, of which Marcuse speaks, was a way toward the liberating of inquiry from the oppressive influence of Christian theology—and all other ideologies for that matter.[11] It began with Descartes' twofold division of substance and terminated in the doctrine of "a value free science." But ironically, in the achievement of this value neutrality, is contained the ultimate subservience of science to national ideologies and vested interests. The process by which science is subjected to interests external to itself is described in the introductory remarks to this paper, and I will not discuss it further here.

Forty years ago the beneficence of science and technology was virtually unquestioned. Utopia was promised in the form of freedom from want and disease. Today this promise has become an illusion. We have a vast trillion-dollar war-science-technology establishment side by side with unmitigated squalor, hunger, and death from malnutritional disease. We are told encouragingly that this is not the fault of science and technology but rather the result of the uses to which it is put. The apology is too easy. In fact, the very neutrality of scientific thinking and technical procedures leaves them vulnerable and prohibits them (scientists and technologists) from reflecting upon the potentialities and uses of their own disciplines. It is the neutrality of science to human concerns and collective human needs, it is the absence of reflection on itself as a social enterprise, that creates its vulnerability and allows its misuse. "Value neutrality" is not neutral. It is a negative value.

There has been for some time a sharp distinction between values and facts; going back to Hume at least, but implicit in much earlier philosophies. In order to sustain this distinction it would be necessary to define each in an exclusive manner which, to my knowledge, has never even been attempted.[12] Facts are said to be empirically denotable, observable things, persons and events. Values are frequently said to be transcendental. But if they were, we could not speak of them at all. So, for values, we are left with written or uttered prescriptions, valuing behavior (protecting, desiring, nurturing, loving)—all forms of behavior, active or verbal. Some say that value is created by labor, that it is contained in the prices of commodities or of a work of art. Others reduce value to a theory of prices, as capital, or as a medium of exchange, coin, currency, or gold. It has been said that found objects, precious stones and metals, possess value. Values are said to be socially and historically determined attitudes, again a kind of behavior. This list of what might qualify as a value is certainly not exhaustive. But it is clear that every item and act enumerated can equally qualify as a fact. Someone might say that friendship is a value, but not a fact. But again, friendship is determined behaviorally. What values, indeed, are not facts? And if it is not possible to hold to the distinction between facts and values then the only reason I can see for clinging to it is to reinforce the popular belief in value-free science and mystify the continuity between desired ends and suitable means.

The problem is that behind its proclaimed value neutrality, the scientific enterprise operates already with a concealed ethic—with implicit values, prescriptions and directives. This hidden imperative demands that scientists (a) not study values because they are not facts, and (b) not reflect on and study the values inherent in their own enterprise—primarily the negative value of value neutrality.[13] Durbin, in his paper, profoundly disagrees either that the situation described here obtains among the "technical community," or that it ought not to. With the bottom of this disjunction I agree. As for the first term, let it be refuted by the recent behavior of scientists who have arrayed themselves against the value neutrality that prevails in their various disciplines. And I cannot concur with Durbin's effort to shift the focus from institutionalized control and distortion of the scientific-technical process to the field of discourse. A "deliberate meaning community" will not be established until "interests" concealed behind discourse are "fully disclosed." This presupposes a community of interest including the larger community of nonscientists. And it must be achieved in the social field. We must settle accounts with the institutional problems first.

Now the very philosophy that Durbin embraces, instrumentalism, has, not insidiously, propagated the doctrines of truth as efficiency in operation, and technique for the sake of technique. Inasmuch as pragmatism and instrumentalism make no claim to truth, however tentative, they are philosophies of surd instrumentality, thoroughly neutralized internally. It is not unexpected then that the unqualified ethical good of Dewey's system is "growth," an objective of which today we must be more than wary. And it is the instrumentalist attitude toward technology that easily equates feasibility with desirability. Pragmatism was, indeed, the philosophy of Sweet Commerce! How bold a philosophy when everything was still to be gained, how fat and timorous when everything was still to be kept.

Positivism, a less sophisticated version of pragmatism, however, has dominated philosophy of science in more recent times. Positivism is a conception of knowledge that denies the possibility of reflective reconstruction of the institutional environment of scientific and technical activity. Actually, positivism is a theory of measurement (not even a theory, rather an account of mensuration procedures) parading as a theory of knowledge. This is established in the positivistic taxonomy of the sciences which reduces meaning to a set of quantitative statements about the observable movements of physical elements; and in its phenomenalistic posture meaning is further reduced to the logical relations among physical language statements minus the physical facts.

Both positivism and pragmatism teach that knowledge is derived from the application of scientific procedures. In each instance the circularity of the argument is obvious: in pragmatism it turns on proving the claim by referring to the results of science; in positivism it proposes a complete identification of knowledge with the procedures themselves, thus excluding the possibility of contextual (social, historical) reflection. The positivist equation of procedure and knowledge does, unlike pragmatism, allow for truth claims. But these are so narrowly tied into the verificationist criterion that they are trivial at best and solipsistic at worst, because there is no attempt to explicate the social and historical context of scientific communication. This positivistic preoccupation with linguistic convenience rules out the possibility of giving an account of the epistemological foundations of science. Accordingly, both philosophies treat science as a neutral procedure for objectifying the neutral elements of intersubjective experience.

If these approaches induce a value-free science and technology, then Popper's more critical philosophy goes even further in that direction. By abandoning the verificationist approach to meaning and substituting systematic falsification, Popper fashions a scientific theory that lends itself more and more to highly efficient (eliminate the bugs), but still unphilosophical, technical application. I think the rejection by Popper of the verificationist apparatus has tended to reinforce in his philosophy that element which was already present in positivism: a conception of science and technology as providing ever more efficient technical instruments for control of the environment. This would not be so bad except that in our time the "environment" encompasses society in addition to nature. The technological control of society entails the technological control of human beings.

Herein we can locate the roots of dehumanization and the spread of technology into the human sphere. In terms of the model introduced at the beginning of this paper, positivism reflects the growing social demand for technical and professional specialization which, when applied to the control and manipulation of human beings, led to the psychological movement called logical behaviorism. And pragmatism in an earlier stage represents the technical requirements of industrial growth as is made perfectly plain in Sidney Hook's useful and revealing *Metaphysics of Pragmatism* (1925). Technical rationality follows the developing possibilities of machinery: adaptability, replaceability, mechanical uniformity, functional repeatability. The machine process is transmitted subtly over time into the theoretical domain of science, redefining its scope and method in terms of technical and instrumental functions, and suppressing its critical and reflective capacities. In the course of this process there appear any number of philosophies of science of kindred orientation: pragmatism-instrumentalism, operationism, conventionalism, positivism and Popperism. In the social productive sphere there is a corresponding development: Taylorism and in general the standardization and routinization of work which today reaches even into the academy, into social and scientific planning, research and development, into social control, behavioral modification, rehabilitation therapy. These last software techniques are deliberately devised for adapting human subjects to their surroundings; while highly sophisticated computerized control systems are imposed to increase efficiency and productivity. To the extent that philosophies of science, particularly instrumentalism and positivism, have provided the rationale for the expansion of technology into the human sphere, they are unlikely to comprehend, or give significant criticism of, this cultural phenomenon.

In conclusion, I will indicate where Durbin and I have arrived at similar conclusions.

1. My paper supports Durbin's criticism of Agassi's proposal that science and technology be held distinct. I agree with Durbin, for the reasons given in my paper, that science and technology are continuous and that this can be shown by pointing out their institutional similarities, a few of which I have tried to do. I also agree with him that a philosophy of technology must be normative.

2. I support Durbin's conclusion that "basic" and "product-oriented" research are indistinguishable. In my paper I have tried to show why, if only in a schematic way.

3. I agree with Durbin that science and technology are not neutral and that because of the nature of class society in the United States (and for other reasons in the USSR, for example) they tend to subserve the interests of the rich and powerful and subvert the interests of the poor and weak. In connection with this, I have tried to show that the proclamation of "scientific neutrality" has often served as a smoke screen for special interests; that it is itself an ideology in its profession of purity from, and antipathy to, other ideologies.

4. I agree with Durbin that his own dismissal of Marxist criticism is, at

best, glib. Marxism is more than a programmatic political doctrine, as I have tried to show in the foregoing. And science and technology must be safeguarded against their ever-growing subjection to class and undemocratic institutional exploitation.

FOOTNOTES

1. Often the working scientist, say in chemical synthesizing, is unaware of the specific need that his work is designed to meet, i.e., whether it is sought for medical application or to expand the CBW arsenal.
2. Benjamin Farrington, *Science and Politics in the Ancient World,* Barnes and Noble, New York, 1965, p. 67.
3. *Precepts,* Chapter VI, edited and translated by W. H. S. Jones, London, Loeb Library, Heinemann, as quoted in Farrington, *ibid.,* p. 65.
4. Otherwise known as Cosmas of Alexandria, *Topographia Christiana* (A.D. 547). English trans. J. W. McCrindle, 1897.
5. Notably, Ernst Cassirer and Jacob Burckhardt.
6. As late as the second half of the fifteenth century, the Florentine Academy under the leadership of Ficino and Pico della Mirandola limited the scope of philosophy to a reconciliation of scholastic and Platonic thought. See Cassirer, *The Individual and the Cosmos in Renaissance Philosophy,* Harper and Row, New York, 1963, p. 2.
7. Farrington, "Democritus, Plato, and Epicurus," in R. W. Sellars, V. J. McGill, M. Farber, *Philosophy for the Future,* Macmillan, New York, 1949, p. 2.
8. For the best concise account of this development see Edgar Zilsel, "The Sociological Roots of Science," *The American Journal of Sociology,* Vol. XLVIII, No. 4, January 1942, 544–562.
9. It was this lack of instrumentation among the pre-Socratics that left two of its most remarkable achievements in a purely speculative condition, i.e., the Anaxagorean hypothesis on the evolution of animal species and Democritean atomic theory.
10. Herbert Marcuse, "Science and Phenomenology," *Boston Studies in the Philosophy of Science,* Humanities Press, New York, Vol. II, 1965, pp. 280, 283. It is important here to notice that Marxist theory of technology is not "monolithic," as Durbin suggests in his paper. Indeed, the Marxist-phenomenological view represented in Marcuse is quite at odds with the more orthodox (and positivistic-progressive) interpretations of many Marxists. Durbin's equation of Marxism with Marcuse's approach is both misleading and too easy, as has been the facile, and often outrageously truculent, dismissal of critical theorists, notably Marcuse and Habermas, in English language literature on technology and values.
11. For a short discussion of the process see my "Science, History and Human Values," Willis H. Truitt and T. W. G. Solomons, *Science, Technology, and Freedom,* Houghton Mifflin, Boston, 1974, pp. 4–12.
12. One can, of course, hold the distinction syllogistically. Any and all absurdities have been so defended. I consider Alasdair MacIntyre's delightful treatment of the matter definitive. See his "Hume on 'Is' and 'Ought'," *Philosophical Review,* 1959.
13. Marx Wartofsky has recommended the institutionalization of a discipline precisely designed to look into these kinds of questions. See his "Is Science Rational?" Truitt, Solomons, *loc. cit.,* pp. 202–210.

TECHNOLOGY AS IDEOLOGY

Kai Nielsen THE UNIVERSITY OF CALGARY

I

Throughout this conference on social responsibility and technology, attended by scientists and philosophers, I have been forcefully struck by a general lack of awareness of the political and ideological dimensions of the cluster of moral problems cast up by the effects of technology in our societies. No matter how we characterize technology, no matter what we say about the relationship between technology and science, the irreversibility of certain general features of our industrial societies and the subtle moral problems posed by that development, it is evident as can be that there are plain ills flowing from the use to which technology is put in advanced capitalist and bureaucratic socialist societies, and that the ending of these ills is primarily a political problem. That is to say, it is a question of the kind of socio-economic order we are going to have. The correction of the ills emerging from this utilization of technology is essentially a matter of there coming into being a more rational and a more humane social order with a radically changed socio-economic system.

Most of the problems attendant on our use of technology are not problems which are inherent in technology. Technology will go in one way under a social order in which profit and capital accumulation are the dynamic elements; technology will go in another way in a social order geared primarily to production to satisfy unmanipulated human needs.[1] To be sure, there will be the common problem of attention to the side effects of developing a given technology, but diverse effects of the utilization of technology will obtain where there are, rooted in very different conceptions of what is a good society, different rationales for when a technology is to be developed, modified or abandoned. There is, we should remind ourselves, no such thing as technology, but different, often closely interlocked technologies, developed and utilized for different purposes by different classes with different and typically conflicting interests. These remarks should be commonplaces. Only ideological mystification by and about technology obscures this from consciousness.

What I think is generally not seen is how this is hidden from us by the pervasiveness of scientism and the technocratic consciousness in our society. What I want to argue is that among its various functions, technology in our societies has come to have a pervasive and indeed a pernicious ideological function which, allied with an instrumentalist conception of rationality and a scientistic conception of knowledge, obscures from us—as all ideology does—the real and indeed manifestly political and economic nature of the problems posed by technology in our society.

In talking of scientism and the technocratic consciousness, I am employing these conceptions in much the same manner as the Frankfurt School and such later but related figures as Jürgen Habermas and Lucien Goldmann.[2] Scientism is the belief that science, and science alone, can give us a genuine knowledge; what cannot be established and sustained scientifically cannot be rationally believed. Science, scientism would have it, is our *sole* legitimate authority in fixing belief. As Habermas puts it himself in his "Why More Philosophy?": "By 'scientism' I mean science's conviction that science must no longer be regarded as one form of possible cognition, but that cognition must be identified with science."[3] It is, as he adds, "the attempt to find reasons for the cognitive monopoly of the sciences. . ."[4] The belief is, to put it simply, that if science cannot tell us something, we cannot know it. However, as is clear in Habermas's work, a critique of scientism is not and should not be a critique of science itself but only a critique of an inflated and indeed conceptually confused conception of the authority and province of science.[5]

Technocratic consciousness is consciousness dominated by scientistic assumptions. It identifies *rationality* with the use of the scientific method and what has been called "instrumental rationality" (the choosing of the most efficient means for attaining whatever ends or objectives one happens to have). It sees technology as applied science and takes the problems of human beings to be problems, inasmuch as they are genuine problems and not emotional harassments felt as problems, which are resolvable, if resolvable at all, by science and technology. We should not, such a consciousness would have it, be romantic about participatory democracy or control from below or paranoid over its absence; rather we should allow, as much as is politically and ideologically feasible, the scientific experts to settle social problems, for they alone, over such complex matters, really know what is going on. Where there would not be too much of a public outcry, these scientific experts should in the various domains of their expertise do the fundamental deciding. We do not need and indeed should not attempt to make real a Millian or Habermasian conception of an informed citizenry deliberating on and debating these issues and then democratically deciding what is to be done; rather, what we should have is a passive citizenry ratifying what has been worked out by experts who alone are in a position to understand what really is at issue.

Jürgen Habermas has argued that technology in cultures such as our own functions ideologically and indeed in a politically repressive manner. It inhibits an understanding of our actual situation and obscures from us the possibilities of altering that situation in a more just and human direction, such that we humans could achieve a greater satisfaction of needs and a fuller realization of our potentialities as human beings. The power and importance of Habermas's claims about technology and ideology notwithstanding, he does not state his case perspicuously. I shall in the next three sections attempt to give an interpretative reconstruction of it and then turn to a critical examination of its exact import for what is to be said and indeed undone about the political and ideological role of technology and the technocratic consciousness.

II

Habermas begins his "Technology and Science as 'Ideology'" with an examination of Marcuse's claim that "what Weber called 'rationalization' realizes not rationality as such but rather, in the name of rationality, a specific form of unacknowledged political domination."[6] (RS 82) "Rationalization" in this context refers to purposive-rational action to efficiently organize means to the achievement of ends or to develop decisive procedures for choosing between alternatives. Rationalization occurs in society when there is an extension into ever further areas of society of such purposive-rational decision procedures; we get, with such rationalization, replacement of traditional authority with a secularized rational authority rooted in science and technology. It is the extensive and effective use of what I have called instrumental rationality. Rationalization, modernization and secularization have always gone together. Habermas further believes that with the advent of state-regulated capitalism as a system-stabilizing device and with the "growing inter-dependence of research and technology which has turned the sciences into the leading productive force," orthodox Marxist analyses of capitalist society are not, as they stand, adequate. Rather, it is his belief that Marcuse's thesis that technology and science take on the function of legitimizing political power is a crucial element in understanding and possibly altering our socio-political situation (RS 100 −101)

In showing why Marcuse regards such rationalization—such a purely instrumental use of reason—as so strategically ideological, Habermas also expresses some conceptions which are distinctive of his own views about rationality and scientism. If we identify rationality with rationalization and instrumental rationality, rationality is then, as I have already noted, conceived of as something which is constituted by the making of efficient choices concerning means where the aims and ends are simply presupposed in a given situation and are not subject to a reflective and critical examination—are not something which can be either rational or irrational. The whole social framework of interests in which strategies are chosen, technologies applied, and systems established is simply removed by conceptual fiat from reflective examination and rational reconstruction because such matters are by definition, given that conception of rationality, beyond the scope of reason.

This is a very high price to pay for taming reason and there is, Habermas argues, a distinct inadequacy in such a conception of rationality. Where rationality extends "only to relations of possible technical control"—to the correct choices of means to achieve whatever purposes we may have—then rationality implies, it is claimed, "a domination of nature or society." Being rational becomes a matter of achieving or maintaining a certain scientific expertise; it becomes a matter of choosing the correct technologies. And this is a matter of scientific know-how and not a matter to be resolved, where several people are involved, by discursive public discussion and reflective assessment by participants treated as equals. So rational choice, in a world where there has been such a rationalization of the conditions of life, comes to making choices in accordance with the discoveries and rationales of scientific experts. In capitalist societies these choices are in turn, in a general way, made in accordance with the interests of the ruling capitalist class. In this way "'rationalization' of the conditions of life is synonymous with the institutionalization of a form of domination . . . ," i.e., the rule of the scientific experts serving a ruling class. (RS 82) The general type of ends or purposes human beings are to have and the general direction they are to take are just assumed or taken as an unquestioned given—a given which, when viewed reflectively, appears to be in certain important respects quite arbitrary. The historical fact is ignored that the utilization of a technology to efficiently achieve certain ends expresses a social decision concerning what the ruling interests—in our society the capitalist class—intend to do with men and things.

Given modernity and this conception of instrumental rationality, the very relations of production of a given society "present themselves as the technically necessary organizational form of a rationalized society." (RS 83) And, while no doubt on reflection people surely know better, in effect these relations are treated as if they were a fundamentally unchanging natural order of things.

Such a conception of rationality has built into it systems-relative criteria of rationality. And this, of course, limits very seriously its capacity to function as a standard to criticize the relations of production of a society. With such a conception of rationality, we cannot develop a "critical standard for the developmental level of the forces of production in relation to which the objectively superfluous, repressive character of historically obsolete relations of production can be exposed." (RS 83) Moreover, such a conception of rationality hobbles social theory very seriously, for operating under its limitations we can hardly develop even a foothold for a theory of social evolution. Yet, as Habermas rightly recognizes, such a theory of social evolution (development) must be the foundation of a critical-hermeneutical theory of society.[7] Rationality so conceived is concerned in its exclusive preoccupation with strategies and means, with making a given system work more efficiently and exposing inefficiencies in the workings of that system; but such a rationality cannot "function as the basis of a critique of the prevailing legitimations in the interests of political enlightenment." (RS 84) We have—now in a distinct Freudian sense of "rationalization"—with such a conception a perfect rationalization for the prevailing system. There are and can be no grounds for rationally criticizing the system as a whole, for such considerations are, for it or any other system, questions beyond the competence of reason. That is, they are beyond the very scope of reason. Only an irrational ideologist hung up with some "total ideology" would try to argue about the legitimacy of a social system as a whole. In this general sense, if such a conception of rationality is correct, there is no rational alternative to accepting the status quo.

The capitalist mode of production, as distinct from earlier modes of production, has a way of challenging traditional societies and providing a rationalization which accelerates modernization and provides capitalist societies with a new basis of legitimation. It works this way: capitalist economic mechanisms make permanent "the expansion of sub-systems of purposive rational action" and by identifying rationality per se with this instrumental rationality (technological rationality), they create a new rational authority replacing or partially replacing the by now vulnerable traditional authority. That is to say, this new rational authority provides a *new* legitimation—a new and less vulnerable ideology—for the current, and no doubt dominant, political system, i.e., bourgeois parliamentary democracy.

III

Habermas agrees with Marcuse that we have in such a situation *ideology disguised as rationality*. Moreover, using this new "scientific" standard of instrumental rationality as the measure—the standard—of rationality *sans phrase*, the power-legitimating and action-orienting traditions lose their cogency. This is particularly evident, Habermas claims, for mythological interpretations and religious world views. (RS 98) The fundamental rationality of the scientific world perspective is used to criticize the dogmatism of traditional interpretations of the world. Habermas regards this crucial instrument of modernization as perfectly legitimate in certain contexts; but he also argues powerfully, as Wittgenstein would, that when those canons of rationality are applied to domains other than scientific ones, one gets the new and artfully disguised ideology of scientism and not a genuinely emancipatory critique. But in its proper spheres such an employment of instrumental reason is indispensable in the attainment of human emancipation and liberation. However, as Habermas himself stresses about the later Wittgenstein, it now appears at least to be the case that in abandoning such a general scientific standard of rationality we have in effect historicized and relativized standards of rationality to the diverse and often incommensurable domains of discourse and forms of life. Habermas wants to avoid what he takes to be such a Wittgensteinian implicature, but it is not clear how or even whether he can.[8] (RS 98–99)

Such considerations aside, and to continue my account of Habermas, one should note that this scientific ideology becomes politically more important with the collapse in advanced capitalism of the laissez-faire ideology of liberal capitalism. This is so because of the pervasive and persistent fact—a fact which could hardly be ignored—of state intervention to secure or try to secure the system's stability. Habermas remarks:

> The permanent regulation of the economic process by means of state intervention arose as a defense mechanism against the dysfunctionalist tendencies which threaten the system, that capitalism generates when left to itself. Capitalism's actual development manifestly contradicted the capitalist idea of a bourgeois society, emancipated from domination in which power is neutralized. The root ideology of just exchange, which Marx unmasked in theory, collapsed in practice. (RS 101)

This ideological collapse meant that the capitalist order had to find a new ideological legitimation. What, following Weber, Habermas has called the rationalization and secularization of society, has rendered the "traditional form of legitimation on the basis of cosmological world-views . . . impossible" and the ideology of laissez-faire, if taken seriously, would make the system too unstable. (RS 102) In this way capitalism came to need a new ideology. Scientism perfectly fills a goodly portion of the bill, and indeed it does become a central element in the ideology of advanced capitalism.

Where the welfare state is well developed, dysfunctions of free exchange are compensated for and an ideology of achievement is wedded to at least a partial meritocracy linked with a suitably minimum level of welfare and secure employment for most of the citizens of that state. So here we have in this managed capitalism of advanced, and we hope late, capitalism a fine wedding of convenience between scientistic ideology and a welfare state ideology. The role of the political system in such a match is to maintain stabilizing conditions for an economy that guards against risks to growth and guarantees social security and the chance for individual upward mobility. "What is needed to this end," Habermas argues, is latitude "for manipulation by state interventions that, at the cost of limiting the institutions of private law, secure the private form of capital utilization and bind the masses' loyalty to this form." (RS 102)

This new welfare state legitimation, Habermas argues, requires help from its partner. It requires, that is, the underpinning of scientism. Government action under such circumstances is directed toward maintaining the system's stability and controlled growth. The thrust of government activity is toward the "elimination of dysfunctions and the avoidance of risks that threaten the system." For this stability to be achieved, "political problems" or "social issues" must come to be regarded as technical problems requiring for their resolution the technical expertise of scientists who will come to function as the new mandarins. (RS 101–103)

We have in such a situation a "scientization of politics"; the crucial thing—given such an ideology and viewed from its vantage point—is to solve technical problems about the society's functioning while avoiding serious discussion of the goals and basic commitments of the society; discussion of the actual rationale of the society and the adequacy of its conception of the good life is, under such an ideology, in effect treated as so much rhetorical guff fit for graduation exercises and the like.

Such a scientization of politics requires for its success "a depoliticization of the mass of the population." (RS 103–104) Political problems are now treated as technical ones. Any actual unmanipulated discussion in the public sphere of the ends of society is very likely to be dysfunctional; for an unbiased, informed discussion of such issues would very likely produce a considerable amount of dissatisfaction among the participants once their political consciousness was raised.

IV

How are the masses of people sold this depoliticization program? Habermas argues, following Marcuse, that this trick is brought off "by having technology and science *also* take on the role of an ideology." (RS 104) *In theory,* conceptions of human welfare get linked with economic growth and scientific rationalization—the so-called modernization—of society, while *in reality* "social interests still determine the direction, function and pace of technical progress." (RS 105) But in late or advanced capitalism, science and technology have become such a key tool in keeping the system intact and such an important productive force, that the fate of these social interests of the ruling class is integrally interlocked with it. This gives a seeming autonomy to science and technology as factors determining the health, i.e., the continued effective functioning, of the system. This makes it appear to be the case that the logic of scientific-technical progress determines the direction of development of the social system. *Viewed from a capitalist perspective,* no rational politics can fail to act in accordance with its technical directives designed to meet the functional needs of the system.

Here we have the full and unfortunately beguiling force of late capitalist ideology. It replaces, in our understanding of ourselves, a model of a reflective understanding with a scientific model of *instrumental reason* masquerading as the very essence of rationality. That is to say, the model of what Habermas calls communicative action (interaction) is replaced by a scientistic model. "Accordingly the culturally defined self-understanding of a social life-world is replaced by the self-reification of men under categories of purposive-rational action and adaptive behaviour." (RS 106) Ideally, where such an ideology would get a full and effective implementation, systems analysis would replace political dialogue aiming at a rational consensus. Habermas realizes, of course, that such a *Brave New World* "has not been realized anywhere even in its beginnings. But it serves as an ideology for the new politics, which is adapted to technical problems and brackets out practical questions." (RS 106–107) In reality, and answering to this fantasy, we have now in the advanced industrial societies less of the "manifest domination of the authoritarian state" and more "manipulation by state power utilizing technical-operational administration." (RS 107) We are increasingly administrated in such a way that our freedom is drastically undermined while it still remains the case that our societies do not show the key features of classical fascism. This is enough to make it a mistake to call our societies fascist societies, but an acknowledgment of this fortunate fact has unfortunately led theorists such as Leszek Kolakowski to be rather too complacent about freedom in bourgeois societies.[9] Having rightly seen the importance of bourgeois liberties and guarantees, Kolakowski, perhaps because of his quite different experiences in Poland, has missed the insidious and typically unobvious ways in which late capitalist societies are managed in the interests of a ruling class.

In the context of arguing that the ideology of scientism is deeply engrained, Habermas remarks that in reality, the "moral realization of a normative order is a function of communicative action oriented to shared cultural meaning and presupposing the internalization of values." (RS 107) But scientistic ideology undermines the effectiveness of such communicative interaction by increasingly supplanting it by conditioned behavior; indeed, the technocratic consciousness is so deeply ingrained in an increasingly large number of people that they have lost a sense of the difference between purposive-rational action on a scientific model and human interaction on the model of dialogue and discussion aimed at rational consensus. "The concealment of this difference" from the consciousness of so many human beings proves "the ideological power of the technocratic consciousness." (RS 107) In late capitalist and probably in all advanced industrial societies, "controlled scientific-technical progress itself" has become "the leading productive force" and the new "basis" for the acceptance by the masses of the social order as a legitimate social order. (RS 111)

Technocratic consciousness is an elusive ideology, for unlike previous ideologies "it does not have the opaque force of a delusion that only transfigures the implementation of interests." (RS 111) It is, however, all the more effective for all of that, for it is "more irresistible and further-reaching than ideologies of the old type." (RS 111) Both in the popular consciousness and for the technocrats themselves, scientism comes through as being "the scientific attitude" and in reality no ideology at all requiring a moralistic "justification." The recognition of the need for discussion and struggle in the public sphere so that people can attain a mature political consciousness and a genuinely communicative ethic is so muted that people in reality have no political education at all. And with this ideology's veiling of practical problems, it not "only justifies a particular class's interest in domination and represses another class's partial need for emancipation, but affects the human race's emancipatory interest as such." (RS 111)

It is Habermas's conviction that this new scientific ideology is less vulnerable to a critique of ideology than are classical ideologies. With it there is no "rationalized, wish-fulfilling fantasy"—no drummed-up conception of the good life and man's station and his duties. Indeed it is difficult, though still imperative, for us in our cultural milieu, even to recognize that this scientific ideology is an ideology, though manifestly not only an ideology. Moreover—and this is linked with our difficulty in seeing it for what it is—it does not in the manner of other ideologies project an image of the good life in accordance with which it can, as could the laissez-faire ideology of liberal capitalism, be criticized in a number of different dimensions. Yet this scientific ideology does something which has been the task of all ideologies, and which warrants calling it an ideology, namely, it works "to impede making the foundations of society the object of thought and reflection."[10] (RS 112) It distorts our understanding of ourselves and our institutions.

Positivism has played an important function in making this scientific ideology effective. Scientific (logico-empirical) knowledge, positivism claims, alone is genuine knowledge; it is not only that metaphysics is impossible but that there cannot even be any knowledge of good and evil or of human destiny. The very question form "What is the meaning of life?" is in reality not a vehicle for a genuine question. When it is asked by someone, its role in discourse is that of a pseudo-question expressive and evocative of attitudes but all the same a "question" to which, even in principle, no answer can be given. Even where, as on Russell's, Stevenson's and Hägerström's accounts, morality is taken seriously, it is still a noncognitive matter; there is, on their accounts, not even the possibility of attaining knowledge of foundational moral and political norms. That is to say, these norms are not statable in propositions which could be either true or false. Indeed, on their accounts, the very belief in such norms— the belief that there could be such norms—is a belief in an illusion. Their existence or reality is quite impossible. Morality, including political morality, is finally a matter of decision and commitment rather than of knowledge. I take it that this is what Habermas means when he says that on such positivist or neo-positivist accounts there is a "repression of 'ethics' as such as a category of life." (RS 112)

Habermas believes that these positivist conceptions, in one form or another, pervasively infiltrate the consciousness of people in advanced industrial societies, many of whom have never heard of positivism. Such a general positivistic view of the world deeply affects their self-images and thus their actions. "The," as he puts it, "reified models of the sciences migrate into the socio-cultural life-world and gain objective power over the latter's self-understanding." (RS 113) Where such an ideology is in effect, there is no awareness of a distinction between the practical and the technical, all forms of political and cultural argument are regarded as "mere ideology," and there is a massive depoliticization of people—the mass of the population in these societies just do not think in political terms.

This scientific ideology—utilized in one form or another both by advanced capitalist social systems and bureaucratic socialism, e.g., the USSR and the German Democratic Republic—seeks to reduce to our interest in technical control a distinct fundamental human interest which determines a distinctive form of knowledge, namely, our interest in mutual understanding and communication without domination or distortion in something like a public sphere which can affect our collective destinies. (RS 113) Marx sought, as Habermas himself seeks, to show how it is possible for people to bring under rational control structural changes in society and indeed of society. This is a central element in what it is to have a critical theory of society. However, under the sway of a scientistic ideology, the dominant twentieth-century cultural consciousness has reduced and deformed this critical conception into something quite different—something which is quite without its critical power. What, on such a deformation, counts as a critical science of society is a conception of the technical control of society on the model of a control of nature. This, he argues, is a scientistic distortion of what critical theory is.[11]

Building on some "predictions" of future control techniques of Herman Kahn's—personality alteration through drugs, genetic control of individuals and the like—Habermas projects a negative utopia in which we would have something like the realization of a "cybernetic dream of the instinct-like self-stabilization of societies" in which men would make their history with will (with planned technical control) but without consciousness (reflective self-understanding). (RS 111)

V

In our interest in emancipation and in the attainment of a genuinely critical theory of society that would both project and guide us toward the attainment of a rational society, how does Habermas propose that we react to this scientistic ideology? Presumably (a) by exposing scientism as an ideology in the way he, Marcuse, Horkheimer and Adorno have done and (b) by making it clear that reason, rationalization and rationality are in effect being *persuasively* defined if they are identified with instrumental rationality. (RS 118 – 119) Besides instrumental or technological rationality there is as well a critical and normatively committed conception of rationality.[12] This other conception of rationality is rationality in the sense of critical awareness. For it to be achievable in cultures such as ours there must exist a culture pattern in which there is "unrestricted discussion, free from domination, of the suitability and desirability of action-orienting principles and norms in the light of the socio-cultural repercussions of developing sub-systems of purposive-rational action. . . ." (RS 118 –19) There must be institutionalized a genuine political education issuing in the public sphere in undominated, unmanipulated human beings, self-consciously engaging in repoliticized decision-making processes. (RS 119) Rationalization in accordance with such a conception of critical rationality would not necessarily lead to a better functioning of the social system. It would instead, I should think, though Habermas does not actually say this, lead to the dissolution or transformation of certain social systems, e.g., both late capitalism and bureaucratic socialism. And in achieving this, "furnish the members of society with the opportunity for further emancipation and progressive individuation." (RS 119)

It is, however, certainly possible to be skeptical about this Habermasian conception of a fuller conception of rationality going beyond instrumental rationality. We can, on such a conception, assess not only the means for the attainment of ends but we can assess even our most fundamental ends themselves. But how is that done? What tests of truth or objectivity do we have here?

That to worry about this need not be just another expression of scientism and objectivism can be seen by the way such worries are pressed by such nonpositivists as Philippa Foot and Alasdair MacIntyre.[13] Moreover, to be rational persons on Habermas's account, we not only would have to be able to do our sums, recognize good evidence and unsubstantiated claims and be good at ascertaining the most effective means for attaining our ends, but we must also be emancipated human beings with an enlightened consciousness and with an understanding of what it is to control our own destinies. To be so enlightened we must be able to see through the idols of our tribe so that we are no longer entrapped in our historically and culturally given illusions and ideologies. A thoroughly rational person will have what Habermas calls a coherent total consciousness.

This is a very tall order, and again, skepticism about whether we have anything like an adequate understanding of what we are talking about here is not unreasonable. We do not, even within our culture, have any considerable consensus about who are and who are not the enlightened and emancipated human beings who see through the cultural traps of their time and place and have some more reasonable conception of what a more rational and human social order would look like, together with some plausible account of how it can be achieved. We do not even agree about exemplars here. Would Chomsky or Kissinger, or neither, come close to being one? Or is it Alan Watts or Thomas Merton? Different subcultures within our culture have rather different and conflicting conceptions of enlightenment and emancipation. Even if we limit our reference group to the educated elite we would get sharp disagreements about such people as Chomsky and Kissinger. We also tend to think of an ideology as something the other fellow has and we are, understandably enough, blind about the way our own outlooks are subject to ideological distortion. And we do not agree about what the illusions of our age are. Is it such an illusion to be caught up by religious commitments or beliefs? Is it to have some faith or at least hope for a socialist future or to believe in the essential soundness of our own bourgeois social order—our pluralist democracies? Is it to believe that we have a common humanity on which a humanly adequate morality could be raised?[14] Reflective and informed people deeply disagree over such matters, and it is not even evident that they agree on a method for achieving a consensus. It is, moreover, a naive bit of scientism to think we could apply something called the scientific method which would, if intelligently used, give us a determinate answer to these staggering questions.

This lack of *de facto* consensus about what is the case here and about how we would decide what is the case suggests that our key conceptions are essentially contested. This means that these conceptions, like what is "true art" or "true religion," are not conceptions concerning which there exist core criteria which would provide us a basis for agreement.

Utilizing his theory of communicative competence and his consensus theory of truth, Habermas would no doubt respond that under *ideal* conditions, where there would be no communicative distortion, people would achieve consensus and that under these conditions, disagreements concerning such conceptions would disappear or at least be drastically reduced. On one reading, these remarks are little more than truisms. Read in that quasi-tautological way or even in a more substantive way, they are not unreasonable. But what needs to be done is to show, particularly for such morally weighted notions as emancipation and enlightenment, that—where the "we" cuts across both cultures and classes—we have some rational basis for agreement about what is emancipation and enlightenment and what it is to suffer from false consciousness.[15] Where false consciousness is widespread, consensus about these conceptions would hardly be "a true consensus" or "a rational consensus." In cool moments, where we are aware of the causes of our beliefs and the consequences of acting on them, we still sometimes are at sea about emancipation and enlightenment. Habermas has done little with such strictly normative conceptions and until he or someone working in the same manner does, we should remain skeptical about their claims concerning critical rationality.

It would, however, in the present circumstances and for our arguments about scientistic ideology be a mistake to build too much on this skepticism. Indeed, we should be skeptical about that skepticism. Consider the concept of emancipation. While it is indeed a problematic concept which cries out for elucidation, is it really so problematic, so essentially contested, that there is nothing substantial which can be established about what emancipation is in virtue of which we could make substantively true nontruistic statements about emancipation which would give us a wider basis for rational agreement than we now have?

Let us ask what we are talking about when we say of a human being that he or she is emancipated. We try to show that this person has a good self-understanding and a good understanding of others. Such a person knows his (her) own motives and the motives of others and the effects of acting on them and can assess these motives. Such a person will be an *emotionally mature* and *autonomous* person with a firm sense of *self-identity* and *self-control*. Such persons will be impartial persons capable of *fairness* and *objectivity*. They will be informed and knowledgeable and they will *see through the ideology* of their own society and the *ideologies* of other societies with which they come in contact. They will have a good understanding of *human needs,* and in the light of that understanding they can understand and assess the importance of their own desires. They will have an incisive understanding of the evils in their world and an understanding of human aspirations for a better social order. They will, as well, have an understanding of what divides and what integrates human beings and of what *human solidarity* and *freedom* consist in, and they will have some understanding of the conditions for attaining them. They will, in cultures such as ours, have a *reasonable* understanding of how and where scientific procedures are relevant to the realization of human problems and they will utilize them where relevant. They will be *self-reflective* about their ends and choose what they choose after a *cogent* and *objective* examination of the alternatives and with (a) a good understanding of what their preferences are and the weight they attach to their various preferences, and (b) a knowledge of the causal conditions of these preferences and the probable consequences of the various alternative policies embodying those preferences should they be adopted. This, of course, implies a very thorough *self-understanding*. Finally, emancipated human beings will not take dogmatic stances and will criticize all forms of *dogmatism* while remaining sensitive to the hidden and typically unconscious dogmatism and ideological distortion in antidogmatic posturing. They will also be aware of the way that a focusing on the dangers of dogmatism and ideological commitment can serve irrationally to block commitment and *praxis*. They will be sensitive to all ideological influences and they will engage in dialectically ramified but still *rationally controlled* reflection on the *ends of life* and will, in response to this reflection, develop their powers—their varied and distinctive capacities as human beings.[16]

In spite of the fact that the words italicized in the above paragraph are all terms which are in many contexts of their use employed to express essentially contested concepts, there is enough content in the above characterization of what it is to be an emancipated person to give *some* tolerably unproblematic content to that notion, such that with an understanding of the boundary conditions, the empirical facts, and with a utilization of this characterization of emancipation, we can make a cluster of true or false statements about the comparative emancipation of a Chomsky, a Kissinger, or a Bellini. It is an exaggeration typical of liberal thinking, to claim "emancipated human being" functions in the same rationally irresolvable way as "true art" or "true religion."

However, even if I am wrong about that and the analogy is closer than I have allowed, this does not vindicate what I have characterized as the technocratic consciousness or show that technology does not have the hidden ideological role Habermas alleges. It would not show, for example, that there were not genuine reflective and critical interests in emancipation distinct from, and not reducible to, our scientific interests and interests in technical control. It would not show that scientific mandarins were uniquely competent, by virtue of their scientific knowledge and authority, to solve the problems of life for us. It would show rather that the resolutions of such problems were more context-dependent and more relativized than I, and many others, believe to be the case. Indeed, certain rationalistic—and it is hoped rational—expectations of ours would be dashed but this situation is still far from the situation in which "anything goes," for the context still has in a certain way an objectivity of its own, and in such a situation we can make certain judgments of better or worse.

If I am near to the mark in this qualified and partial defense of Habermas's account of technology as ideology, then a number of important considerations receive corroboration. It is a mistake to argue, as Galbraith and others have, that the technological order is autonomous and self-augmenting. It is rather in some considerable measure a creature of ruling-class interests and its autonomy is an ideological mystification.[17] We should not see technology as something neutral but as, in William Leiss's words, an "organized human activity for practical purposes" which results in a "specific allocation of the world's material resources" with "*specific* possibilities now being planned for the immediate future."[18] We should come to see whose interests these utilizations of technology reflect, and it is by understanding these interests that we can assess the moral cogency of different uses of technology. This is not a scientific task but a social and political task requiring informed moral reflection for its rational resolution. We must not look for a technical or scientific resolution of these issues but a political one which, we hope, will also have a nonideological, i.e., rational, moral grounding. Looking at our societies and the possibilities for technical innovation open to us, both with the present socio-economic order and with historically possible alternative ones, we should be asking whether, using the technology rationally achievable within the frameworks of these various social orders, our needs will find their most nearly adequate satisfaction.

It is in the context of thinking about alternative social systems and their comparative viability that we should assess the development of technologies. It is not a question of a Luddite or romantic rejection or devaluation of science or technology.[19] Both are key human activities essential for a viable life in twentieth-century industrial societies. But they are not the measure of all knowledge or human rationality and they should not determine the value of our common life as social beings.[20] Technology by itself cannot be our savior and it need not be our destroyer or our master. Whether, as a useful instrument of the ruling class, it does either of these last two things depends on the outcome of the various political struggles for a transformed socio-economic order which will engage us in the next several decades, whether we like it or not.

FOOTNOTES

1. This is skillfully shown by William Leiss, "The Social Consequences of Technological Progress: Critical Comments on Recent Theories," *Journal of Canadian Public Administration*, Vol. XIII (1970), pp. 246–262; and Bob Eccleshall, "Technology and Liberation," *Radical Philosophy Eleven* (Summer, 1975), pp. 9–14.

2. Jürgen Habermas, *Knowledge and Human Interests* (Boston: Beacon Press, 1971), translated by Jeremy J. Shapiro. Lucien Goldmann, *The Human Sciences and Philosophy* (London: Jonathan Cape, 1969), translated by Hayden V. White.

3. Jürgen Habermas, "Why More Philosophy?," *Social Research*, Vol. 38, No. 4 (Winter, 1971), p. 650.

4. *Ibid*.

5. This is very clearly expressed in Sara Ruddick's critical notice of *Knowledge and Human Interests*, in *The Canadian Journal of Philosophy*, Vol. II, No. 4 (June, 1973), pp. 545–569.

6. Habermas is principally concerned to examine here Marcuse's essay, "Industrialization and Capitalism in the Work of Max Weber." An English translation occurs in Marcuse's *Negations: Essays in Critical Theory* (London: Allen Lane, Penguin Press, 1968). Habermas's remarks occur in the sixth essay of his *Toward a Rational Society* (Boston: Beacon Press, 1968), translated by Jeremy J. Shapiro. Page references are given in the text under the code item RS. On this general topic see as well William Leiss, *The Domination of Nature* (Boston: Beacon Press, 1974), pp. 199–212; and Hans Peter Dreitzel, "Social Science and

the Problem of Rationality," in *The Politics and Society Reader*, ed. by Ira Kratznelson, *et al.* (New York: David McKay, 1970), pp. 360–377. These two essays are crucial for a rounded understanding of the issues pursued by Marcuse and Habermas in this section.

7. See the preface to his *Legitimation Crisis*, translated by Thomas McCarthy (Boston: Beacon Press, 1975). See also his *Theory and Practice*, p. 14.

8. See as well his discussion of Wittgenstein in his *Philosophischepolitische Profile* (Frankfurt am Main, Germany: Suhrkamp Verlag, 1971), pp. 141–146. He makes it very plain there that he believes that if science and critical philosophical reflection cannot "go beyond the dimension of actually living language-games in which they are rooted," critical reason is in reality undermined. (See p. 146 in that text.) For some of the general issues involved here, see Jürgen Habermas, *Zur Logik der Sozialwissenschaften* (Frankfurt am Main, Germany: Suhrkamp Verlag, 1970).

9. This is evident in his rather pitiful response to E. P. Thompson. Miliband and Saville refer to Kolakowski's response as "in some ways a tragic document." I think pitiful is the more accurate characterization. Leszek Kolakowski, "My Correct Views on Everything," *The Socialist Register*, 1974, pp. 1–20.

10. Habermas makes us see how seductively and perniciously systems theory does this in his discussions with Luhmann. Jürgen Habermas and Niklas Luhmann, *Theorie der Gesellschaft oder Sozialwissenschaften* (Frankfurt am Main: Suhrkamp Verlag, 1971).

11. Paul Lorenzen shows how, between positivism and Habermas, there is a dispute about what is to count as critical theory. He also provides some reasonable grounds for siding with Habermas on this issue. Paul Lorenzen, "Enlightenment and Reason," *Continuum*, Vol. 8, No. 1 (Spring-Summer 1970), pp. 3–11.

12. This conception of rationality is articulated most explicitly in *Theory and Practice*, Chapter 7, and *Toward a Rational Society*, pp. 118–119.

13. Philippa Foot, "Morality and Art," *Proceedings of the British Academy*, Vol. LVI (1970), pp. 1–16; and Alasdair MacIntyre, "The Essential Contestability of Some Social Concepts," *Ethics*, Vol. 84, No. 1 (October 1973), pp. 1–9.

14. Alasdair MacIntyre and E. P. Thompson argue powerfully for this last conception. See Alasdair MacIntyre, "Notes from the Moral Wilderness—I," *New Reasoner 7* (Winter 1958–59), pp. 90–100; "Notes from the Moral Wilderness—II," *New Reasoner 8* (Spring 1959), pp. 89–98; and "Breaking the Chains of Reason," in *Out of Apathy*, ed. by E. P. Thompson (London: 1960); and E. P. Thompson, "An Open Letter to Leszek Kolakowski," *The Socialist Register* (1973), pp. 1–100.

15. See the essays by Paul Lorenzen, "Enlightenment and Reason," Kurt Jürgen Huch, "Interest in Emancipation," and Jürgen Habermas, "Summation and Response," all in *Continuum*, Vol. 8 (Spring-Summer 1970). See also Albrecht Wellmer, "Communication and Emancipation: Reflections on the 'linguistic turn' in Critical Theory," *Stony Brook Studies in Philosophy*, Vol. 1 (1974), pp. 74–120.

16. Paradigmatic examples of this are the essays by MacIntyre and Thompson cited in footnote 14.

17. J. K. Galbraith is a captive and an effective purveyor of this mystification. J. K. Galbraith, *The New Industrial State* (Boston: Houghton Mifflin, 1967), Chapter 2. See William Leiss's criticism of Galbraith in his "The Social Consequences of Technological Progress," *Journal of Canadian Public Administration*, Vol. XIII (1970), pp. 249–250.

18. *Ibid.*, p. 255.

19. This is clearly argued for by T. B. Bottomore, *Sociology as Social Criticism* (New York: Pantheon Books, 1974).

20. Alasdair MacIntyre, "Notes from the Moral Wilderness—I," *New Reasoner 7* (Winter 1958–59), pp. 90–100; and "Notes from the Moral Wilderness—II," *New Reasoner 8* (Spring 1959), pp. 89–98.

HUMANIZATION OF TECHNOLOGY: SLOGAN OR ETHICAL IMPERATIVE?

Edmund Byrne INDIANA UNIVERSITY–PURDUE
UNIVERSITY AT INDIANAPOLIS

How are we to humanize technology? Technology obviously can affect human beings negatively, so few would deny that there might be some need to "humanize" it. But there is little agreement about what such an endeavor might entail beyond paying some attention to "human values" along with the mainstream concerns of technology. To be determined, then, is how much attention, and to which values. How do we decide, like the Little Prince, which ones are most deserving of our attention? The easy answer, long heard on Madison Avenue, is simply to equate value with interest—and then, if necessary, create the interest. But a value founded only on interest is as flighty as a day at the stock exchange. So, in a world of finite resources, it is crucial that we think beyond interests to needs. But in a society such as ours where commerce is king, a determination of needs is not a very high priority; and, as a result, there is in the land a shortage of experts on how to decide what a human being really needs in order to be fully human.[1] Yet this we must learn if we are ever going to be serious about humanizing technology. For, a world full of people in need will no longer endure the privilege of interest artificially created and sustained. Thus, the basic question to which humanization of technology points is this: What, if anything, do human beings really need that technology might somehow deprive them of?

The answer to this question on the part of most proponents of technology, and even of some proponents of its humanization, tends to be: Nothing that more technology can't eventually restore a hundredfold! So it is not surprising that, to date, the humanization of technology movement has done more for certain corporate images (skillfully advertised) than it has for people. Yet in the very process of co-opting the concept of humanizing technology, the corporate interests in question have inadvertently called attention to at least one affirmative answer to our basic question: Technology can and does deprive human beings of personal and professional responsibility, and it is poorly equipped to restore it.

It is, of course, by no means obvious that any such charge is justified or, if justified, is rightly laid at the door of technology. For, to begin with, not everyone's ox is ordinarily gored at the same time or in the same manner. So as a result, what Garrett Hardin has called "the tragedy of the commons" may strike not collectively but only selectively.[2] And accordingly, so long as there are prophets available to say that this may reasonably be hoped, the old Malthusian zero-sum game will undoubtedly continue to be played by players who seek to prove their fitness by surviving. In other words, so long as man's ego feeds on symbols of superiority over other men, it seems unlikely that "To the victor belong the spoils" will be replaced in human consciousness by stern warnings to the effect that pride may precede a fall. For, even the gentleman who had just completed half the distance of his fall from a tall building was reportedly then of the opinion that everything was just fine—so far!

What follows from these seeming jeremiads is not doomsday minus a lifetime or two, but rather a suggestion that technology, like fate, does not treat everyone alike. And accordingly, no individual group, however perspicacious in its own eyes, is in a position to declare that technology has been humanized, even, I dare say, in its own regard.[3] So, as a start, one might honestly recognize that some people's lives are negatively affected by machines, and that for just that reason they may not appreciate the subtleties of scholarly rationalizations to the contrary.[4] But in order to come around to any such uncommon insight, one must further learn that there is more to "assessing" technology than merely testing a machine or attempting to estimate how long it will take people to accept what the machine will produce. The "more" that is here at issue, furthermore, is not encompassed by any merely quantitative systems approach, however subtle and sophisticated that may be. Nothing less than responsible concern for all reasonably possible consequences will suffice, whether such is presently deemed feasible or not.[5] In either case, this is an ideal toward which assessment, or evaluation, of technology must consciously, and conscientiously, move.[6] This being done, a systematic appreciation of human needs, including a need for responsibility, would be incorporated into one's analysis and, indeed, would be made in some fashion controlling. To show that this is eminently reasonable requires an adequate account both of (1) the teleology of man-machine relations and of (2) the ethical considerations implicit therein. With regard to the first, it must be recognized that man-machine teleology is not confined to the level of machine design and operation but necessarily includes that of organizational policy and planning. With regard to the second, it must be shown that ethical considerations are engendered on both levels, especially the latter. What follows is, essentially, an attempt to support these claims.

I. BEYOND "HUMAN FACTORS"

That man-machine teleology includes the level of organizational policy and planning would seem obvious from the viewpoint of management science and of the economics of technology.[7] But to the engineer concerned with system design, human goals and purposes tend to be relevant only insofar as human factors have an effect, especially a negative one, on a system's efficiency. Thus, the propaedeutic task of opening the way to an ethics of technology involves showing that considerations of design are subservient to and dependent upon those of management goals. Much of this work has in effect already been undertaken by such scholars as Lewis Mumford, Jacques Ellul, Victor Ferkiss, Nobert Wiener, and many others.[8] Here it will suffice to restate the basic thrust of their conclusions schematically by way of a kind of phenomenological comparison between the concepts of *cyborg* and *prosthesis*.

As will be seen, these concepts are models of the man-machine relationship in the sense that each is "an ordered set of assumptions about a complex system."[9] But included in the assumptions of each model are conflicting interpretations as to whether the requirements of machine or those of man are controlling. Given this conflict of interpretation, it will be argued that neither model can be adequately tested without reference to management goals.

Consider, as a point of departure, what an assembly-line worker is reported to have said about his job on the line:

> I don't like to work on the line. . . . You can't beat the machine. Sure, maybe I can keep it up for an hour, but it's rugged doing it eight hours a day, every day in the week, all year long. It's easy for the time-study fellow to come down there with a stop watch and figure out just how much you can do in a minute and fifty-two seconds. . . . But they can't clock how a man feels. . . . I like a job where you feel like you're accomplishing something and doing it right. When everything's laid out for you and the parts are all alike, there's not much you feel you accomplish. The big thing is that steady push of the conveyor—a gigantic machine which I can't control.[10]

The "time-study fellow" here referred to has since learned to talk about employee responsibility in the form of "job enrichment" and the like. But for the most part he, or what is now in fact a variety of specialists on man-machine relations, attempts with his particular set of tools to evaluate how effectively any given man-machine system is doing or would do whatever job is assigned to it. This evaluation might take the classic form of comparing present, ongoing man-machine performance against some desired standard, productivity or whatever, as is done by means of the various time-study techniques which may all be subsumed under the title of Taylorism.[11] Other, more recently developed approaches study animals for clues to making better machines (bionics), study machines for clues to making better men (cybernetics), or study men's use of machines in order better to coordinate the work of man and machine together (ergonomics).[12] Each of these and other related approaches takes it for granted that machines will play an increasingly important role in human endeavors, and so strive to render the relationship between man and machine as palatable and especially as productive as possible. They also have in common, by way of corollary, a kind of professional indifference to such broad normative questions as whether, to what extent, and under what circumstances it is desirable for human beings to be conjoined with machines. Thus, to the extent that man-machine specialists even bother to speculate beyond their carefully circumscribed programs of research, they tend simply to assume that in the future men and machines will and should be getting together even more than they already are today.[13]

This widely held version of the doctrine that more is better thrives best in minds that see no need to seek alternatives and hence would not appreciate the Socratic dictum that the unexamined life is not worth living. But there are alternatives—indeed, to use Robert Theobald's expression, alternative futures—which are in fact latent in the alternative ways in which futurists tend to speak about an evolving merger between man and machine. This postulated merger, or symbiosis, can look very different depending on whether the human being is thought of as a component of a machine system or the machine as a component of a human system. Any such model that gives the machine priority I call a cyborg, and one that gives the human being priority I call a prosthesis.

Each of these concepts has a fairly well-established meaning within one and the same circumscribed field of discourse, namely, that of bioengineering. But the concept of cyborg encapsulates the perspective of the engineer; that of prosthesis, the perspective of medical specialists. Each may be a synonym of the other with regard to the *explicandum,* but not with regard to the explication. And in this difference there is, so to speak, all the difference in the world.

The notion of a cyborg, in the first instance, refers to a system of human and machine components combined and coordinated in such a way as to utilize the capabilities of each toward the accomplishment of what neither can do alone. The word "cyborg" itself was coined by combining the two words, "cybernetic" and "organism," and abbreviating them to the two syllables, cyb and org.[14] The word "cybernetic" comes from a Greek word meaning to control or govern; and, as used today, it refers to a science that combines engineering and neurology to study ways of automatically controlling or regulating process both in machines and in human beings. A cyborg, then, is a kind of inevitable side effect of the cybernetic endeavor in that it involves a system consisting of both human and machine components, all of which are coordinated and controlled toward the accomplishment of a pre-set task. In this sense, a semi-automated assembly line might be thought of as a cyborg. A more obvious example would be that of an astronaut ensconced in his space suit and meticulously plugged into the artificial environment of his space capsule, all of which is elaborately regulated and controlled on earth by an even more complex system of interrelated men and machines.[15] In the broadest sense, then, any system of machinery the design and proper functioning of which calls for some continuing human input (what engineers call "man-in-the-loop") may be thought of as a cyborg.

The concept of a prosthesis, in the first instance, refers now as it has for centuries to a device or instrument whereby some function of the human organism that has been impaired is at least partially restored. In this sense, a cane; a pair of eyeglasses; a brace, support or corset; a hearing aid are all examples of prostheses or, as they are also called, prosthetic devices. Also appropriately included under the heading of prostheses are such sophisticated artificial devices as joints, arteries, pacemakers, hands, feet, arms, legs and so on.[16] By way of extension, one may also speak of any medical apparatus whatever as being a prosthesis in relation to man, however momentarily it is actually in contact with, or a component of, the human biological system. In this sense, the heart-lung machine, the renal dialysis machine, the encephalograph, the cardiograph, and countless others may all be thought of as prosthetic devices, inasmuch as they all have as their common function to contribute in some way to the well-being of the human organism. Still more broadly, one may think of an entire hospital complex, with all of its staff and facilities, or, for that matter, of the entire health care delivery system of a given nation or even of the world as a whole, as being prosthetic. In the widest sense of all, one may say that any machine or set of machines, whether medically related or not, is prosthetic if and to the extent that it supplies for some inadequacy of one or more human organisms.

As may be noted from this brief explication of terms, "cyborg" and "prosthesis" represent markedly different models of the man-machine relationship. "Cyborg" stresses man's inferiority to the machine; "prosthesis," the subservience of the machine to man. "Cyborg" orients a relatively undifferentiated human being or group of human beings to the otherwise unattainable requirements of a machine system; "prosthesis" directs the capabilities of certain machines to the limitations or handicap of an otherwise independently valuable human being. Thus, at least to the extent that language is revelatory of reality, the cyborg model tends to give priority to the needs of a machine; the prosthesis model, to the needs of a human being. And thus, the former may be said to "dehumanize" human beings, the latter to "humanize" machines.

In summary, then, in the case of a prosthetic relationship the machine compensates for a deficiency in the human organism, whereas in the case of a cyborg relationship the human organism compensates for a deficiency in the machine. As a sign of the former, if the organism could function well on its own, there would be no need for the prosthesis. As a sign of the latter, if the machine could function well on its own, there would be no need for the human component.

Implicit in these remarks is a suggestion that we are dealing with two incompatible ways of thinking of the goal-directedness of a man-machine system. But such is not necessarily the case. When viewed as a prosthesis, the end or goal of a man-machine system is presumed to be given (call it human well-being) and the means are determined accordingly. When viewed as a cyborg, means are also selected with a view to an end; but the end itself is comparatively arbitrary, and, though determinate once selected, remains at least in principle subject to change. Thus, as a direct consequence of the higher degree of freedom attributed to a cyborg, it is possible to view the cyborg as itself having a prosthetic goal.[17] It may therefore be contended that, though the prosthesis is by definition directed toward human well-being, the cyborg may be so directed by choice.

It is, then, precisely at this point that mere phenomenology must give way to the reality principle. For, as more than one advertising agency has discovered for its corporate client, there is no point in suggesting that one's favorite technology is arguably dehumanizing when it is possible to portray it instead as being a notably humanizing factor in society. This may be, and indeed has been, done in various ways, such as by focusing attention either on some great benefit allegedly to be derived from a suspect technology, or on the absence of notable harm therefrom, or, still more irrelevantly, on some wholesome, altruistic human beings who just happen to be associated with that technology. What is left unattended by such diversionary public relations is, of course, the whole range of basic questions that might arise out of a thorough cost-benefit analysis, especially where such analysis is carefully honed to determine whether those who bear most of the costs are even approximately the same as those who reap most of the benefits.[18] But any such course of investigation might well lead, on occasion, to the embarrassing discovery that the Emperor's new clothes are only original equipment that comes with the model. And, what could be even more embarrassing, the Emperor in question might even be one who has written over the entrance to his palace: "Let no one ignorant of prosthetics enter here."

The point here, as Aristotle saw after only thirty years under Plato, is that there are more things in the world than definitions dream of. In particular, not even an enterprise such as the health care industry is necessarily directed to human well-being merely because it describes itself as being dedicated to what I have associated with prosthesis. For, as numerous analysts of health care delivery in the United States have concluded, in effect, there might be far more similarities between a hospital patient and our unhappy assembly-line worker than any neat distinction between teleological models would allow.[19] Nor will the discrepancies in question be eliminated by simply improving the systematization of the system.[20] For, what is essentially being contested in the ongoing debate over health care in America is whether patients or professionals are intended to be the principal beneficiaries of the system.[21]

In other words, the weakness of some human beings may be exploited to increase the power of others. And in the process of selecting means to that end, be it called human well-being or whatever, those to be exploited will be evaluated as would any other proposed tool on the basis of efficiency, accessibility, durability, cost, and other related factors.[22] Thus, for example, one who wanted to build a pyramid in a labor-intensive economy may well have estimated a need for w soldiers, x whips, y slaves, and z blocks of stone. This comparatively primitive approach to industrial planning, involving what Lewis Mumford calls a megamachine, has given way of late to considerably more mechanized approaches.[23] And in the new and essentially different context of automation, the powerless human being is finding more and more frequently that he does not even qualify, in Kantian terms, to be used as a means to an end, to say nothing of being the end toward which the means chosen are directed.

As the foregoing is intended to illustrate, then, one and the same man-machine system may be evaluated very differently depending upon how the teleology thereof is construed. On a broader scale, the same can be said about any technology, since a technology, at least when viewed as a system, inescapably involves man-machine relations. What tips the normative scale in either direction is the dimensions of one's teleology. It may be assumed that any operator of a machine can ordinarily do more with than without the machine, whether the latter be viewed as correcting a deficiency or augmenting a capability. What really matters, however, is not the internal ends designed into the man-machine system but the externally intended ends toward the attainment of which such a system is meant to contribute.[24] It is these external ends, in turn, which are usually appealed to to determine the value of the component system.

The importance of external ends in evaluating technology is perhaps most clearly exemplified by the quasi-sectarian rhetoric of military budgeting, whose proponents have given more weight to such concepts as "balanced forces," "counterforce" and "deterrence" than to the hard realities of performance capability.[25] But analogous and, to some extent, overlapping considerations are common to the marketplace, where competitive strategy not infrequently tends to determine product selection and design.[26] From the viewpoint of the potential consumer, in turn, a given technology is seldom evaluated (except in the case of "impulse buying") just on the basis of technical performance. Rather, a given technology is evaluated in terms of its overall impact upon the techno-social system into which it would have to be inserted. Such evaluation could be, and in some instances has been, disastrously short-sighted, especially with regard to a technology which is unquestionably effective in and of itself though not necessarily in relation to any particular use to which it might be put. But to the extent that no consideration peripheral to a system's maintenance (e.g., novelty, prestige, pressure from competitors) is allowed to be controlling, people generally get the kind of technology they want, especially if the people in question happen to be in positions of power.

To this extent, at least, people do tend to evaluate a technology not just in terms of what it can do, but in terms of what it can, or is likely to, do *for them;* thus, they favor the technology that complements or, better, augments their interests; they do not favor the technology that threatens their interests. This being the case, they can usually be counted on to manifest various sorts of discretion and selectivity. At the extremes, they may either reject out of hand or openly welcome a new technology for no better reason than that it is new. But choice of technology is not usually so simple and straightforward, except in perhaps metaphysical and hortatory statements for or against technology in general. What is far more often the case is that a proposed technology is viewed somewhat ambivalently, by virtue of such considerations as the following: on balance, its advantages do not clearly outweigh its disadvantages; projected costs threaten to exceed benefits; alternative technologies are available and none can be shown to be clearly preferable to any other; undesirable consequences of utilization can be anticipated and are seemingly inevitable. This ambivalence toward technology may be said to constitute a fundamental characteristic of our times, especially where undesirable consequences are discerned. Indeed, this ambivalence may even be identified as the one central moral question of our technological age: to approve or not to approve, to have or not to have, to use or not to use an available or attainable technology.

What a question such as this asks us to determine is, essentially, how to decide when, where, and under what conditions a technological development may be considered an asset rather than a liability.[27] Many subsidiary questions are, of course, involved in this one basic, though complex, question. But the most important of these, I think, is this: Is the present or proposed technological development aimed primarily at satisfying (1) special (e.g., vested economic or political) interests, or (2) clearly demonstrable human needs? It has, of course, been customary for centuries in marketing circles to foster "needs" that correspond with products that one is prepared to manufacture and distribute. But this mercantile approach to a definition of need proves not to be persuasive when technology on which many people have become dependent shows unmistakable signs of being obsolescent. This broad-ranging problematique involves many different kinds of technology, if not all sooner or later, but the general nature of the problem seems especially well illustrated in our times by the example of the American petroleum industry's response to the "energy crisis."[28]

As described by industry advertising in the United States, the essence of the crisis is really very simple: America needs more petroleum, whatever the cost; so the industry is responding by an all-out effort to produce it. What this means in terms of technology, of course, is increased emphasis on equipment to get more oil out of older fields, to find and exploit new fields which were heretofore comparatively inaccessible in terms of technology, cost or politics, and to expedite importation of foreign oil.[29] As a result of this assessment of the problem, we find the once stalled Alaskan pipeline now well underway, many heretofore taboo government-owned lands now available for exploration and exploitation, and a marked increase in support for both offshore exploration and construction of coastal superports for tankers. Complementing these developments in the political arena is a campaign to offset increased costs by way of deregulation of natural gas, increase in the price of gasoline to the consumer, modification of emission standards, and sufficiently competitive improvements in automobile fuel economy (why only now, and so easy to do?) to challenge fuel-economizing foreign cars and to forestall introduction of alternatives to the gasoline-consuming motor vehicle. Moreover, this campaign almost succeeded in persuading Congress to approve increased federal taxes on gasoline to fund research and development of alternative energy technology. This latter would undoubtedly continue to be concentrated in the areas of coal and nuclear power, since these have come to be increasingly under the control of the same oil industry that has in the past controlled most aspects of our energy policy, including that of research and development.[30]

These reactions to OPEC's success at establishing a pricing cartel are, of course, well understood by the OPEC leaders themselves. Indeed, it is even more likely that their currently posted price for a barrel of oil is determined by the known current cost of producing a barrel alternatively from shale. It may therefore be assumed that, when the cost of such an alternative approximates that of OPEC petroleum, the latter's price will be adjusted accordingly. In the meantime, analysts are attempting to assess the staying power of the OPEC cartel in light of such divergent interests as those of an OPEC country like Venezuela, whose oil reserves are near exhaustion, and one like Saudi Arabia, which still has abundant reserves.[31]

In the meantime, what remains peripheral to these high-level calculated risks, of course, is any marked interest, at least for the time being, in a serious effort at developing energy technologies that are in no way dependent upon, or even under the control of, the petroleum industry.[32] Individual and even many corporate interests would seem at first glance to have everything to gain from such research and development, which have already been effectively stalled for at least several decades too long. But, thanks in large measure to the efforts of the petroleum industry itself, dependency on petroleum and petroleum-fueled technology has come to pervade almost the entire network of America's economic life and lifestyle, from the wrapper for my lunch to the suburb from which I carry it to work in my gasoline-powered automobile on a concrete-and-steel expressway.

Lest this point still not be obvious, consider how extensive has been this American dependence on petroleum. Natural gas has become the principal source of space, especially residential heating. Derivatives of oil (petrochemicals) are used in the manufacture of numerous synthetic materials, such as nylon, which, for all its advantages, requires high-energy consumption for its production and is a nonbiodegradable substitute for such biodegradable natural products as cotton, silk and wool.[33] DDT and other chemically related biocides, which are by-products of American petroleum research, have created as many problems as they ever solved. It is well known how Rachel Carson's *The Silent Spring,* and then in time both state and federal government agencies, came to recognize the harmful consequences of using these products in our own country.[34] Now similar concerns are being articulated, with increasing insistence, in the Third World as well, especially among younger scientists in the developing nations, who recognize, as their leaders have not, that education and socio-economic improvement are far more reliable approaches to the elimination of malaria, as has been proven in practice in Israel, Malaysia, Taiwan and California.[35] Then, too, America's enormous military consumption of petroleum, not only as fuel but in the form of napalm, herbicides and other destructive agents, supports the claim that our government has even come to think of petroleum as an instrument of international problem-solving.[36]

Most oil consumed by Americans, however, is still in the form of gasoline; and nine-tenths of all gasoline consumed in the United States is consumed as fuel in motor vehicles, where it produces not only horsepower but also noxious emissions which still defy even dedicated efforts at neutralization. In addition to the increasingly serious health problems which follow from concentrating both people and oversized automobiles in urban areas whose modernity is tainted by the absence of adequate mass transportation, the petroleum complex is responsible for excessive misallocation of resources. Motor vehicles (over a hundred million of them already in 1968) require one-fifth of all the steel and two-thirds of all the rubber consumed in the United States, along with large amounts of such other materials as glass, chrome, mercury, copper and sulfur.[37]

Thus has the American petroleum industry locked us into excessive dependence on the production, refinement, distribution, and consumption of a substance which is both nonrenewable and increasingly limited in supply. Versatile investors, no doubt, will "get out" of petroleum in time for the next energy era. But in the meantime this disproportionate concentration shall have been the direct or indirect cause of countless deleterious effects, especially in the automotive industry and its many satellites. The United States, accordingly, cannot continue to develop solely or even primarily on the basis of a petroleum-fueled technology. So it will either change over to a different energy technology or it will begin to break down. In either case, those whose livelihoods are directly or indirectly dependent upon the petroleum industry—and that includes most Americans, to some degree—may well experience professional and personal breakdowns of their own.

In a word, human beings will be affected negatively by virtue of their particular relationship to a complex technology. Yet there is nothing in all of this that human factors analysis, or anything like it, would even be prepared to consider, to say nothing of doing anything about. Thus are we driven by the very seriousness of the present crisis—which, of course, extends far beyond the borders of the United States—to begin to recognize philosophically as well as institutionally that the teleology of man-machine relations is preeminently a question of managerial responsibility.

II. TOWARD HUMAN RESPONSIBILITY

The foregoing considerations notwithstanding, it hardly needs mentioning that, except for the symbolic ritual of electoral politics, the concept of managerial responsibility has traditionally been given a very narrow, if not altogether meaningless, interpretation. For, in spite of numerous laws, in this country and abroad, that are aimed at regulating business and industry, a corporate executive who is not prone to fraud or embezzlement is rarely held personally responsible for anything except satisfactory aggrandizement of corporate profits.[38] Such, at least, would seem to be a legitimate interpretation of John Kenneth Galbraith's account of the corporate "technostructure,"[39] which tends to prevail over such alternative accounts as Ferdinand Lundberg's finger-pointing at "the rich and the super-rich."[40] Nor would one yet be likely to find very wide support for the proposal now being explored to introduce ethical considerations into investment decisions.[41] Maintaining the kind of respect for white collars that was manifest throughout the post-trial Watergate sentencing, our decision-makers prefer to find scapegoats on the man-machine level of the laborer. In short, our ethics of technology tends to be confined to the operational, or first-order, teleology of short-run goals, and seldom encompasses the managerial, or second-order, teleology of long-run goals.[42]

So accustomed have we become to this pre-system approach to allocating responsibility that we seldom even notice that it is inadequate and, arguably, unjust except when, as in the case of America's energy problem, we are abruptly confronted with a "glitch" that is so glaring as to force us to reconsider not simply a system-component but the system as a whole. This, in turn, suggests a need to distinguish between operational and managerial glitches. (The operational glitch usually points to no more than a flaw in first-order design which may be corrected by repair, improvement or replacement of some particular subsystem. But the managerial glitch points to a flaw in the second order design which may be corrected only by repairing, improving or replacing the entire system to the ends of which the subsystem contributes.) But such distinctions notwithstanding, people—like the semiconscious characters in a Chekhov play—tend if at all possible to limit their remodeling instincts to first-order considerations rather than acknowledge the need for a new second-order system.

In other words, what is especially noteworthy about America's energy problem is only the dimensions of its possible consequences, not its uniqueness. Such technological overdevelopment is a common and well-known characteristic of the evolutionary process, and has resulted in the extinction of innumerable species whose particular set of endogenous tools proved inadequate when the environment to which they were adapted changed.[43] Thus any species, once evolved, has a vested interest in the maintenance of the kind of environment in which it has been successful—a principle, in other words, of stability, conservatism, synchrony, system maintenance.

In historical times, and on the cultural level, the vested interests of some human beings have often led to a technological overdevelopment for the sake of system maintenance. And on occasion such overdevelopment may result in a glitch of some notoriety and seriousness, as was the case, for example, when an American pilot of a high-flying reconnaissance plane was shot down over the Soviet Union and placed on trial. So also in the case of the Cuban Missile Crisis, of various disasters associated with storage and shipment of poison gases, or with the premature distribution of an inadequately tested drug such as thalidomide, or with the malfunction of an inadequately tested system during a manned or unmanned space flight, and, most recently, all the electronic skullduggery referred to for short as "Watergate," not to mention the demise of American influence in southeast Asia in spite of considerable technological efforts to achieve the contrary.

What such notorious glitches tell us about technology is, first of all, that human beings learn how to use an available technology far more quickly than they learn how to decide whether, to what extent, and for how long that technology ought to be used at all. Said glitches would seem to suggest the insurmountable inadequacy of the purely "technological fix." But, if so, they have not notably discouraged men from entrusting the achievement of their ambitions to yet other technological devices. For example, during America's recent involvement in southeast Asia we came to see communications technology, among others, made the basis for what has all too appropriately been entitled an "electronic battlefield."[44] Indeed, it is in large measure due to confidence in a fully implemented electronic battlefield that the Nixon administration believed it could withdraw American ground forces from Vietnam without significantly diminishing the efficacy of our country's military "presence" there. But, as subsequent events have shown, the American approach to problem-solving in southeast Asia may well go down in history as a prime example of a glitch in the doctrine of technological fixing.[45]

To be learned from all of this, secondly, is that people tend to pass judgment on any given technological quest for power largely on the basis of their views about the goals thereby to be pursued. For proponents of those goals, for example, the end does tend to justify the means, however noxious those means may be. Thus, if there is or is ever to be anything like an ethics of technology, its greatest challenge is to learn how to surmount the bias of those who may deign to speak for it.

It is perhaps in order to evade "humanizing" considerations such as the above that the all-systems-go technophile chooses to canonize the cliché that technology (like science) is in and of itself "value-neutral," and just happens to get used for ends that are deemed either good or evil. The obvious (and, for technocrats, convenient) corollary to this cliché is that, when an embarrassing if not catastrophic glitch happens to surface, one may call for stricter regulation of the consumers rather than the purveyors of technology. Such a view suggests either insufficient consideration or willful disregard of the social, political and economic realities that determine the context and orientation of technological endeavors. For, in a word, it is not generally the case that technology is just incidentally developed and then fortuitously used, after the fact, in support of power; far more commonly, power tends to mandate the development of whatever technology it deems can contribute to its own maintenance, if not aggrandizement, and to discourage whatever may not.[46]

To the extent that this statement of the relationship between technology and power is correct, it is perhaps unnecessary even to ask what motivates power to technologize. For, on this reading, power need not have any reason to technologize beyond itself, beyond its own maintenance and expansion. But people on occasion do tend to have alternative aspirations; so the power motive is generally reconciled with people's aspirations, insofar as possible, by way of ideology. It has been claimed in recent years, especially by one of the most ideological of technocrats, that ideologies are extinct, at least in developed countries.[47] But this claim of being free of ideology is but a special form of the general tendency on the part of people of whatever persuasion to think of their own ideas as being wholesome, benevolent, constructive, and of those of others as being somehow sinister, deleterious, destructive. From a more epistemological point of view, an in-group tends simply to think of its own views as representing truth, or *the* truth, and of others' as being, at best, ideologies or, if malevolence is suspected, even conspiracies. Such cultural bias (which, to be sure, might take the form of an inverse cultural bias) does not provide a very solid basis for distinguishing between "good" and "bad" technologization of power. A more adequate basis, it seems to me, is the philosophy of man which a given technologization presupposes.

There are, with respect to technologization, two basic kinds of philosophy of man: one defensive, the other supportive. By a defensive philosophy of man I mean one which thinks of man primarily as a hostile and destructive being that must be carefully watched and vigilantly guarded against. By a supportive philosophy of man I mean one which tends rather to think of man primarily as the earth's most valuable resource—one which, if properly supported, cannot help but increase and multiply the amount of good in our midst. When made the basis for technologization, the former produces a technologization of despair, the latter, a technologization of hope.

A technologization of despair is one which is based on the cyborgean belief that human beings in and of themselves are capable of making only minimal if not counterproductive contributions to goals deemed worthy of attainment. Such technologization accordingly aims at maximum mechanization of all means to the ends pursued, including the control and, if necessary, the suppression of attempts at human intervention. A technologization of hope is one which is based on the belief that it is only or at least primarily in and through the contributions of human beings that any really worthwhile and lasting good can be accomplished. Such technologization accordingly aims at mechanizing only the processes which can supplement and complement human pursuit of humanly desirable goals, the most comprehensive of which is the prosthetic goal of human well-being as such.

Although no rigorous mapping seems possible in these matters, there is good reason for suggesting that one's approach to technologization depends very much upon such psycho-social factors as the following: (1) the degree of well-being desired for one's most preferred interest group; (2) the degree of commitment to that desire; (3) the extent to which that desire is being or is deemed capable of being fulfilled; and (4) the seriousness of threats (real or imagined) to such fulfillment.

The interest group in question may be a single individual, a family (nuclear or extended), a community (clan, tribe, village, ethnic or racial group, commune or whatever), a corporation, a society or nation-state, a particular geographic region or continent, or even mankind as a whole. Each of the four factors calls for a subjective estimate, which, however, may be based on whatever objective information is or is adjudged to be pertinent. Depending on which of the factors are given greatest emphasis, one's attitude toward technologization may be favorable either to a technologization of despair (e.g., an emphasis on the first and fourth factors) or to a technologization of hope (e.g., an emphasis on the first and third).

As the foregoing suggests, the primordial policy goal that should be implicit in all planning is to identify genuine human needs and try to figure out how they might be given priority among human interests. This being stated, what can we identify as a genuine human need?

It seems obvious that nutrition is a universal need among human beings, and as such a human need. But perhaps neither sexual pleasure nor even a minimum of sleep is a universal need. What, then, of shelter? Artificial as distinguished from organic means of transportation? Membership in some sort of cultural group? Membership in a country club? As these questions are meant to suggest, as soon as one moves beyond the obviously physical or organic on to the psychosocial and cultural level, universality seems to give way rapidly to selectivity. It would, for example, be foolish to say that everyone needs to participate in a fertility rite; but it would be just as foolish to claim that everyone needs a raincoat and an umbrella.

In other words, it is perhaps where needs leave off, at the edge of the physical or organic, that culturally determined interests begin. But some interests may very properly be designated common interests, in the sense that they are in some fashion common to all human beings. Primordial among these common interests are undoubtedly what Abraham Maslow identified as constituting a hierarchy of "needs."[48] Developed to account for certain unexpected data in industrial psychology and now being applied rather broadly, Maslow's theory amounts to a claim that humans take an interest in their needs sequentially, beginning with that which is most basic, namely, survival, and rising to that of self-actualization. Whether all this applies universally, outside of the industrial setting, is a question that need not be resolved here. For, even allowing for the possibility of discrepancies both between and within groups, it does seem reasonable to say, with Maslow, that above and beyond the basic needs for survival and security, human beings need esteem and self-actualization, or what I prefer to identify, respectively, as professional and personal responsibility. So, even granting that there are surely culturally determined ways of ascertaining whether and to what extent one's needs or interests in this regard have been satisfied, they remain for all that transcultural.

With thoughts such as these in mind, I assume (1) that every need contributes to or brings forth an interest (necessity the mother of invention); and (2) that every interest manifests or contributes to a need (invention the mother of necessity). Telescoping these two assertions into one, it may be claimed that, phenomenologically speaking, an interest is a presumptive need. Beyond the phenomenological, however, there remains the gnawing problem of ascertaining whether and to what extent any alleged need is in accord with human nature. This problem, once deemed very simple, has now become almost insoluble, largely because of the way in which the theory of evolution has come to dominate considerations of human nature in general and of human use of technology in particular.[49]

For centuries man's understanding of and expectations for himself have been conditioned by one or another absolutist notion of human nature. Whether contrasted with superhuman, even divine, natures (as was rather customary early in the game) or with subhuman natures, as the present generation seems to prefer, human nature is a theoretical construct which serves to separate and even isolate man from his environment. Even when thought of evaluatively, as in the expression "just human nature" or as in Hobbes's view of "man a wolf to his fellow man" or as in Nietzsche's "human, all too human," the assumption has been that there exists somehow a distinct and describable entity that is both causal and constrictive with regard to human behavior.

Curiously enough, human nature is seldom thought of as being adequate to the demands that are placed upon human beings. Rather must it (human nature) be saved or redeemed in some fashion, whether by drawing upon divine grace or the forces of the cosmos, by finding a charismatic leader, or by building a machine or two. In any event, human nature, precisely inasmuch as it is deemed to be definable, suggests, or indeed constitutes, bounds to what a human being is capable of without external support of some kind from somewhere. What these bounds may be is quite diversely articulated. But differences of detail aside, the very common effort to distinguish, i.e., single out, man from everything else simply points to the still more basic truth that human beings are inseparably connected to a greater and richer world within and beyond themselves.

Some people, of course, have been so taken with the chemistry of consciousness that they attempt to locate the greater and richer world in a realm beneath ordinary experience. Others, usually thought of as mystics, have pointed beyond the ordinary. But all these "heads," be they psychedelic or monastic, eventually get around to telling us that the world they see is the world we all see, only they see it better. Generally speaking, they claim to like our world better as they imagine it to have been before humans came along to modify it, at times almost beyond recognition. And whether one thinks of them, in this regard, as "wilderness nuts" or whatever, it would be folly indeed to disregard the particular message which they in their own way help to keep alive in the world, namely, that we are tied to and totally dependent upon the natural environment which not only bore the first of our species but bears each of us as well.[50]

That there are unbreakable ties between things "in nature" and man has been noted by the most primitive of peoples as well as by the anthropologists who chose to interpret such "savage thinking" as totemism.[51] Sophisticated theories aside, there is something hauntingly important about the primitive's sense of at least selective oneness with nature. For, unless you take very seriously the sports world custom of selecting a team mascot, modern man has lost not only this, but just about any other, sense of belonging to nature. On occasion, to be sure, lava still flows too far, mine walls still collapse on miners, dams still fail to hold back impounded water, or the deteriorating carcass of some strange form of marine life is washed ashore to detract from one's early-morning swim. But, such distractions aside, modern man has been encouraged to believe the lethal myth that he is as free of nature's humbling constraints as is the butterfly of its once imprisoning chrysalis. That this freedom, like that of the butterfly, is attributed to our advanced technology only serves to thicken the plot. For, it is not unheard of that even mighty flying machines, like Icarus, fall out of the sky, or that whole cities, like Atlantis, are suddenly reclaimed by the earth. Indeed, it is at least conceivable that still mightier machines may be coming to supplant us, as Samuel Butler before and many science fiction writers since have forebodingly fantasized.[52] But in the meantime we "run" our machines, not altogether unlike the nature-doomed innocents on the supposedly unsinkable *Titanic,* or the sociopolitically preoccupied passengers on Katherine Anne Porter's *Ship of Fools.* Discounting such signs of vulnerability in machines as well as in man, technocrats and technophiles see very positive images of themselves in their machines; and so they feel free to discount as shortsighted the sort of fear for man's good image that burdened William Jennings Bryan in the so-called "monkey trial" of a teacher named Scopes and, incidentally, of another named Darwin.[53] Momentary glitches aside, so the tacit argument would claim, we makers and manipulators of machines are eating the pudding that proves, so the burden of proof must be on the nay-sayers and second-guessers. These latter curiosities, it is expected, will eventually disappear once and for all, just as surely as has the whooping crane and the Inca and all other weaklings weeded out in the unemotional struggle that ever sees the fittest survive to glory in their ordeal-proven fitness.

Such, in brief, is the legacy of social Darwinism, which transformed Darwin's meticulous study of our animal origins and heritage into a brazenly antibiological and at least tacitly antihuman ideology to legitimate the industrial will to power.[54] By no means defunct, this ideologization and ultimate mechanization of the evolutionary view of things is perhaps most brazenly propounded by Richard R. Landers in his guided tour through the "dybosphere."[55] It is rather cleverly argued by Bruce Mazlish, who claims that neither Sigmund Freud nor Jerome Bruner went far enough when they pointed out how human beings have learned to acknowledge their oneness with nature, with animals, and with one another. They must learn to acknowledge, as indeed they are doing, their oneness with machines. For, as Mazlish has it, it is no longer possible even to think of man apart from machines.[56]

One must admire the epistemological daring of such a claim, for it is patently false at least on the level of overt awareness. But it illustrates very well indeed the increasingly common device of interpreting all man-machine relations as being prosthetic, by simply accentuating the beneficial aspects of the machinery in question.[57] But, as at least one science-fiction writer has suggested, even prosthetized man may use his supplementary capabilities for other than universally edifying purposes.[58] Which is to say, in effect, that however mechanized man may become, if a mechanized man is to remain recognizably human, he shall have to make an independent choice or two along the way.[59] And however such choices come to be made, and whatever problem they may concern, the underlying issue will ever remain whether human beings wish to continue being human. What this means in practice is known to anyone who has ever seen a terminal patient refuse to be "helped" by exotic biomedical technology.[60] Its broader ramifications may be suggested by way of some concluding reflections on the ethical and legal import of certain seminal ideas of two biologists, René Dubos and Garrett Hardin.

Dubos asserts in his *So Human an Animal* that "in practice people need what they want" and that what they want depends less on biological than on social requirements.[61] The major thrust of his argument is to the effect that if human beings are to survive they must rediscover the importance of satisfying inherited biological needs even at the cost of transforming or, if necessary, doing away with cultural demands that are antibiological. This may be done, thinks Dubos, with the help of scientific information, provided that science can be directed to formulating and resolving basic questions about who we are and where we are going. In this way, he anticipates, we may move beyond biologically harmful technological "cultures" to a universal "civilization" that will somehow perpetuate primitive oneness with nature by means of scientifically legitimated institutions.

Implicit in this view is the assumption that human beings are capable of creating—or, in Gabor's expression, inventing—their own future, regardless of the cultural conditions and resulting cultural bias to which they are inevitably subject in the present. In short, Dubos assumes with Gardner Murphy that human beings are free—that, in addition to their biologically inherited "first nature" and their socioculturally imposed "second nature," they have also at their disposal a kind of "third nature" of spontaneity, ingenuity and creativity.[62] Unfortunately, there is much in history that might better not have happened at all, however creative may have been the historical agents that brought it about. Michaelangelo may indeed have enriched us with his ingenuity; but then there were Genghis Khan, and Adolf Hitler, and possibly the man next door. Similarly, one technological development might increase food production; but another might take arable land out of cultivation indefinitely. The same invention might help either to maintain or to destroy a culture. Thus does the mixed impact of technology lead us to the suspicion that the freedom to choose is more a burden than a blessing, especially when the kinds of choices that one must make almost certainly will have consequences beyond anyone's ability adequately to predict or control. But existential anguish alone is no substitute for adequate safeguards. Nor, at the other extreme, is the exculpatory determinism of one who claims impotence in the face of events which he has not single-handedly set in motion.

The principles elaborated at Nuremberg with regard to medical experimentation on human beings are applicable always, not only to war crimes as such but to any situation where human beings may be harmed without their informed consent.[63] Such, at least, is the standard which has evolved in most courts of law as well as in agencies such as the NIH which regulate funded medical experimentation requiring human subjects.[64] This doctrine of informed consent is by no means a panacea for every possible abuse of the laudable desire to advance human knowledge; but at least it is available, and it does provide a court with an entirely appropriate basis for awarding damages demonstrably undergone. Analogously, causes of action lie for damages to an individual resulting from faulty products, by virtue of such legal theories as that of implied warranty. And, of course, our courts have been finding for plaintiffs in numerous cases involving proven and causally determinable damage to the environment.[65] It seems clear, however, that a procedure such as litigation for damage already done is a minimally effective instrument for dealing with the kinds of problems that can conceivably arise from almost any technology if it is not adequately monitored.

Thus the need not only for prior review and regulation but also for the elaboration of policy on the basis of which such review and regulation shall be carried out. Prior review and regulation can be handled in part by means of such legal remedies as declaratory judgments, injunctions and restraining orders. Policy considerations are better arrived at, at least in principle, by way of a balanced mixture of legislative enactment and appellate review. As regards evaluation of technology, this ongoing process ought to have as its principal goal to enable society to assure itself that any creative endeavor supported in its behalf is fully *responsible*.

That creativity must be made responsible does not mean, in case this is not obvious, that those who create (or invent) should have responsibility attributed to them arbitrarily by anyone in need of a scapegoat. Intended here is what Garrett Hardin has appropriately called the cybernetic concept of responsibility, which he associates with Charles Frankel's proposal that "a decision is responsible when the man or group that makes it has to answer for it to those who are directly or indirectly affected by it."[66] On the basis of this definition of responsibility Hardin elaborates what amounts to two axioms for a systems approach to the evaluation of technology:

1. that "the morality of an act is a function of the state of the system at the time the act is performed . . .";
2. that "[w]henever the state of a system needs to be taken into account before an act can be approved or disapproved, an administrative agency is needed."[67]

This line of reasoning, as Hardin himself notes, can readily end in a cul-de-sac if the administrative agency in question cannot itself be held responsible; and in the absence of adequate information about how, and with regard to which, that agency makes decisions, quality control by way of overseeing will be ineffective at best.[68] Many, if not most, regulatory agencies of our federal (to say nothing of state) government such as the FCC, the FTC, the AEC, and perhaps already even the EPA, are often indistinguishable, in terms of personnel as well as results, from the vested interests which they have been established to regulate.[69] But an agency such as the FDA, for all of its inadequacies, does manage to give some meaning to its charge to keep us from ingesting what is not both safe and effective.[70] Much more is required, however—not necessarily in the form either of more laws or of more regulatory agencies, but in the form of more citizen involvement in the process whereby existing agencies decide what is in our best interest.

Citizen involvement in and of itself, of course, in the absence of enlightened appreciation of what is truly at stake in a given case, can do as much harm through zeal as might an ineffectual regulatory agency through neglect. Accordingly, it is absolutely essential to the humanization of technology, as here understood, that appropriate governmental and citizen interests be adequately complemented with duly issue-oriented professional responsibility on the part of scientists and engineers. Such, it would seem, is one of the principal conclusions arrived at by geneticists at the recent Asilomar Conference, who in effect chose to take seriously a cybernetic concept of responsibility on the level of peer evaluation.[71]

What remains to be seen is whether professional self-regulation of the sort agreed to at Asilomar will encourage others in analogous situations also to take responsibility even for unintended consequences that enter only peripherally into their planning for circumscribed goals.[72] For, one of the most prevalent consequences of modern industry-subservient technology is frustration of those most human of all human values, personal and professional responsibility. If these are ever effectively destroyed, then technology will indeed have become an end in itself and Cyborg will be its name. But if we do in fact learn how to require of ourselves and of others rational yet adequate responsibility for the foreseeable consequences of our actions, it may yet be possible to take Marshall McLuhan's dictum that our environment is a "programmed happening" not as a warning but prosthetically.[73] To settle for any less will be in effect to say to posterity that in our time the humanization of technology was mainly a manipulated slogan, not an ethical imperative.

FOOTNOTES

1. This is not meant to imply that no attempts are being made to develop such expertise. See, for example, Kurt Baier and Nicholas Rescher, eds., *Values and the Future*, New York: Free Press, 1969; Environmental Protection Agency, *The Quality of Life Concept: A Potential Tool for Decision-Makers*, Washington: EPA, 1973.

2. "The Tragedy of the Commons," *Science* 162 (Dec. 13, 1968) 1243–48, and widely reprinted thereafter, e.g., in *The Everlasting Universe*, eds. L. J. Forstner and J. H. Todd, Lexington, Mass.: Heath, 1971, pp. 174–93.

3. Such is, in effect, one message that can be heard in some critiques of the MIT/Club of Rome approach to designing a "world system." See H. S. D. Cole, et al., *Models of Doom: A Critique of the Limits to Growth*, New York: Universe, 1973; Ervin Laszlo, ed., *The World System*, New York: Braziller, 1973; David L. Sills, "The Environmental Movement and Its Critics," *Human Ecology* 3 (January 1975), 1–41, esp. 19; Charles Susskind, review of *The Limits to Growth*, *Ecology Law Quarterly* 2 (Fall, 1972), 879–88.

4. Involved here are both physical and economic injury. With regard to the former, Bureau of Labor statistics show that between 1961 and 1970 the manufacturing injury frequency rate rose from 11.8 disabling injuries per million man-hours worked to 15.2, an increase of 29 percent. According to the National Safety Council, there are 2,300,000 disabling injuries and 14,200 deaths per year from accidents on the job, resulting in $9.3 billion in lost wages, insurance, medical expenses and property damage—and there are indications that more accurate figures would be ten times higher. See Dan Cordtz, "Safety on the Job Becomes a Major Job for Management," *Fortune* 86 (November 1972), 113. Thus the manifest need for the Occupational Safety and Health Act of 1970, PL 91-596, S. 2193, 91st Congress, which went into effect April 28, 1971. In the first three years of OSHA, 172,000 inspections led to 115,000 citations involving 592,000 violations. Alexander J. Reis, "Three Years of OSHA: The View from Within," *Monthly Labor Review* 98 (March 1975), 35–36. Indirect or economic injury may arise, for example, from use of inappropriate technology, a point often made by critics of W. W. Rostow's stage theory of development. See John M. Culbertson, *Economic Development: An Ecological Approach*, New York: Knopf, 1971, pp. 257–91; Jagdish Bhagwati, *The Economics of Underdeveloped Countries*, New York: McGraw-Hill World Univ. Library, 1966, pp. 225–30; Stephen Enke, *Economics for Development*, Englewood Cliffs, N.J.: Prentice-Hall, 1963, pp. 189–206. By way of contrast, see Charles P. Kindelberger, *American Business Abroad*, New Haven and London: Yale University, 1969; Andrew Shonfield, *The Attack on World Poverty*, New York: Vintage, 1962.

5. The assessment done for NASA by the American Academy of Arts and Sciences verbalized such broad concern by declaring that its goal was to enable NASA to carry out its activities "in such a fashion that the net total of secondary effects can be optimized—within reasonable limits"—Raymond A. Bauer, *Second-Order Consequences*, Cambridge: MIT, 1969, p. 199. But under the circumstances this meant little more than searching for horses after the barn door was opened. Yet to be achieved is consistently responsible assessment *before* a new technology has become a political and economic necessity, as was acknowledged by the Panel on Technology Assessment of the National Academy of Sciences, *Technology: Processes of Assessment and Choice*, Report to the U.S. H.R. Committee on

Science and Astronautics, Washington: U.S. Government Printing Office, 1969. See also Donella H. Meadows, *et al., The Limits to Growth,* New York: Universe 1972, pp. 146–55.

6. The word "ideal," as here used, is appropriately defined by Russell L. Ackhoff as "an objective which cannot be obtained in any time period but which can be approached without limit," specifically by an ideal-seeking system, i.e., "a purposeful system which, on attainment of any of its goals or objectives, then seeks another goal and objective which more closely approximates its ideal." Ackhoff, "Towards a System of System Concepts," in *Systems Behaviour,* eds. J. Beishon and G. Peters, London: Harper & Row for The Open Univ. Press, 1972, p. 87.

7. See Edwin Mansfield, *The Economics of Technological Change,* New York: Norton, 1968; Herbert A. Simon, "The Science of Design: Creating the Artificial," *The Sciences of the Artificial,* Cambridge and London: MIT, 1968, pp. 55–83; Michael Shanks, *The Innovators: The Economics of Technology,* Baltimore: Penguin, 1967; Robert W. Campbell, "Strategic and Operational Decision Making," *Soviet Economic Power,* 2nd ed., Boston: Houghton Mifflin, 1966, pp. 53–82.

8. The ideas of Mumford, Ellul and Wiener are perceptively introduced, along with those of several other well-known analysts of technology, by William Kuhns, *The Post-Industrial Prophets,* New York: Weybright and Talley, 1971; Harper Colophon, 1973. Those of Ferkiss and others, by Donald W. Shriver, Jr., "Man and His Machines: Four Angles of Vision," and by Paul T. Durbin, "Technology and Values: A Philosopher's Perspective," *Technology and Culture* 13 (October 1972), 531–76.

9. Meadows, *et al., op. cit.,* p. 20.

10. Robert W. Guest, "Men and Machines: An Assembly-Line Worker Looks at His Job," in *Modern Technology and Civilization,* ed. C. R. Walker, New York: McGraw-Hill, 1962, pp. 99–100.

11. Frederick W. Taylor, *The Principles of Scientific Management,* c. 1911, New York: Norton, 1967. See also Taylor Society, New York, *Frederick Winslow Taylor: A Memorial Volume,* 1920, Easton, Pa.: Hive, 1972; Samuel Haber, *Efficiency and Uplift,* Chicago and London: Univ. of Chicago, 1964.

12. The literature in each of these areas, especially that of cybernetics, is, of course, extensive. The following are especially helpful introductory surveys: Alphonse Chapanis, *Man-Machine Engineering,* Belmont, Cal.: Wadsworth, 1965; O. G. Edholm, *The Biology of Work,* New York: McGraw-Hill World Univ. Library, 1967; F. H. George, *Cybernetics and Biology,* Edinburgh and London: Oliver & Boyd, 1965; Lucien Gerardin, *Bionics,* New York: McGraw-Hill World Univ. Library, 1968; Arthur Porter, *Cybernetics Simplified,* London: English Universities, 1969; John F. Young, *Cybernetics,* New York: American Elsevier, 1969.

13. See, for example, Stafford Beer, *Cybernetics and Management,* New York: Wiley, 1964; Edwin G. Johnsen and William R. Corliss, *Human Factors Applications in Teleoperator Design and Operation,* New York: Wiley, 1971. There are, however, some notable exceptions to the rule, e.g., Norbert Wiener's *The Human Use of Human Beings,* Boston: Houghton Mifflin, 1950, and *God and Golem, Inc.,* Cambridge: MIT, 1966, and Anthony G. Oettinger's *Run, Computer, Run,* Cambridge: Harvard, 1969. The underlying value question here at issue, of course, involves weighing costs and benefits of automation. See Ben B. Seligman, *Most Notorious Victory,* New York: Free Press, 1966; George Terborgh, *The Automation Hysteria,* New York: Norton, 1965; "The Office of the Future," *Business Week,* June 30, 1975, 48–84.

14. Manfred Clynes, Foreword to *Cyborg: Evolution of the Superman,* D. S. Halacy, Jr., ed., New York: Harper & Row, 1965, p. 8.

15. An early but comprehensive treatment of the multiplicity of systems that must be coordinated will be found in Siegfried J. Gerathewohl, *Principles of Bioastronautics,* En-

glewood Cliffs, N.J.: Prentice-Hall, 1963. The managerial problems inherent in such coordination are lucidly analyzed by Leonard R. Sayles and Margaret K. Chandler, *Managing Large Systems*, New York: Harper & Row, 1971.

16. See Harold M. Schmeck, Jr., *The Semi-Artificial Man*, New York: Walker, 1965; Donald Longmore, *Spare-part Surgery*, Garden City, N.Y.: Doubleday, 1968.

17. For example, though most of Marshall McLuhan's observations about media as extensions of the nervous system suggest a cyborg, he tends to think of media prosthetically, even when complaining about such overloads on the system as war in the global village. A deliberate usage of the prosthetic model is John McHale, "Doctor Jekyll and the Bride of Frankenstein," AAAS Convention, Philadelphia, 1971. Inversely, a minority opinion among prosthetics specialists has it that artificial devices might eventually function even *better* than the natural organs which they replace. See Schmeck, *op cit.*, p. 176 n.

18. See E. J. Mishan, *Technology and Growth*, New York: Praeger, 1970; Morton Mintz and Jerry S. Cohen, *America, Inc.*, New York: Dell, 1971; Walter J. Hickel, *Who Owns America?*, New York: Paperback Library, 1971; James Ridgeway, *The Politics of Ecology*, New York: Dutton, 1970; Peter d'A. Jones, ed., *The Robber Barons Revisited*, Boston: Heath, 1968; Philip A. M. Taylor, ed., *The Industrial Revolution in Britain: Triumph or Disaster?*, Boston: Heath, 1958.

19. The following are typical of the kinds of issues raised: Jerome Tuccille, *Here Comes Immortality*, New York: Stein and Day, 1973; Gerald Leach, *The Biocrats*, rev. ed., Baltimore: Penguin, 1972; Allan Chase, *The Biological Imperative*, Baltimore: Penguin, 1971; Dannie Abse, *Medicine on Trial*, London: Aldus, 1967; New York: Crown, 1969.

20. See Editors of *Fortune*, eds., *Our Ailing Medical System*, New York: Harper & Row Perennial, 1970; Research and Policy Committee of the Committee for Economic Development, *Building a National Health-Care System*, New York: Committee for Economic Development, 1973; Edward P. Luongo, M.D., *American Medicine in Crisis*, New York: Philosophical Library, 1971.

21. Barbara and John Ehrenreich, *The American Health Empire*, New York: Vintage, 1971; Ed Cray, *In Failing Health*, Indianapolis and New York: Bobbs-Merrill, 1970.

22. See, for example, David A. LeSourd, *et al.*, *Benefit/Cost Analysis of Kidney Disease Programs*, Washington: HEW/Public Health Service, 1968, esp. pp. 53–60.

23. *The Myth of the Machine: The Pentagon of Power*, New York: Harcourt Brace Jovanovich, 1970, pp. 236–99.

24. These broader reaches of purposiveness are developed with the assistance of cybernetic theory by Karl Deutsch, *The Nerves of Government*, New York: Free Press, 1963, pp. 91–93, and by Stafford Beer, *op. cit.* (n. 13).

25. See Alexander P. de Seversky, *America: Too Young to Die!* New York: McGraw-Hill, 1961, pp. 95–110; *F.A.S. Public Interest Report* 27, No. 2 (February, 1974). See also *ibid.*, 24, No. 10; 25, No. 4; 27, Nos. 3 and 5.

26. The computer industry provides an excellent example in this regard. See "A Tyro Challenges IBM in Big Computers," *Business Week*, May 12, 1975, 65–68; "Hewlett-Packard: Where Slower Growth is Smarter Management," *Business Week*, June 9, 1975, 50–58; William Rodgers, *Think: A Biography of the Watsons and IBM*, New York: Signet, 1970. See also "Corporate Planning: Piercing Future Fog in the Executive Suite," *Business Week*, April 28, 1975, 46–54.

27. See Raphael G. Kasper, ed., *Technology Assessment*, New York: Praeger, 1972; Lynton K. Caldwell, ed., *Science, Technology and Public Policy: A Selected and Annotated Bibliography*, Vol. 3, Bloomington, Ind.: Indiana Univ., 1972, pp. 59–66.

28. For numerous examples from other times and places, see David S. Landes, *The Unbound Prometheus: Technological Change and Industrial Development in Western Europe from 1750 to the Present*, London: Cambridge, 1969.

29. William C. Uhl, "Offshore, The Oil Hunt Gets Tough," and Werner Bamberger, "Rigs that Drillers Swear By," *The New York Times*, March 31, 1974, Section 3.

30. Hearings on Fuel and Energy Resources Before the House Committee on Interior and Insular Affairs, 92nd Congress, 2nd Sess., ser. 42, pt. I, Washington: U.S. GPO, 1972, pp. 58–59, 140–41, 301–07, 371; Morton Mintz and Jerry S. Cohen, *America, Inc.*, New York: Dell, 1971, pp. 227–57; Peter R. Odell, *Oil and World Power*, Baltimore, Penguin, 1970, pp. 23–43; Harvey O'Connor, *The Empire of Oil*, New York: Monthly Review Press, 1962; Robert Engler, *The Politics of Oil*, New York: Macmillan, 1961. As these studies tend to show, government policies influenced, if not determined, by the petroleum industry spill over from questions of energy into other areas not exclusive of military intervention. See also in this regard Wesley Marx, *Oilspill*, San Francisco and New York: Sierra Club, 1971, pp. 132–39.

31. "OPEC: The Economics of the Oil Cartel," *Business Week*, January 13, 1975, 77–81; *F.A.S. Public Interest Report* 27, No. 1 (January 1974); "Why OPEC Begins to Abandon the Dollar," *Business Week*, March 31, 1975, 18.

32. To the contrary, the American oil industry, among others, is busy expanding and solidifying its operations in the oil-producing countries. See "Building a New Middle East," *Business Week*, May 26, 1975, 38–54. Meanwhile, others talk of alternatives: Sidney E. Rolfe, "Whatever Happened to Project Independence?" *Saturday Review*, January 25, 1975, 25–28; Glenn Seaborg, "Finding a New Approach to Energistics—Fast!" *Saturday Review*, December 14, 1974, 44–48; Dietrich E. Thomsen, "Power from the Salton Trough," *Science News* 106 (July 13, 1974), 28–29; Victor K. McElheny, "Hydrogen—A Way Out of the Energy Crisis?" *The New York Times*, May 12, 1974, Section 3; John H. Douglas, "Coal: The Stopgap Fuel—Maybe," *Science News* 104 (July 7, 1973), 10–12; Arthur M. Squires, "Clean Power from Coal," *Science* 169 (August 28, 1970), 821–27; Peter E. Glaser, "Solar Energy—An Alternative Source for Power Generation," *Futures* I (June 1969), 304–13; "Power from the Sun: Its Future," *Science* 162 (November 22, 1968), 857–61.

33. Barry Commoner, "The Origins of the Environmental Crisis," Keynote Address before the Council of Europe, Second Symposium of Members of Parliament Specialists in Public Health, Stockholm, Sweden, July 1, 1971, pp. 14–15, Table IV.

34. Frank Graham, Jr., *Since Silent Spring*, Greenwich, Conn.: Fawcett Crest, 1970.

35. M. Taghi Farvar, *et al.*, "The Pollution of Asia," *Environment* 13 (October 1971), 10–17.

36. Steven Rose, ed., *CBW: Chemical and Biological Warfare*, Boston: Beacon, 1968; J. B. Nielands, *et al.*, *Harvest of Death: Chemical Warfare in Vietnam and Cambodia*, New York: Free Press, 1972.

37. Virginia Brodine, "A Special Burden," *Environment* 13 (March 1971), 22–24; Howard Lewis, *With Every Breath You Take*, New York: Brown, 1965; Max Carasso, *People vs. Cars*, New York: Autofacts, 1970; Helen Leavitt, *Superhighway—Superhoax*, New York: Ballantine, 1970. For more detailed and technical information, consult Committee on Environmental Information, 438 N. Skinker, Saint Louis, Missouri.

38. Thus the entirely predictable dissatisfaction with management of most American automobile manufacturing companies for not having anticipated the impact of the energy crunch on their industry. See Marylin Bender, "The Energy Trauma at General Motors," *The New York Times*, March 24, 1974, Section 3; "What's Wrong at Chrysler?" and "What's Right at American Motors?" *Forbes*, December 1, 1973, 28–33. (In June, 1975, Chrysler announced it was abandoning the large car market entirely, and rumors had the company being sold to *Volkswagenwerke*.) For a detailed analysis of the problem of allocating responsibility *internally* among multiple managers, see Sayles and Chandler, *op. cit.* (n. 15), pp. 298–321.

39. Galbraith, *The New Industrial State,* Boston: Houghton Mifflin, 1967; New York: Mentor, 1968, pp. 46–108. See Morton Mintz and Jerry S. Cohen, *op. cit.* (n. 18), *America, Inc.,* New York: Dell, 1971, pp. 59–108.

40. New York: Lyle Stuart, 1968; Bantam, 1969.

41. Charles W. Powers, ed., *People/Profits: The Ethics of Investment,* New York: Council on Religion and International Affairs, 1972; "Focus on Social Responsibility: Reflections on Law, Morality and Equal Justice," *Trusts and Estates* 114 (April 1975), 209–12+; Philip I. Blumberg, "Introduction to the Politicalization of the Corporation," *The Record* 26 (May 1971) 369–85.

42. An important exception to this exclusion of the managerial or strategic level from ethical considerations was Anatol Rapaport's *Strategy and Conscience,* New York: Harper & Row, 1964. See Andrew Wilson, *The Bomb and the Computer,* New York: Delta, 1968, pp. 177–83, and Charles Hampden-Turner, *Radical Man,* Garden City, N.Y.: Anchor, 1971, pp. 361–410; Hans Jonas, "Technology and Responsibility: Reflections on the New Tasks of Ethics," *Social Research* 40 (Spring 1973), 31–54. The dissipation of managerial responsibility here at issue was foreseen by Norbert Wiener as a likely consequence of the increasing automation of the office. See his commentary on C. P. Snow's "Scientists and Decision-Making," in *Computers and the World of the Future,* ed. Martin Greenberger, Cambridge: MIT, 1962, p. 26. Compare, however, Sayles and Chandler, *op cit.* (n. 5), esp. pp. 173, 201, 278–281, 306 ff.; R. R. Ritti, *The Engineer in the Industrial Corporation,* New York: Columbia Univ., 1971.

43. Loren Eiseley, *The Invisible Pyramid,* New York: Scribners, 1970, pp. 16–17.

44. *Investigation into Electronic Battlefield Program.* Hearings Before, and Report of the Electronic Battlefield Subcommittee of the Senate Subcommittee on Preparedness Investigation. Hearings Nov. 18, 19, 24, 1970, 91st Congress, 2nd Session., Washington: U.S. GPO, 1970. Report, 92nd Congress, 1st Sess., Washington: U.S. GPO, 1971.

45. Andrew Wilson, *op. cit.,* pp. 201–10. In this connection, the Pentagon Papers hardly need additional commentary. They may be supplemented, however, by Noam Chomsky, *American Power and the New Mandarins,* New York: Random House Pantheon, 1967; "On Changing the World," *Problems of Knowledge and Freedom,* New York: Vintage, 1971, pp. 53–111; David Halberstam, *The Best and the Brightest,* New York: Random House, 1972; Wilfred G. Burchett, *Vietnam: Inside Story of the Guerilla War,* New York: International, 1965; Harold L. Wilensky, *Organizational Intelligence,* New York: Basic Books, 1967, pp. 24–34, 188–89.

46. See Amitai Etzioni, *The Active Society,* New York: Free Press, 1968, pp. 197–222; Paul Dickson, *Think Tanks,* New York: Atheneum, 1971; Adam Yarmolinsky, *The Military Establishment,* New York: Harper & Row, 1971, pp. 237–323. "Power," of course, here means the resultant of a vector of forces which taken together may be thought of as an "Establishment." This is well illustrated by the classic case of the emergence of NASA in the 1960's: Enid Curtis Bok Schoettle, "The Establishment of NASA," *Knowledge and Power,* ed. S. A. Lakoff, New York: Free Press, 1966, pp. 162–270. See also Jerome R. Ravetz, *Scientific Knowledge and its Social Problems,* Oxford: Clarendon, 1971.

47. See Chaim I. Waxman, ed., *The End of Ideology Debate,* New York: Simon and Schuster Clarion, 1969; Paul E. Sigmund, Jr., *The Ideologies of the Developing Nations,* New York: Praeger, 1963.

48. *Eupsychian Management: A Journal,* Homewood, Ill.: Irwin-Dorsey, 1965. See also Maslow, "Criteria for Judging Needs to be Instinctoid," in *Human Motivation: A Symposium,* ed. M. R. Jones, Lincoln: Univ. of Nebraska, 1965.

49. The theory of evolution formed the backdrop to Marx's *Capital* which in fact was dedicated to Darwin. As with Marx, so do current writers find it congenial to express value preferences with regard to industrial development by taking a stand on the chicken-or-egg

question of whether tools or brains came first in time. See Clifford Geertz, "The Impact of the Concept of Culture on the Concept of Man," in *New Views of the Nature of Man*, ed. J. R. Platt, Chicago and London: Univ. of Chicago, 1965, pp. 93–118; Lewis Mumford, *The Myth of the Machine: Technics and Human Development*, New York: Harcourt Brace Jovanovich, 1967, pp. 22–54.

50. See John Passmore, *Man's Responsibility for Nature*, New York: Scribners, 1974, pp. 73–126, 173–81; David L. Sills, *op. cit.* (n. 3), 22–24. With regard to the chemistry of consciousness, see William Braden, *The Private Sea*, Chicago: Quadrangle, 1967; Alan W. Watts, *The Joyous Cosmology*, New York: Vintage, 1962; R. E. L. Masters and Jean Houston, *The Varieties of Psychedelic Experience*, New York: Delta, 1966, esp. pp. 151–83. For a more alimentary view of our dependence on nature, see Donald E. Carr, *The Deadly Feast of Life*, Garden City, N.Y.: Doubleday, 1971.

51. See Claude Lévi-Strauss, *Totemism*, Boston: Beacon, 1963; *The Savage Mind*, Chicago: Univ. of Chicago, 1966.

52. Butler, "The Book of the Machines," in *Erewhon*, 1872; New York: Airmont, 1967, pp. 143–65. The modern classic along these lines is undoubtedly Isaac Asimov's *I, Robot*, New York: Gnome, 1950, and its now famous laws of robotics.

53. Scopes v. Tennessee, 152 Tenn. 424, 278 SW 57; 154 Tenn. 105, 289 SW 363.

54. See Richard Hofstadter, *Social Darwinism in American Thought*, rev. ed., Boston: Beacon, 1955.

55. *Man's Place in the Dybosphere*, Englewood Cliffs, N.J., 1966. See also Harold Sackman, *Computers, System Science, and Evolving Society*, New York: Wiley, 1967; John David Garcia, *The Moral Society*, New York: Julian, 1971; Dennis Gabor, *Inventing the Future*, New York: Knopf, 1971, esp. ch. 8, "Men and Machines."

56. "The Fourth Discontinuity," *Technology and Culture* 8 (January 1967); reprinted in *Perspectives on the Computer Revolution*, ed. Z. W. Pylyshyn, Englewood Cliffs, N.J.: Prentice-Hall, 1970, pp. 195–207. Compare, however, Laurence H. Tribe, "Technology Assessment and the Fourth Discontinuity: The Limits of Instrumental Rationality," *Southern California Law Review* 46 (June 1973), 617–60.

57. This approach is, of course, commonplace among advocates of more technology, e.g., Herbert A. Simon, *The Shape of Automation for Men and Management*, New York: Harper & Row, 1965. But humanist observers, including Karl Marx with regard to industry, Marshall McLuhan with regard to electronics, and Norman Mailer with regard to Project Apollo, have attempted as much. More cautious affirmations include Loren Eiseley, *The Invisible Pyramid*, *supra*, and Lewis Mumford, *Technics and Civilization*, New York: Harcourt, Brace, 1934. See also, with regard to nineteenth-century American writers and statesmen, Leo Marx, *The Machine in the Garden*, Oxford: University Press, 1964.

58. Michael Crichton, *The Terminal Man*, New York: Knopf, 1972.

59. By thus identifying freedom of choice as the operationally distinguishing characteristic of a human being, I associate myself somewhat more closely with concerns of the legal profession than with those of academicians who have wondered prolifically if machines can think, as in Alan Ross Anderson, ed., *Minds and Machines*, Englewood Cliffs, N.J.: Prentice-Hall, 1964. I readily acknowledge the influence of the existentialists at this point, as well as the related, though more scholastic, stance of Mortimer J. Adler, *The Difference of Man and the Difference It Makes*, New York: Holt, Rinehart and Winston, 1967. See also the theological approach of Harold E. Hatt, *Cybernetics and the Image of Man: A Study of Freedom and Responsibility in Man and Machine*, Nashville and New York: Abingdon, 1968.

60. See Marya Mannes, *Last Rights*, New York: Morrow, 1974; Elisabeth Kübler-Ross, *On Death and Dying*, New York: Macmillan, 1969.

61. New York: Scribners, 1968, pp. 170 ff.

62. Gardner Murphy, "Three Kinds of Human Nature," *Human Potentialities*, New York: Basic Books, 1958, pp. 15-25.

63. Jay Katz, ed., *Experimentation with Human Beings*, New York: Russell Sage Foundation, 1972 pp. 283-321 ff.

64. *Ibid., passim* and esp. pp. 523-608 and 690-92.

65. See Earl F. Murphy, *Man and His Environment: Law*, New York: Harper and Row, 1971; Norman J. Landau and Paul D. Rheingold, *The Environmental Law Handbook*, New York: Friends of the Earth/Ballantine, 1971; and Malcolm Baldwin and James K. Page, Jr., eds., *Law and the Environment*, New York: Walker, 1970, valuable both for its discussion of changes within the legal profession and for its annotated bibliography, pp. 375-412; John E. Bryson and Angus Macbeth, "Public Nuisance, the Restatement (Second) of Torts, and Environmental Law," *Ecology Law Quarterly* 2 (Spring 1972), 241-81. See also Christopher D. Stone, *Should Trees Have Standing? Toward Legal Rights for Natural Objects*, Los Altos, Cal.: Kaufman, 1974.

66. Hardin, *Exploring New Ethics for Survival*, New York: Viking, 1972; Baltimore: Penguin, 1973, pp. 102, 103. This cybernetic, or systems, concept of responsibility, which focuses on environmental impact of decisions, is only remotely related to that presented for purposes of allocating responsibility within corporate management by Sayles and Chandler, *loc. cit., supra* (n. 38). Compare, however, the recent trend in law towards managerial liability, with the resulting development of, and issues surrounding, managerial liability insurance. See "Should Shareholders be Personally Liable for the Torts of their Corporations?" *Yale Law Journal* 76 (May 1967), 1190-1204; John S. Morrison, "Factors That Limit the Negligence Liability of a Corporate Executive or Director," *Univ. of Illinois Law Forum* 1967 (Summer) 341-50; William E. Knepper, "Corporate Identification and Liability Insurance for Corporation Officers and Directors," *Southwestern Law Journal* 25 (May 1971), 240-63; "Officers and Directors: Identification and Liability Insurance—An Update," *Business Lawyer* 30 (April 1975), 951-67. See also Emil F. Sos, Jr., "Liability of Engineer for Defective Design," *Cleveland State Law Review* 19, (January 1970) 184-93.

67. Hardin, *op. cit.*, p. 134. See, however, Panel on Technology Assessment, *op cit.* (n. 5), pp. 37-38, 65.

68. Hardin, *op cit.*, pp. 135-40.

69. See Mintz/Cohen, *op cit.* (n. 18), pp. 294-313; William H. Rodgers, Jr., *Corporate Country: A State Shaped to Suit Technology*, Emmaus, Pa.: Rodale, 1973; Joseph C. Goulden, *The Superlawyers*, New York: Dell, 1973. This alleged inefficacy of administrative agencies is compounded by similar complaints with regard to Congress' benign approach to its oversight function. See Ferdinand Lundberg, *The Rich and the Super-Rich*, New York: Bantam, 1969, pp. 584-678; Panel on Technology Assessment, *op. cit.* (n. 5), pp. 9-10, 26-27. Compare, however, *Federal Agencies and the Public Interest: New Directions in Administrative Practice*, Proceedings of the ABA National Institute, *Administrative Law Review* 26 (Fall 1974); Robert E. Jordan, III, "Alternatives Under NEPA: Toward an Accommodation," *Ecology Law Quarterly* 3 (Fall 1973), 705-57.

70. Katz, *op. cit.*, pp. 736-93, 856-82. See also *How Safe Is Safe? The Design of Policy on Drugs and Food Additives*, Washington: National Academy of Sciences, 1974. With regard to citizen involvement, see Panel on Technology Assessment, *op. cit.* (n. 5), pp. 41, 70-71, 83.

71. A working paper on research policy in the field of genetic engineering was adopted in February, 1975, by an international group of molecular biologists meeting at Asilomar State Park near Monterey, California. Then an official summary statement of their report to the Assembly of Life Sciences of NAS was approved by the latter's Executive Committee on May 20, 1975, and subsequently published in *Science, Nature,* and *Proceedings of the National Academy of Sciences*. See *Science News* 107 (March 8, 1975), 148-149, 156; (June

7, 1975), 366; Paul Berg, *et al.,* "Asilomar Conference on Recombinant DNA Molecules," *Science* 186 (June 6, 1975), 991–94. By way of background, see Joshua Lederberg, "The Freedoms and the Control of Science: Notes from the Ivory Tower," *Southern California Law Review* 45 (Spring 1972), 596–614.

72. See Hasan Ozbekhan, "The Emerging Methodology of Planning," *Fields Within Fields,* 1973, pp. 63–73; Harold Gilliam, "The Fallacy of Single-Purpose Planning," *Daedalus* 96 (Fall 1967), 1142, 1143.

73. "Environment as Programmed Happening," in *Knowledge and the Future of Man,* ed. Walter J. Ong, New York: Holt, Rinehart and Winston, 1968, pp. 113–24. See also C. A. Van Peursen, *The Strategy of Culture,* New York: American Elsevier, 1974, pp. 189–90, 192, 196.

WHAT IS TECHNOLOGY?

Robert E. McGinn STANFORD UNIVERSITY

With the increasing recognition of technology as a pivotal force in modern society, philosophers and other scholars have begun to consider problems arising out of the availability and use of new technologies. However, perusal of the emerging body of literature reveals an underlying divergence of opinion on the basic question, "What *is* technology?" A similar lack of consensus characterizes (and vitiates) much of the public discussion of technology that has occurred in recent years. While supporting the elaboration of a comprehensive social philosophy of technology and analyses of ethical and human values issues raised by modern technologies, I believe it important to improve our understanding of the nature of technology per se. What follows is intended as a contribution to this goal.

I propose to answer the question "What is technology?" in two parts; first, by inquiring into the structure of technology, i.e., by answering the question "What kind of activity-form is technology?" and second, by specifying in general terms what constitutes the content of technology. After offering a substantive example of technological activity to illustrate the presence of each of the previously identified structural aspects, I will conclude by showing how the resultant more differentiated concept of technology may be of use in clarifying the controversy over whether technology is "value-laden" or "value-free."

Definitions of technology have tended to fall into two general categories. Either it is defined in terms of one or two salient characteristics thought to be central to it, or as a concatenation of a larger, more disparate set of such characteristics, none of which, it is suggested, can be omitted without failing to capture something essential to technology. Illustrating the first tendency, Singer, *et al.*, in their *History of Technology*, define technology as "what things are done or made" and "how things are done or made,"[1] while Mesthene defines technology as "the organization of knowledge for the achievement of practical purposes."[2] Alternatively, Donald Schon asserts that technology means "any tool or technique, any product or process, any physical equipment or method of doing or making, by which human capability is extended."[3] In my view neither definitional strategy succeeds in conveying an adequate notion of what technology is. My approach here is not definitional but characterological. I propose to treat technology as a *form of human activity,* others of which include science, art, religion, and sport. (It should be added that recent work of Jane Goodall and associates has shown that humans are not the sole practitioners of technology. Not only do chimpanzees use tools—e.g., compressing leaves, inserting them in water-filled tree trunks as sponges—but they also engage in rudimentary tool manufacture: stripping leaves off branches and inserting them into tree trunks to extract termites.) After making an important preliminary distinction, I will identify and discuss eight important aspects of the activity-form technology, some more "internal" to it than others; each, however, worth including in an attempt to understand better what kind of activity technology is. Taken individually, each aspect in its specific content constitutes a necessary condition for an activity to be properly termed an instance of the activity-form technology.

Before proceeding, a preliminary distinction is necessary. On those rare occasions when "technology" *is* used to refer to a kind of human activity, this usage tends to lump together two distinguishable albeit related kinds of activity. Consider the following quote: ". . . a technologist . . . must before long turn back towards the real world, so to speak, and design his motor car, his factory, his computer, [and] his transport system in the real world rather than in the laboratory."[4] Notice that the things the technologist is said to design fall into two different categories. Earlier the author calls them "device[s]" or "machines" and "systems." Here I shall be concerned to characterize not the activity of the person—usually called the "systems design engineer"—who produces designs of various types of socio-technical systems (e.g., factories and transport systems), but rather the activity (taken as a whole) of those individuals, engineers and others, whose efforts issue in material objects of one sort or another. In the sequel, unless otherwise indicated, when speaking of technology as a form of human activity and analyzing its various aspects, "technology" should be understood as referring only to the latter activity, not the former. The activity of making material objects should not be assimilated to the activity of designing and constructing the systems in which those material objects will operate or be used. This, of course, is not to suggest that these two activities ought to be isolated in practice. On the contrary, the criteria governing the making of material objects ought to take more seriously into account those governing the operations of the socio-technical systems in which the objects will be deployed.

The first aspect of technology pertains to the nature of the results or "outputs" of the activity-form: its characteristic outcomes are *material* as opposed to ideational in nature, as, e.g., tends to be the case with religious activity. Thus prayer, although sometimes involving technique, is not an instance of technology. This is not to say, as will shortly become clear, that ideational phenomena play no role in technology. As technology is widely thought to involve the making of things, it might seem adequate to say that technology is a material product-making activity. This, however, would fail to cover some recent and other imaginable activities which would likely be termed instances of technology but which are not accurately subsumed under the material product-making rubric; e.g., plastic surgery, genetic engineering, and a technical process which made a person invisible. I suggest the rubric be broadened to "material product-making or object-transforming activity," where the transformation in the material object is from an initial state or condition to a subsequent one. Evidently one could subsume both elements in this revised rubric under the latter, i.e., material object-transforming, with material product-making viewed as a transformation from the raw materials stage to the final product stage. However, I shall retain the dual rubric.

Some technological activity results in the production of what might be termed "stuff": "raw" materials (e.g., in a certain size or shape) or partially or wholly "cooked" materials (e.g., synthetic fibers or plastics) out of or with the aid of which the final products (objects) of subsequent technological activity will be made (transformed). Setting aside this subset of the material outcomes or outputs of technological activity, we shall call the elements that remain *technics* (if they are material products) and *quasi-technics* (if they are transformed material objects, e.g., people who have undergone plastic surgery).[5] Items of technics thus include not only tools and the usual cars, planes, and computers, but also, e.g., contact lenses, musical instruments, clothes, certain kinds of foods and furniture (viz., those that have undergone some significant transformational process), as well as certain works of art (e.g., pieces of sculpture). Not included under the category of technics as here characterized would be the trimmed tree trunk or block of marble on which the sculptor begins to work, the plastic out of which the contact lens is to be made, or the synthetic crystals with which semiconductors are made.

It is well to point out that "technology" is perhaps most often used to mean the complex of technics—more specifically, modern technics—or a subset thereof: viz., machines, tools, and weapons. In my view this usage is undesirable, for not only does it cast its net too narrowly as regards the physical objects it embraces, but it also clouds the important fact that technology is as much a multifaceted human activity as any other, and is not just "those machines out there," mistakenly viewed as somehow possessing lives and possibly demonic powers of their own.

The second aspect of the activity-form technology concerns a characteristic of the overall process by means of which technics are made; viz., such processes are *fabricative* in nature. By this is meant that at least some of the significant features of the technic are due primarily to the technologist's working of his or her will on the constituent ingredients or parts, rather than their all being primarily the result of the operation of chemical, biological, or physical laws. A practitioner may, of course, *facilitate* the occurrence of interactions among ingredients. Thus the fact that the material outcome possesses properties resulting from the operation of such laws may in a sense be said to be due to the volition of the practitioner, but not *primarily* so as here understood and intended. Not every stage of the technological process involved in the production of a given kind of technics is fabricative, particularly so in the complex processes of modern technology. Consider, e.g., what engineers call the "preliminary design stage." However, taken as a whole, it is still the case that all technological processes producing technics are fabricative. Technologists and ancillary staff fashion, shape, construct, form the technic (or prototype or model of same). On this account I contend that neither rain-making by seeding clouds nor agriculture are tokens of the activity-type technology. Although they both, when successful, generate material output, neither is fabricative. Whether such activities are based on scientific understanding or not, both involve facilitation of the occurrence of chemical and biological processes. Of course, each may be called technological in the weak and here uninteresting sense that their practice involves various technics, the products of prior full-blooded technological activity. Essentially, however, both are in their most sophisticated forms specimens of applied science. In sum, the fabricative character of technology pertains to the nature of the processes in which its issue is brought into being.

The third aspect of the activity-form technology is that of *purpose*. Technology is an explicitly and highly purposive form of human activity. But is there any specific purpose which can be said to be characteristic of technology in general? Edwin Layton claims that "the central purpose of technology" is "design": "adaptation of means to some preconceived end."[6] Not so: rather than being technology's general purpose, design is rather a vitally important *means* to the realization of that purpose, viz., expansion of the realm of the humanly possible; by this is meant the realm of possibilities which may be utilized by human beings, not mere theoretical possibilities.[7] For technology is a quintessentially practical form of human activity. Particular expansions of the realm of the humanly possible are often undertaken in the name of increasing the power of their beneficiaries; either directly, through increasing their capacities to realize goals in the face of obstacles to doing so, or indirectly, through the expansions' provision of vehicles for practitioners' self-realization. The expansion of the humanly possible through technology occurs in at least six different modes.

1. direct extension: providing a direct extension of some already existing human function or capacity, as in the cases of the telescope, the megaphone, and the computer;

2. qualitative innovation: offering a qualitatively new addition to the repertoire of human capacities, as do, e.g., the airplane and the submarine;

3. risk reduction or elimination: enabling one to do something, previously done, but only with the risk of incurring certain costs, without or with significantly reduced risks, as, e.g., in the case of the birth-control pill and the asbestos fire-fighting suit;

4. improvement of performance: offering the ability to do something easier or more efficiently than it would be done previously, as, e.g., in the cases of snowshoes, chainsaws, and solar-energy cells;

5. substitution: enabling one to do Y, previously precluded by doing X, by X's being done technologically, e.g., reading while one's lawn is watered by an automatic sprinkler;

6. increasing the means for expression of the inner life: providing for the aesthetically or otherwise motivated representation of emotions, beliefs, perceptions or other states or conditions of consciousness in external, tangible forms, as, e.g., in the cases of musical instruments, sculpture, perfumes, etc. (This might be viewed as the provision of additional means whereby humans can express their capacity for the representation of consciousness or experience. Thus, this mode of expansion differs from that of direct extension described in (1) above.)

It should be noted that more than one of these modes may pertain to a given technic.

These modes of expansion suggest the following two more general purposes of technology: (1) facilitating successful human adaptation to a possibly threatening environment (natural or human) by augmenting one or another dimension of human power with respect to that environment (e.g., sunglasses used in driving, a well operating in an arid land, or weapons used for obtaining food or self-protection); and (2) aggrandizement of human power even in a context or environment devoid of threat; put differently, in the name of thrival rather than survival (e.g., the development and use of technology in order to increase one's economic power far beyond that required for survival).

There is a claim lurking in the background which should be brought to the foreground. Although difficult or impossible to confirm, it may be that at a deeper level, the most general purpose of technology is to increase human power via the above-mentioned expansion of the realm of the humanly possible. Sometimes the increase may serve human survival interests, on other occasions it may serve human aspirations to self-aggrandizement or domination. On a somewhat more metaphysical and speculative level, recall the early Sartre's claim that human life in all its aspects, including, e.g., sexuality, is a fruitless attempt to escape from the anguish of ontological contingency by becoming God, the being at once *pour-soi* and *en-soi*. I think it worth conjecturing whether technology itself might finally be one of man's attempts to deny and transcend his finitude and its associated vulnerability, and to become Godlike by acquiring more and more power in all its dimensions. While urging consideration of this conjecture, I do not mean to deny that one or more of the relatively more mundane levels of purpose or modes of expansion might not be operative in the development of particular technologies. The question remains however: is technology's purpose to be understood, following Marx, as a human means of successfully adapting to a recalcitrant natural or social environment, or, alternatively, in the spirit of Nietzsche, as motivated by survival *only in special cases,* but more generally and profoundly as a vehicle for the expression of the alleged insatiable "will to power"?[8] It would appear then that a full understanding of the most general or deepest level of the purpose of technology, if such there be, must await further understanding of human nature itself.

Now we may ask: Is the activity of, say, the organic chemist an instance of technology? After all, much of his/her work involves the synthesis of new compounds and seems, intuitively speaking, more fabricative than either agricultural or cloud-seeding activity. However, even if the synthesizing activity of the organic chemist could not be ruled out as technology by either of the first two aspects of technology, there is a sense in which it might still be disqualified on the basis of the third. When synthesizing compounds, the organic chemist, *qua* organic chemist, is interested in the structure, composition, properties, and reactions of the compounds made, not in the fact that the compounds synthesized may be put to some mode of technological purpose, to aid in expanding the realm of the humanly possible. Only insofar as the production of a material substance (chemical or otherwise) for one or another technological purpose is a goal informing the practitioner's activity, rather than simply the achievement of scientific understanding through the discovery of knowledge, does the activity remain eligible for the appellation "technology." To the extent that a technological purpose is operative the practitioner is not acting *qua* organic chemist but *qua* technologist. Thus one and the same specific activity may be either an instance of the activity-form of science or one of technology, depending on the nature of the purpose with which the activity is pursued. But can one and the same specific activity be simultaneously an instance of both activity-forms? Consider the following: even the synthesis of new compounds by the organic chemist in order, say, to develop a new contraceptive pill, fails to be fabricative in the sense in which technology is. The activity of compound-synthesis is sometimes fabricative, but only in a weaker, second-order, indirect sense: besides facilitating the interaction of materials, the organic chemist may be said to manipulate or control, the *environment* within which the laws of chemical composition and reaction, as it were, operate to determine the synthesized compounds. But in technology the material outcome itself is fabricated by the practitioners(s).

Thus, if one reserves "technology" for activities which are fabricative in the strong sense and which are not disqualified by any of the other aspects discussed, then it follows that the work of our organic chemist is, although akin to, not a bona fide instance of the activity-form technology. To the extent that this synthesizing activity is felt to be an instance of technology, this may reflect the family resemblance that obtains between the sets of aspects or features of the activities of our chemist and those of paradigmatic technological activities. If, however, one has no quarrel with the weak, second-order, indirect sense of "fabricative," then, other necesary conditions being satisfied, one is liable to maintain that it is possible for a specific activity to be legitimately viewed as being *simultaneously* a token of the activity-types science and technology. This possibility, however, does not in any way negate the important differences between these two activity-forms.

The fourth aspect of the activity-form technology is *resources*. Technology is a resource-based, resource-expending form of human activity. In the course of making its products or transforming its objects, technological activity utilizes a variety of resources, some of which are precisely its prior products: technics and extracted raw and processed materials. In addition, technological resources include information, available energy, nonprocessed and, on occasion, nonextracted materials, people (witness the construction of the Egyptian pyramids and the Alaskan pipeline), and, more recently, capital, a resource increasingly needed to acquire other resources.

The fifth aspect of technology is *knowledge*. Technology is based upon, utilizes, and generates a complex body of knowledge, part of which may reasonably be called specifically technological knowledge. Evidently, knowledge might be viewed as of a piece with the technological resources noted in the preceding section. However, since technology is often thought to differ from science precisely on the ground that it, unlike science, has essentially nothing to do with knowledge worthy of the name, it seems worthwhile to individuate it as an integral aspect of technology.

The knowledge in technology is not wholly reducible to nonintellectual (e.g., neuromuscular) skills or to independently developed scientific knowledge. There seem to be three components to the body of technological knowledge. First is the knowledge of how to do certain things by making or utilizing certain material products or by transforming certain material objects. Consider the following: there was once a time when it was not part of this component of technological knowledge that bodies located in different planes could have their movements coordinated by a system of linkages or interlocking gears. The same is true of the making, controlling, or extinguishing of fire. Granted, the discovery and application of such possibilities may well have involved certain skills, mental, physical, or mental-physical (e.g., perception). However, it seems likely that the processes of coming to "know how to do" that have occurred in these cases were bound up with grasping certain generalizations of the form "if P (is done), then Q (happens or can be done)." Moreover, such acquisition of knowledge may be coupled with insight into how such results can best be achieved and into the designs,[9] however rudimentary, for achieving them. If it is seen that throwing sand or water on fire can, under certain circumstances, be effective in controlling or extinguishing it, this knowledge, although until relatively recently wholly nonscientific (in the sense of being grounded in the understanding of underlying scientific laws), is not simply an instance of nonintellectual "know-how." Increases or improvements in "knowing how to do" are not only compatible with the achievement of new intellectual knowledge but are often undergirded by such knowledge.

The second component is the knowledge of the resources, especially energy and materials, used in technological activity, both of their general properties and those of the resources used on a particular occasion, the more so if those selected (or rejected) have been tested in some way or other. If so, the resultant test data—part of the activity's information resources—may also be viewed as part of this component of technological knowledge. Indeed, such data can be correlated and used for predictions at what might be termed a pretheoretical level. The engineer's prediction of matters as apparently simple as the pressure drop in turbulent flow through a pipe is based on such a procedure even today. There is, in fact, a body of "theory" on how to construct such correlations.[10]

This brings us to the third component of technological knowledge, consisting of knowledge of the methods used in reaching the desired result of the activity. It is artificial to maintain that in all cases this knowledge is nothing but a species of nonintellectual know-how, particularly for modern engineering, many of whose practitioners have intellectual knowledge and understanding of a host of sophisticated methodologies, e.g., that of "parameter variation."[11]

In sum, there is a broad spectrum of kinds of knowledge integral to technology, one which has evolved over historical time: from inchoate, atheoretical praxis to skilled, purposive correlation and planning methods for taking and organizing data in the name of some form of improvement or optimization. That much technological knowledge of the kinds distinguished above has traditionally[12] arisen from praxis or inductively from acquaintance and testing, rather than deductively from scientific laws, does not negate its status as bona fide knowledge.

The sixth aspect of technology is *method*. Technology incorporates and proceeds in accordance with certain methods. Method would seem to have always been an element of technology in several ways. First, method has been present in the process of determining whether the production of a technic in one way should yield to a different, possibly better way of making it. The traditional method for doing so has been that of trial and error, while in modern times there are different, more formal methods used in making such determinations. On a more microlevel, method has always been present in technology in the more or less careful selection of materials and in the development and employment of techniques used to acquire, prepare, and formally or informally test them as well as to do likewise to the result of the technological activity. Method is involved in technology in a third sense, in that technological activity generally proceeds through its various stages in a nonarbitrary order, one not dictated primarily by the law of nature, but by the exigencies of fabrication and the desire to realize a successful outcome. From initial conception of desired goal to final testing of material outcome, there is a multileveled methodicity in the procession of technological activity; whether in the manufacture of stone axes with its rudimentary methodicity, or in the development of windmills, water wheels, propellers, and space capsules with the complex of more sophisticated methods involved in making those technics.[13]

The seventh aspect of technology is the *sociocultural–environmental context* within which the activity unfolds. This aspect might be called an informing and enabling rather than a constitutive one, although it is inseparable from technology and no less important than the preceding aspects for understanding it. With one kind of possible exception, technology always unfolds in a sociocultural context. The possible exceptions involve early practitioners, some of whom may have made their tools in isolation and from scratch, i.e., such that their methods, knowledge, resources, and purposes were not provided for them by a social unit or cultural tradition. In all other cases, such provision makes the sociocultural context a condition for the possibility of technological activity. In the exceptional case, it is evidently the environmental context that shapes technology, affecting the products produced, purposes pursued, resources utilized, and methods employed.

Technology depends on sociocultural context in a second way: the economic, political, ideological, and social interests of various parties may have directive effects on technological practice.[14] Consider the effects of economic conditions on research into energy and automative technologies, of international politics on weapons technology during the Cold War, of claims to ideological superiority and fears of losing prestige on the development of aerospace technology needed to reach the moon, and of social unrest on research into optical recognition devices used to keep nonresidents from entering apartment buildings. More strictly cultural factors can exert similar directive pressures. The effort invested in sophisticated biomedical technology is not unrelated to Western attitudes toward death and identity. If we awoke tomorrow imbued with Schopenhauerian world views, would we remain as supportive of and interested in, say, research in prosthetic and cryogenic technology aimed at increasing human longevity?[15]

A third connection between technology and its sociocultural context lies in the fact that the making and operation and use of technics frequently and increasingly take place in settings which may reasonably be termed sociotechnical systems—bounded complexes of interacting social and technical elements—e.g., research and development laboratories, armies, factories, hospitals, communications and transportation systems, etc. Indeed, much of modern technology, both its practice and the use of its products, can occur only within the context of such large-scale sociotechnical support systems.

The eighth and final aspect of the activity-form technology is, like the seventh, more informing than constitutive, but is equally inseparable from technology: that part of the *practitioner's mental set* which relates to any of the preceding aspects, as well as to the meaning and significance of his or her activity. In what relations do practitioners of technology stand to its results, purposes, knowledge, resources, methods, and sociocultural-environmental contexts? Of concern here are both cognitive and affective phenomena: those of perception (including observation and prehension), thought, and reasoning, as well as those of feeling and emotion. It would be surprising indeed if the appropriate mental sets of Yir Yoront makers of stone axes (a totem in their society, indeed one present in their archetypal world view), Samurai warrior makers of swords, the craftsmen in George Sturt's wheelwright shop in Farnham, England, in the 1880s and 1890s, and today's General Motors automotive engineers were not significantly and interestingly different.

Here are two examples of technology-related mental sets which are, it would seem, worlds apart. The following rich passage is from Marx's essay "On James Mill" and presents the author's romantic conception of the complex, self-conscious mental set of the unalienated practitioner of technology:

> Supposing that we had produced in a human manner; each of us would in his production have doubly affirmed himself and his fellow men. I would have (1) objectified in my production my individuality and its peculiarity and thus in my activity both enjoyed an individual expression of my life and also in looking at the object have had the individual pleasure of realizing that my personality was objective, visible to the senses and thus a power raised beyond all doubt. (2) In your enjoyment or use of my product I would have had the direct enjoyment of realizing that I had both satisfied a human need by my work and also objectified the human essence and therefore fashioned for another human being the object that met his need. (3) I would have been for you the mediator between you and the species and thus been acknowledged and felt by you as a completion of your own essence and a necessary part of yourself and have thus realized that I am confirmed both in your thought and in your love. In my expression of my life I would have fashioned your expression of your life, and thus in my own activity have realized my own essence, my human, my communal essence. In that case our products would be like so many mirrors, out of which our essence shone.[16]

Compare the mental set evoked by Marx with that of the individual quoted in the middle of the following passage. Speaking of the Italian passion for embellishment, Luigi Barzini recalls:

> I remember examining an Isotta Fraschini motor many years ago in New York, with an eminent American automotive engineer. He looked at everything for a very long time, then shook his head, looking very puzzled. He asked me: "Can you tell me why parts that do not work are as polished and well finished as those that work? This seems to me to be an irresponsible waste of labor and money." I did not explain to him that one could not restrain Italian engineers, designers, and workers from beautifying everything they made. In fact, most Italians thought that a pleasant appearance was as important as (and often more important than) usefulness. *"L'occhio,"* they always said, *"vuole la sua parte."*[17]

Evidently there are myriad technology-related mental sets. Among their many elements the following would seem worth investigating: What is the practitioner's attitude toward aesthetic considerations in technological activity, both in regard to the environmental effects of resource acquisition as well as vis-à-vis the final product? What is the nature of his conception of the beautiful? Is the practitioner a perfectionist in technological activity? Are personal identity and self-esteem intimately bound up with his work? If so, in what ways? Does the practitioner perceive work as unfolding in a larger social context, as contributing to the strengthening or weakening of a way of life? Does he feel in any way personally responsible for what is done with his products? What are his conceptions of efficiency, of human progress, of the good life—if he has such notions? What are his animating values? Finally, what is the practitioner's concept of nature and of relationships between man and nature? Some of these elements will be at least in part reflections of or responses to cultural milieus. Others, perhaps even more difficult to identify, will be substantially the expressions of individual idiosyncrasy. Investigation of practitioners' technology-related mental sets may contribute to a more complete elaboration of the thought component of technology. If ignorance or neglect of this component has contributed to the mistaken view of technology as less of a human activity than, say, religion or art, as less intellectual an enterprise than, say, science, perhaps attention to its affective subcomponent will help combat the equally mistaken view of technology as essentially a matter of bloodless precision.

To summarize our characterization of technology: it is a form of activity that is fabricative, material product-making or object-transforming, purposive (with the general purpose of expanding the realm of the humanly possible), knowledge-based, resource-employing, methodical, embedded in a sociocultural-environmental influence field, and informed by its practitioners' mental sets. Technology can be differentiated from, as well as related to, other forms of human activity (e.g., science) by comparing it with them regarding differences in substance in the general aspects identified and discussed above, particularly the initial six: its "outputs," the adverbial quality of its processes, its purpose(s), knowledge, resources, and methods.

Before turning to the second part of my answer to the question "What is technology?", here is a fascinating example of a person actually engaged in the activity of technology in one of its fields: canoe technology. I have selected from a detailed account of this man's work,[18] passages which reveal the presence of the above-mentioned aspects of technology, each of the aspects taking on particular substance in the example.

The person in question is one Henri Vaillancourt, of Greenville, New Hampshire, one of the four or five remaining professional makers of bark canoes. To save time and avoid belaboring the obvious, I shall minimize comments on the connections between the passages and the aspects:

[Technics]: Henry makes several different kinds and styles of bark canoes, making all but one of their parts from basic raw materials obtained from the woods.

[Purpose]: He became passionately interested in Indian life, and "the aspect of it that most attracted him was the means by which the Indians had moved so easily on lakes and streams through otherwise detentive forests. He wanted to feel—if only approximately—what that had been like. His desire to do so became a preoccupation. . . . He formed an ambition, which he still has, to make a perfect bark canoe, and he says he will not rest until he has done so." He has visited almost all the other living bark canoe-makers, and has learned from the Indians. He has returned home believing, though, that he is the most skillful of them all.

[Resources]: Virtually all the material resources he uses are obtained directly from nature: from white birch, cedar, black spruce, and white pine trees, to porcupine quills. He has, however, taken to substituting plain asphalt roofing cement for Indian gum made from the pitch of white and black spruce. The only resources of technics he uses are an ax, a froe, an awl, and a crooked knife—no power tools, nails, screws, or rivets. The energy used is that of his body. The only capital he requires is that needed to purchase his tools. He obtains that by selling his canoes (as well as paddles and snowshoes). Apart from the companies that make those tools, Henri's technological activity does not rely in any crucial way upon institutional resources.

[Knowledge]: Henri's mastery of canoe technology is based on a substantial body of knowledge of several kinds. "In 1965 he laid out a building bed, went out and cut bark and saplings, and began to grope his way into a technology that had evolved in the forest under anonymous hands and—as he would learn—was too complex merely to be called ingenious." When he finished it he destroyed it with an ax. "It was a piece of junk," he explained. After visiting the few remaining Indian practitioners in Canada he found his master through the printed sketch and word: Edwin Tappen Adney. "Adney was an American who went to New Brunswick in the 1880s and built a bark canoe under the guidance of a Malecite [Indian] . . . he recorded everything the Malecite taught him. For the next six decades, he continued to collect data on the making and use of bark canoes. He compiled boxes and boxes of notes and sketches, and . . . made models of more than a hundred canoes, illustrating different tribal styles, differences within tribes, and differences of design purpose. A short, low-ended canoe was the kindest to portage and the best to paddle among the overhanging branches of a small stream. A canoe with a curving, rocker bottom could turn with quick response in white water. A canoe with a narrow bow and stern and a somewhat V-sided straight bottom could hold its course across a strong lake wind. A canoe with a narrow beam moved faster than any other and was therefore the choice for war. Adney so thoroughly dedicated himself to the preservation of knowledge of the bark canoe that he was still doing research, still getting ready to write the definitive book on the subject, when, having reached the age of eighty-one, he died." The book was compiled and published by the Smithsonian Institution in 1964. Vaillancourt has also acquired extensive knowledge of his *materials:* "You split cedar parallel to the bark," he commented. "Hickory you can split both ways. There are very few woods you can do that with." He makes his canoes in the shade because "in direct sunlight the bark becomes less flexible, and it needs all the elasticity it can retain during the building process." "If the month is June or early July, the bark will almost pop off by itself. . . . Winter bark is really tight. I'm telling you. You need a special tool, a type of spud, to get it off." Like the Indians, Vaillancourt will spend long periods of time roving birch-filled woods, carefully looking for a perfect tree. Moreover, Henri has intimate knowledge of the Indian *methods* and has mastered the skill of using the crooked knife, his finishing tool. His knowledge relative to canoe-making is thus in significant part but by no means wholly nonintellectual skill, even if all of that knowledge is ancillary to a practical technological goal.

[Method]: There is a definite general order in which the bark canoe is assembled which differs substantially from that of the canvas canoe. "In the latter, the canvas comes last. It is stretched across an essentially completed and rigid frame. The Indian, on the other hand, began the assembly with bark. He rolled it right out on the building bed, white side up, and built it from there." There is also method in the sense of technique in many of the stages: when putting the ribs in with a mallet as the canoe moves close to completion, Henri says, "you work from the ends toward the middle, putting the ribs in. You pour very hot water on the bark to keep it flexible at this stage." Method is involved in testing samples of bark in the woods: "[After unloading his canoe last night, Henri] went straight to one of the big birches and took a sample of its bark. Along the grain he bent the sample, steadily applying pressure with his fingers as the bark formed the shape of a U. Gradually he increased the pressure until the bark cracked. It was pretty good bark, he said. It had good relative elasticity. Its layers did not tend to separate." He is presently experimenting with abrasive bone tools of the Indians which had been replaced by the steel tools of the white man.

[Sociocultural-environmental context]: On the environmental context side, there is the obvious connection between the nature of the Indian's forest-cum-lake environment and the character of the canoe technics they built. In Henri's case, the initial choice of bark as his basic material was a product of two facts: he was too poor to afford processed materials like canvas, aluminum, or fiberglass, and the woods around his town were packed with white birch. On the sociocultural context side, two points are noteworthy: Originally in paying the seams of their canoe hulls, the Indians used the gum of the pitch of white and black spruce which they obtained by the culturally transmitted method of tapping trees, analogous to the collection of maple sap. "They boiled the pitch and strained it, and at tribal campgrounds they had communal pitchpots where anyone could go for sealant to touch up a canoe." Secondly, Henri, as noted above, although eschewing power tools, has switched to asphalt roofing cement for his sealant, because collecting and boiling spruce pitch involves unnecessary tedium, not clearly an Indian cultural attitude.

[Mental set of the practitioner]: Elements of Henri's mental set in relation to each of the foregoing aspects are implicit in the above. He is self-consciously a perfectionist, a sentinel for the tradition of bark-canoe-making, aesthetically inclined, and deeply attached to each of his canoes. When he travels he visits his canoes and loves to get them back to "the yard" for repairs. He decorates them with Indian emblems and symbols using porcupine quills. "I've never seen a bark canoe that wasn't graceful," he says. William Morris, take heart!

Turning to the second part of my answer to the question "What is technology?", having dealt with the general structure of the activity-form, I should like to examine what I call "the content of technology." To do so, I need to make clear what is meant by *a technology* (as opposed to, say, several technologies) rather than by speaking of technology as a form of human activity or of "technology as a whole," whatever that may mean. What are some examples of technologies in this singular/plural sense? Canoe technology, aerospace technology, contraceptive technology, pollution control technology, and tunnel technology. But what kind of a thing is "a technology"? Although they are related, there is a subtle but important difference between a technology of a certain kind and the *process* of manufacturing the associated kind of technic. For example, to speak of what canoe technology is, per se, is not to speak of the process of making canoes, whereas to speak of the practice of canoe technology does entail talk about the process of making canoes. By "canoe technology," at a given point in time, we mean the complex of knowledge, resources, and methods used in making canoes. In general, by a (particular kind of) technology, we shall mean the complex of knowledge, resources, and methods involved in the production (transformation) of the associated technics (quasi-technics) of that kind. We may now indicate what we mean by "the content of technology." As a first approximation we shall say that the content of technology is the set-theoretical union or sum of all particular (kinds of) technologies, each of which is a complex of the kind just discussed. However, since technologies change, new ones coming into existence, old ones lapsing into desuetude, the content of technology is evidently a function of time. Thus we should speak of the content of technology at a particular point in time. Moreover, a given kind of technic may be made differently in different societies at the same time. Think of contemporary bicycle technology or of the fact that the Bic Company currently manufactures and sells a nineteen-cent disposable razor only in Greece. Thus the dependence of the content of technology on a societal variable should also be made explicit. One may then speak of, say, medieval Chinese technology or of British technology in 1800. Finally, the content of technology at a particular point in time in a given society is the union of all technologies practiced in that society at that time.

What is the relationship between technology as a form of human activity and technologies? Technology, as a form of human activity, is variously incarnated or embodied in the practice of each technology. Technologies may be individuated or related by characterizing the substantial differences or similarities in technics, specific purposes, methods, bodies of knowledge, resource bases, contexts, and practitioners' mental sets associated with the different incarnations or embodiments of the activity-form technology in the practice of the technologies under consideration.

We conclude by making use of this more differentiated concept of technology in clarifying the controversial issue of whether technology is "value-laden" or "value-neutral."

Technology: "Value-laden" or "Value-neutral"?

There are at least five ways in which technology is a value-laden enterprise.

1. If, following Kurt Baier,[19] the value of a thing consists in its ability to confer a benefit on someone (by whom or in whose behalf it is used), then to the extent that it is successfully practiced, technology is inherently value-laden. For, as our discussion of the general and specific purposes of technology indicated, technics, the primary products of technological activity, are made precisely in order to fulfill value-laden purposes, whether of practitioners or of those in whose service they labor. Thus the (noneconomic) *value* of a technic generally reflects the *values* of those for whose purposes it is made and used.

2. Technology is also value-laden in a sense similar to that in which, according to Neitzsche's analysis, *Wissenschaft* is. For Nietzsche, science, a different form of human activity, constitutes an essentially Socratic optimistic and would-be triumphant approach to dealing with the problematic nature of the human condition, as opposed to, say, the consolatory approach implicit in the tragic world view.[20] Science, Nietzsche claims, places something akin to an absolute value on truth and assumes that life can be made meaningful by understanding it through rational, intellectual means.[21] As a different way of appropriating the world, technology, with its animating Promethean *Geist,* is also optimistic and assigns a kind of categorical value to "technological progress." The predominant spirit informing post-Renaissance technological activity, a spirit liberally fueled by its remarkable successes, assumes that, *ceteris paribus,* the human condition can only be ameliorated and rendered more meaningful by ongoing technological progress (i.e., improvements in the knowledge or resource or methodological sectors of various technologies).

3. Technology is value-laden insofar as use of resources to advance its "cutting-edge" may preclude their use in the more prosaic work of ameliorating people's everyday lives. There is, of course, nothing in the nature of the activity of technology per se which necessitates such value-laden trade-offs, although the categorical value widely ascribed to technological progress makes such conflicts more likely to arise in the modern world of Spaceship Earth.

4. Another extrinsic dimension of technology's value-laden character arises from the institutionalization of modern technology, i.e., from the institutional locus of its practice in modern society. Technology's products and the values they serve are increasingly determined not by practitioners' values and specific goals but externally, by those directing the institutions controlling the resources requisite for modern technological practice.

5. Finally, technology is value-laden in the young Marxian sense that its products are generally expressions, objectifications, or crystallizations of at least some of the individual preferences, tastes, idiosyncrasies, world views and cultural values of their designers or makers (or those for whom they labor). For example, preferences for or against ornamentation, unalloyed functionality, maximum efficiency, durability, or convenience, character, organicity or simplicity, may be and often are reflected in technics and make them expressive of a range of evaluative attitudes at best tangentially related to the central specific purpose the technic is intended to serve. The quotation concerning the Isotta Fraschini motor illustrates this point nicely.

To sum up, I have tried to answer the question "What is technology?" by distinguishing between the structure and the content of technology; i.e., between the various aspects of its structure as a form of human activity and the various technologies, the practice of each of which embodies the activity-form technology, the union of which makes up what I called the content of technology. After illustrating the presence of these structural aspects in a particular technology, that of the bark canoe, I made use of this more differentiated concept of technology in elucidating the value-laden character of technology. It remains a task for another occasion to elaborate and analyze the structure of the other branch of technological activity to which I referred at the outset: the design of sociotechnical systems in which technics are used. It is appropriate that this task be deferred temporarily since, in response to environmental, consumer, technology assessment, and accountability movements, the designers and operators of sociotechnical systems are beginning to incorporate steps which, if widely adopted, will lead to significant changes in this branch of technological activity.[22]

FOOTNOTES

1. C. Singer, E. J. Holmyard, and A. R. Hall, eds., *A History of Technology* (Oxford University Press: Oxford, 1954), Vol. I, p. vii.
2. Emmanuel G. Mesthene, *Technological Change* (Mentor: New York, 1970), p. 25.
3. Donald A Schon, *Technology and Change* (Delacorte: New York, 1967), p. 1.
4. John J. Sparkes, "Technology and General Education," *Prospects,* Vol. 4, 1974, p. 94.
5. Below, when a statement is made about technics, the parallel statement about quasi-technics will not ordinarily be affixed.
6. Edwin T. Layton, Jr., "Technology as Knowledge," *Technology and Culture,* Vol. 15, Number 1, January, 1974, p. 37.
7. I am indebted to Professor Robert J. Baum for this refinement.
8. This expression is not used here in a pejorative sense.
9. To some readers it may seem that the notion of *design* has been given short shrift in this essay. Perhaps, it may be argued, design should have been accorded the status of one of the individual aspects of technology. I prefer, however, to think of design as a stage in the process of making (many) technics or their prototypes in which knowledge, methods, and resources are drawn upon and utilized to formulate plans for the fabrication of technics. Design has emerged as a vital stage of most processes for making technics only in the modern era. To claim that design has always been part and parcel of technology, including, say, ancient technology, is to reduce "design" to a point at which it amounts to little more than a conception of a specific technological purpose and an image of a desired technical outcome.
10. See, e.g., P. W. Bridgman, *Dimensional Analysis* (Harvard: Cambridge, 1922), and S. J. Kline, *Similitude and Approximation Theory* (McGraw-Hill: New York, 1964).
11. Walter G. Vincenti, "The Air Propeller Tests of W. E. Durand and E. P. Lesley: An Example of Engineering Research by Parameter Variation" (unpublished paper).
12. In modern times an increasing portion of the resource component of technological knowledge, particularly as regards material resources, has become more specifically scientific in nature.
13. Methods used in the practice of modern technology include a variety of qualitatively distinct procedures and techniques: e.g., descriptive geometry, design procedures, comparative systems studies, testing, similitude techniques, order of machining and assembly, and parameter variation.
14. See, e.g., Nathan Rosenberg, "The Direction of Technological Change: Inducement Mechanisms and Focusing Devices," *Economic Development and Cultural Change,* Vol. 18, No. 1, Part 1, October 1969, pp. 1–24.
15. One would like to see serious research aimed at articulating various aspects of the cultural milieu which have affected the development of specific technologies, research complementing that devoted to the more familiar politico-economic directive factors.
16. Karl Marx, "On James Mill," in *Early Texts,* David McLellan, Transl. and ed. (Blackwell: Oxford, 1971), p. 202.
17. Luigi Barzini, *From Caesar to the Mafia* (Bantam: New York, 1972), p. 269.
18. John McPhee, "The Survival of the Bark Canoe," *The New Yorker,* February 24, 1975, pp. 49–94.
19. Kurt Baier, "What is Value? An Analysis of the Concept," *Values and the Future,* K. Baier and N. Rescher, eds. (Free Press: New York, 1971), pp. 33–67.
20. Friedrich Nietzsche, *The Birth of Tragedy* (Vintage: New York, 1967), pp. 95–96.
21. Nietzsche, *On the Genealogy of Morals* (Vintage: New York, 1969), p. 151.
22. I am indebted to my colleagues S. J. Kline and W. G. Vincenti for detailed critiques of an earlier draft of this essay.

SHIFTING FROM PHYSICAL TO SOCIAL TECHNOLOGY

Joseph Agassi BOSTON UNIVERSITY AND TEL AVIV UNIVERSITY

INTRODUCTION

The following essay was read to an audience well prepared for both a holistic approach and for participatory democracy. The holistic approach is that which endeavors to see each detail, each act, however small, not in isolation but in relation to the main features of its settings and environment. The chief holistic point of this essay is that technology is in part concerned with machines, in part with humans, and only the artificial act of isolation—at times laudable, at times lamentable—distorts an item of technology to look as if it were merely physical technology or merely social technology. So this chief point of the present paper was well received, yet I elaborated on it more than I would have done had I been more familiar with my audience. The second point was that of participatory democracy. Following E. M. Forster *(Two Cheers for Democracy)* and K. R. Popper *(The Open Society and Its Enemies)* I consider human fallibility and the resultant need for criticism ample justification for democracy. But I find participatory democracy the only one that works; this because of the fact that unless people learn to exercise their democratic rights regularly and unless they acquire and develop some experience of political responsibility, democracy is likely to be a mere formality. And in participatory democracy I envisage both centralized government and industrial democracy and open schools (i.e. democratically run self-governed ones). It was curious to find a conference intent on participatory democracy questioning it after hearing a paper which defends it. I suppose the cause of this is the absence of a detailed defense. But I cannot do participatory democracy any justice in a paper not devoted to it. Indeed, each of the three main fields of participatory democracy, decentralized government, industrial democracy (including job redesign, etc.) and open schools, each of these deserves extensive independent studies.

The question of the present study is, however, more practical in its thrust: How can we quickly bring about significantly more participatory democracy?

I. PHYSICAL AND SOCIAL TECHNOLOGY INTERTWINE

Technology has many aspects—physical, psychological, sociological. At times the physical side is prominent simply because the physical side of a technological problem reigns supreme. Thus, when we want to protect England from air raids by building a barrier that will prevent any metallic object from going through, the social and other aspects of the problem are taken for granted, and one immediately strains one's imagination and thinks of force fields that should do it, and, of course, finds the problem insoluble. The very insolubility of the physical aspect of the problem makes it look as if the problem were purely physical. Yet suppose the physical problem were solved. There would arise, at once, all sorts of social problems, organizational problems, etc. For example, the British would want to be able to go through the barrier, and for that they would need certain arrangements to bring it down at prearranged secret gaps in time. Their frequency would depend on the cost of lowering and reraising the barrier, the importance of the missions, etc. There would soon be moral problems: a stray pilot has failed to return but is not reported missing; should we permit the risk of his hitting the barrier or should we permit the risk of a stray enemy rocket going through? The cost of the barrier would decide, even, whether it be raised day and night or nights only.

This is the stuff science-fiction is made of. And the difference between science-fiction and fiction about current technology is merely in this. Science-fiction assumes at the very start that a certain problem in physical technology which is clearly beyond our grasp, whether in principle or in mere fact, is solved, and proceeds to other aspects of the technological problem; i.e., it proceeds to the social, economic, political, and psychological problems involved in the implementation of the solution to the physical technological problem. Indeed, this, according to Isaac Asimov, is what generally characterizes science-fiction; and he calls it social science thought experiments. We can just as well see in fiction and nonfiction similar examples of social science experiment, where the physical technology has offered a gadget and its implementation raised other problems. There are stories where the obstacle to implementation was popular prejudice, whether religious or social. And, both sociologists and social anthropologists observe, solving the social side of the implementation of a new gadget may change the face of a society. I shall discuss this later on.

In parallel to the solution, real or imaginary, of the physical aspect of a technological problem, which is so paramount as to make us oblivious of the lesser, social aspects of the problem, we can have the solution, real or imaginary, of the political aspect of a technological problem, which is so paramount as to make us oblivious of the lesser, physical aspect of the problem. When the political solution is imaginary, it is called utopian. The difference between a utopia and an Asimov-type science-fiction is just here: the utopian writer solves a political problem and only works out some physical technological corollaries to his solution, whereas the science-fiction writer goes the other way round. It is therefore not very surprising that, whereas the utopian sets his society in a surrounding of a fairly primitive physical technology, the science-fiction writer puts his advanced physical technology in politically primitive societies. No one knows that better than Asimov. In one of his science-fiction works a secret document of supreme technological importance is the long-lost Constitution of the United States. In his Foundation trilogy, Selden is the echo of Karl Marx and Galactic History is a large-scale replica of the Marxist view of history. His *Caves of Steel* and *Naked Sun* depict two extreme societies, one overcrowded, poor planet and one almost uninhabited, anomie-ridden, rich planet; and both allude to a third, happy medium somewhere far away—clearly Asimov's ideal is a Europe whose happy medium is between Russia and America.

That physical technological developments play an important role in almost all utopias has been recently overlooked due to the fact that Plato's *Republic* is now viewed as the blueprint for all blueprints of all utopias; but it is Sir Thomas More's *Utopia* itself where physical technology flourishes for the first time. And, need one say, in Marx's communist regime, that utopia to end all utopia, physical technology reigns supreme and wipes out all problems of social technology, sociological, political, even organizational and psychological; and it does so by eliminating all economic problems through the creation of affluence. But since affluence is a relative matter, never an absolute one, the Marxian utopia is the wildest of them all. What it has done to the modern world is to give it the illusion that we need no utopias and that the physical technological aspect of any technological problem is all that matters. Any reader of that pathetic breed of literature called Soviet science-fiction knows its poverty, just as any person familiar with the opposite literature, the protest science-fiction of Russia, from Zamyatin to Abram Terz, has felt it: science-fiction and Marxism do not mix in a communist regime half as well as they mix under capitalism so-called.

So much for fictitious solutions of the political aspect of technological problems. One might wish to add that the fictitious solutions can afford to be large-scale, unlike the real solutions we have thus far attempted. But this is a gross error. First, we have the large-scale political solutions known as the political revolutions. Besides this, however, we have no less large-scale solutions in terms of foreign aid, in terms of planning large cities like Brasilia, in terms of attempts to solve large-scale political, economic, or social problems. The most important of these are largely aspects of one urgent large-scale problem of today, that of bridging the gap between the rich and the poor parts of the world.

The first large-scale solution that comes to mind is that of economic foreign aid. It had many aspects, from physical technology to the psychology of sex. The bold—and I am afraid materialistic—solution was the hypothesis that foreign economic aid will do as foreign aid. It did not do. The problem is more intricate, and includes that of changing people's values, especially attitudes to child-bearing, and to the position of women in society.

Traditionally, we were told, technology is value-free: given some ends and circumstances, the technologist tries to forge the means to achieve these ends in those circumstances. But, as Jarvie observes, just as technology changes circumstances, it also changes ends. Already operational rescarch studies thirty years ago claimed that ends are often impossible, perhaps even inconsistent, and must be reshaped before any physical step can be recommended or even before analysis begins.

But let us leave the large-scale. What inventors know and readily tell us, but what neither social scientists nor theoreticians of technology have paid any attention to, is a very simple fact. Given that in our society women were trained to prefer technical ineptness to technical competence, and given that in our society the household was the province of women, the inventor who wished to sell gadgets for the household must invent highly simplified ones. In this very contention there is a hornet's nest hidden. First, simplified means simplified for the user. Second, you can simplify the use of a machine for a user by making all servicing of the machine, however easy and simple and routine, a matter of repair. To reduce cost of repair, then, you must make servicing more infrequent, perhaps at the cost of increasing the initial cost of the purchase of the machine. I do not mean to say that household implements are nowadays much simpler than other instruments, nor do I know as an empirical fact that attempts to simplify implements for household use was the major impetus in the development of contemporary gadgetry. Evidently, the philosophy of planning the worker's job for him, especially Taylorism which was meant to break down jobs to minute components and leave the thinking out of the worker's task, this philosophy interacted in a strange way with the development of implements and tools other than those used in industry. In particular, industrial society gave rise to a new type of citizen, much more at home in the world of gadgets than ever before.

The altered attitude to gadgets is, indeed, just the tip of an iceberg. When a European writer puzzles over his American counterpart's use of the typewriter the way he uses the quill, his puzzle can only be resolved by revealing the great difference between the attitudes of the two over a wide range of affairs.

Let me end this broad survey with one of Edison's few failures—although it is quite commonplace now. He invented a gadget for voting: every senator can have a small set of buttons on his desk, and his action on them should record his vote; in a matter of seconds the final outcome of a voting process can stand clearly before the Speaker's eye or on a large board for all to see. Such machines, as I say, are extremely common these days. Yet for mere political reasons Edison could not implement it. And even today, many a time, a senator who has this machine and can use it prefers to vote by a roll call.

Let me conclude, then, by repeating my first claim: technology has many aspects, but we single out, in each technological specification or task, the most obviously immediately problematic aspect of it, and classify the given technological specification or task accordingly. Thus, if the social aspect is the more problematic, we call the specification social technology and the task social engineering—whether planning a school or an educational system, whether planning a library or an evening of entertainment or even an industrial concern. When the specification or task is more evidently physically challenging we have on our hands physical technology and physical engineering. Traditionally this is better known as technology and as engineering *tout court;* and other kinds of technology and engineering—social, agricultural, environmental—are known as planning. Technology is really planning theory, as opposed to engineering; but there are almost no social, political, economic, etc., theories of planning, as opposed to just theories or mere practice. It is, indeed, theories of practice, as opposed both to theories of what there is and to mere practice, that need and demand and require our attention.

The important corollary to this is rather obvious. As long as technology was viewed as physical technology only, or biological, for that matter, its moral credentials were *prima facie* unquestionably excellent: it was man's mastery over nature that was its business. And so, social technology immediately smacked of man's mastery over his neighbor, and so its moral credentials could not be seen but as at least *prima facie* very poor, and perhaps also irredeemably very poor. But if all technology, including physical technology, is also necessarily social to some extent, then the credentials of all technology, physical and social, are questionable. Marx deplored the use of the machine for exploitation, yet he was sure that once the social impediment was eliminated, technology could be purely physical and so with no exploitation. Today we are less optimistic and wish to plan and design and supervise our industrial venture as social engineering: we want industrial democracy, job satisfaction, the humanization of industrial relations, etc. We want to plan what Marx thought comes easy (after the revolution): we want to eliminate job alienation by direct planning.

II. WHY PHYSICAL TECHNOLOGY IS DOMINANT

Seeing, then, that technology is a complex of physical and social problems, we can ask, Why is technology viewed mainly as physical technology, i.e., whose difficulties are largely physical; why, when social technology was introduced was there so much objection to the very concept? The answer is complex, but clearly it has its roots in materialism, i.e., the philosophy that all problems will be solved with the sufficient increase of man's control over physical nature. Of course, this is not the whole answer, since experience had to support the philosophy for it to survive. And it was, indeed, spectacularly supported by experience. The reason, it seems, was twofold. First, the social organization permitted the implementation of many solutions to problems of physical technology, old and new. Second, the reforms in the social and organizational parts of the system necessitated for the implementation of technological innovation to begin with, were made, the reformers being quite unaware perhaps, by the very introduction of materialism. This is an oversimplification, since some reformers were more aware than others and since the political system was not so fortunate as the social and economic and organizational systems: the Marxist view of the French Revolution and the Napoleonic Wars as the spread of both materialism and liberal capitalist governments is an oversimplification, to be sure, yet it contains an important kernel of truth. But the chief disaster caused by the unbalanced introduction of technology as the merest physical technology is, of course, the ecological calamity in the developed countries plus the population explosion of the undeveloped ones.

Once we realize that equating technology with physical technology means the oversight of social and political technology, then we should not be so surprised at the great tremors to society and even to our physical existence that technology causes: old-fashioned technology is too short-sighted for words.

Even before the technological disasters were so pressing as they are these days, it was clear that the high premium put on physical technology and the disregard for social technology are painfully expensive. Thus, one of the obvious blessings of technology is easy travel. In spite of the enormous organizational developments that involved easy travel, easy travel was, still is, described in histories of technology as mere physical technology, with the organizational aspect receiving no more than a glance of acquaintance, at times also a nod of approval. But Ellis Island, passports, visas, and quotas—all these and other limitations on travel—are organizational afterthoughts, poor parts of one large solution which we artificially divide into the physical, the economic, the political. Call it what you will, it is pointless to shorten the time of a transatlantic flight or increase its comfort, when the international airport is becoming increasingly impossible to land in, to pass passport and custom control in, to leave by ground transportation; the meeting of relatives and friends; not to mention economic constraints that make travel inaccessible to most people today.

This case can serve us as a mere metaphor for all sorts of developments, and so we can say this. When technology still centers on the transatlantic flight even though the trouble is increasingly at the flight terminals, merely because the technology of the flight is largely physical but that of the flight terminals is largely social, then it is clear that diminishing returns have set in, and that we must reallocate our resources of research technology.

Can we do this? Can research resources at all be allocated? If so, how and at what price? Who, then, is going to engineer this reallocation? This, I think, is today's pressing and urgent problem. Unless we solve it well and fairly quickly, then possibly the problem will soon radically alter; perhaps there will even be no problem to solve soon enough: the future of life on this planet may itself be at stake.

The questions at stake are not easy. They relate to diverse areas or fields of inquiry, most of them virgin soil. To begin with, can we direct interest of researchers? This is a very difficult question. Many people, expert and inexpert, are convinced that tastes are not given to control without severe damage. Yet, even with no controls, there are vast uniformities of tastes—all sorts of tastes—even in the arts. Why Bach and Beethoven, why the Beatles and Joan Baez are popular among vast portions of the population is partly due to the high or peculiar quality of these popular artists, partly because tastes do develop in certain directions under certain conditions. They do not develop with natural necessity, to be sure, and each new step much depends on the existing artistic achievements and the traditions they are embedded in and their accessibility to the audiences in question. Thus, the fact that the sociology of tastes exists and can be developed is hardly contestable; and this includes the taste for research and thus the tastes for particular types and areas of research. We know that incentives for research in terms of position and money can do that much and no more. But we also know that incentives can create traditions and that traditions can do much more. They can, in particular, affect tastes of young aspirants who are far from being allured by a mere secure job or by merely a few extra thousands of dollars a year. We do not know yet how much time such a process takes. From the day you raise the salary of existing, hardly able, researchers by a few thousands, to the day a young ambitious person decides to follow his footsteps because of the resultant tradition—not because of the initial high salary, but because of the youth's dreams of glory—some time has to elapse; but we do not know how long and depending on what. But we do have instances to this effect and we can study them.

We also know that just as high artistic tastes must be sustained by real art, so high research has to be sustained by high aspirations. For example, after the Russians had sent the first Sputnik and beat the Americans in the space race, there were many new federal and private funds to support any program in the philosophy and the history of science. The result was negligible as no money supported any viable program: most philosophers and historians of science worship physical technology. Here, perhaps, is the place to mention the high position of the uninspired technological researcher, i.e., the physical technologist, the fact that so many aspire to emulate him. Thomas S. Kuhn's *The Structure of Scientific Revolutions* opens with the observation that ever so many social scientists aspire to emulate the natural scientist who, as Kuhn claims, is in fact a humdrum physical technologist. The humdrum technologist, the small puzzle-solver, Kuhn has labeled "normal scientist." Kuhn joins the crowds who worship him by offering the theory that the growth of science depends on him. The normal scientist is, no doubt, a product of the incentive system we have in research since World War II. I shall not ask how come a historian of science such as Kuhn takes the most recent developments in the area and views them as eternal. And I would only mention in passing that Kuhn claims that normal science must have scientific periodicals, even though he knows full well that Greek science, normal or abnormal, flourished with no print and no periodical publications.

The question, to repeat, is, can we alter the incentive system so as to make normal science obsolete? To make humdrum physical technology the abnormal rather than the normal, that is. The question is, Will the change of the incentive system make a great difference fast? The question is, How can we alter the incentive system? The question is, Who should decide on such matters? The question is, Do we have the time for all this?

The starting point to all these questions, I feel, is the question I am posing here: Why is physical technology dominant? For, if there is a strong reason for this dominance, then unless we find it and consider it nothing will help. If, in objective reality, physical technology offers mankind the highest incentive, whereas social technology matters only to those with a deep sense of social equity, for example, then I do not think physical technology can ever cease to be dominant. If, however, as I contend, the social conditions of Western Europe of the Industrial Revolution offered the strong incentive for physical technology; if in the nineteenth and early twentieth century, political reasons, and bad ones, stood behind the preference for physical technology over social technology; if the post-World War II preference for physical technology and economic aid and economic growth were the spillover of traditional errors, then we might benefit from arguing against these traditional errors, as Edward Goodman does, for example, and D. V. Segre or, less well, Kenneth Galbraith and Ed Mishan. The realization that present circumstances do not favor economic growth in rich countries, the view that the improvement of the quality of life is not the same as the increase of the real income of the average family, the view that the quality of working life cannot be measured by income—in other words, the current refutations of all materialism—must be the starting point of the reform movement.

It follows, then, that the revolution must be engineered by those more concerned with the truth and with the future of humanity rather than by those concerned with approval and their own high position in accord with the present incentive system. This does not mean that they will be the ones to propagate it or execute it. There is little doubt that unless this revolution will be more democratic—in its aspirations as well as in its institutions—than the last revolution, of Enlightenment and Industrialization, there will be no way to implement it. I mean, of course, participatory democracy. How then can we envisage the New Enlightenment, what instruments of expansion can it have, and will it expand fast enough to prevent the oncoming cataclysm? I do not know, but I intend to examine these questions as best I can.

III. THE DYNAMICS OF RAPID SOCIAL CHANGE

There are three views of rapid social change, the functionalist, the Hegelian, and the Weberian.

The functionalists view any society as a set of interlocking institutions, as best interlocking as possible in the given conditions. Social change, then, is impossible to effect from within. Whether an external factor, then, causes a large change or small is a matter of external interactions, and functionalists have not studied their dynamics. Nonetheless, the functionalist might and often does assume that a system would resist all external causes of change as best it can or else readjust as fast as possible. Thus Japan first attempted a total isolation, and only when this failed did it accept extensive industrialization while attempting to retain as much of its earlier structure as possible.

It is not clear whether functionalism can be so extended to cover the quick adjustment to external changes. But if not, then we can leave it now as useless for our purpose. If we accept the extension as some functionalists have proposed, however, then it is not clear whether the extension is refutable. I think, anyway, it is very hard to explain the large explosion in Western Europe, such as the Italian Renaissance, the Reformation, etc., along such functionalist lines. For, functionalism is an equilibrium theory with an addition concerning the quick return to equilibrium, not a dynamic theory of rates of series of changes or of growth. So much for functionalism.

Hegelianism views all changes as gradual and views all revolutions as a transition from one phase to another. This is the first law of dialectics, so called: quantity turns into quality. However continuous the motion of two vehicles on the road is, and however small is the added velocity of the vehicle behind, sooner or later it will catch up with the vehicle in front—the moment of revolution—and in no time it will be ahead: the moment of victory.

The Hegelian theory of revolutions, thus, seems to me to be analytically true of all continuous processes. It seems to me that it suggests that all revolutions are slow, but some of them seem to be fast because of the oversight of the slow processes that build them up. Now the vehicle behind may be far behind or not, it may have a high or a low excess velocity, it may accelerate at a high or at a low rate of acceleration, the acceleration may be sudden or not—all these are left utterly untouched by the analytic version of the Hegelian theory, and some of these refute any given synthetic version of it. Nevertheless, it is important to see what small changes add up to a revolution, what not. So much for the Hegelian theory for the time being. Its value, I think, is in its pointing out that revolutions must pass certain hurdles, certain points of no return. But not all revolutions. Whether the transition from the accent on physical technology to the accent on social technology is such a revolution, whether it can be made rapidly or must be slow, these questions remain to be discussed later on. For the time being, we can say, the Hegelian theory has given us all the help it can.

The Weberian theory is as static as the functionalist theory, except for Weber's rider, i.e., his theory of the charisma. Max Weber views a society as a collection of interacting ideal types, where a type is more or less an institutionalized individual, be he a capitalist or a bureaucrat or a religious functionary. The charisma theory comes to explain the formation of a new type: a charismatic leader invents a new type and enough people follow him to make his invention stick.

This does not help much. We have charismatic leaders who excited whole populations and failed to make their impact stick. We have hardly charismatic leaders who had very marginal impact on some margins of society which, however, became dominant later in time—through a very slow process or expanding in one moment in time like wildfire. Examples abound, and I shall offer only one which pertains here.

The reason Pestalozzi preached vocational education is complex; he died with hardly an impact; yet he offered an important ingredient in the making of poor Switzerland rich through the idea of vocational training, and he did so long after his death. The expansion of vocational education in Switzerland took like wildfire once it took.

Open schools or free education had all the charismatic leadership it needed, from Homer Lane to his latest successors. The idea has still not become a broad movement, yet we can hardly hope to evolve genuinely participatory democracy without it.

And the question is, Why is free education not a wild success? Why some ideas fail despite charismatic leadership is very clear; the idea may be poor. Timothy Leary, a great charismatic leader, did not know that the allegedly psychedelic drugs are not psychedelic at all, and in addition they can kill. His movement spread like wildfire, but the fire had nothing to feed on. Were Aldous Huxley's theory of the drugs true, the movement would have succeeded quite spectacularly. Nonetheless, the spread of the movement is very remarkable and deserves study, especially for those who notice that some movements, especially in education, should spread rapidly and do not have much success.

The remarkable aspect of the psychedelic movement was the fact that it spread against all resistance from the Establishment. The movement was almost entirely incoherent, and the coherent part of it was inconspicuous if not clandestine. Such movements often produce charismatic leaders, but these are inessential, as was Timothy Leary or any other possible leader of the movement. What creates such movements has barely been studied, but clearly a few factors are essential. First, the movement answers a widespread need. Second, it does so by opening avenues of activity—to novice and to old hand alike; indeed, often such movements are cheap in that they offer easy avenues of action to almost perfect strangers on condition that they convert: this makes it easy to have quick conversions in chain reactions. Third, and most important, the movement is not much dependent on institutions of the established order.

This point deserves much more notice than it has received. Whoever is in charge always appoints people whose chief responsibility is to impede that progress that may lose the one in charge his power. Officers have sergeants to do it, bureaucrats have secretaries and doormen to do it. And the result is that if a reformer has to see the authorities in order to implement innovations that may render too many established people jobless, then the doorman is there to delay his move.

Examples abound. But more interesting is the case where people voluntarily give up positions of strength on the conviction that their services are no longer necessary. But the conviction has to be clear and widespread. It is hard to doubt that the brains of the French aristocracy created the French Revolution, that the British people and government gave up the British Empire, etc. But these are exceptions.

It seems to me that were it a general law that any employee whose services are no longer required automatically gets pensioned off at full salary, then much of the force of reaction would be broken. I know of no strong argument against such legislation.

I would go further: any innovation that requires retraining should lead to the choice between retraining and full pension. I shall not elaborate.

These suggestions are very useful but not sufficient. The call for social experimentation—in schools, on the job, etc.; the call for funds to support experiments, etc.; the legal defense of reasonable experiment in case it fails—all these too are essential and can be passed on local, state, and federal level almost unobtrusively after reasonable rational debate. Let me mention one small example, yet one which offers a strong argument for such legislation. There are complaints against almost any university instructor; the instructor who is cowardly and conservative is automatically protected against complaint; unlike the instructor who cares and experiments, who is almost always penalized as result of the complaint. This cannot go on, and can be altered by simple legislation—whether in existing institutions or by alternative new ones. Here, incidentally, I see the root of the ill-success of all free schools, i.e., those which experiment in democratic self-rule: their success is not so spectacular to spread into the system, and in the system there are too many teachers whose interests dictate opposition to it. The same analysis, I think, applies to industrial democracy, the reform of industry, job-improvement, reform of work, sociotechnical systems, etc. And, I think, common to both these and some other movements is that they are the means to develop a system of participatory democracy. Also, there must be organs to pool experiences in social technology and discuss the empirical and political and intellectual and moral aspects of the goings-on, so that the implementation of the new democracy should be most democratic. And so, indeed, what we need is a movement—self-sustaining but able to draw the support from the establishment.

Let me end by noticing the current situation. The movement is there, in ecology, in the concern for poverty, for urban renewal, etc. People of the movement have distinguished themselves by the dubious battle on the slogan that there are better causes than the moon shot. The question, Is this true, was shown irrelevant by the argument that aborting the moon program would have cost as much as proceeding with it. Keynesian economists argued, further, that the moon program was not necessarily at the expense of other programs. All this to no avail: the movement opposed the moon program until it was swept by the enthusiasm justly accorded the moon landing.

This shows how poor, intellectually, the movement of concern for social technology is. But the concern is there. What is needed is engineering some brain power to go with it. This should provide enough incentive for enough thinkers and social inventors to start the ball rolling in the direction of increasingly rapid social planning and innovation.

To conclude, let us ask how far can anything we can do today save the world from its impending catastrophe? The question naturally splits into at least three parts, how can we affect the Western industrial world, the totalitarian tightly organized countries, communist or not, and the rest. Clearly, what matters most for the rest is the problem of economic growth plus democratization; for, as this problem fails to be solved more and more countries become totalitarian. The possibility of developing an increasingly democratic system to go with an increasingly industrial system is so unstudied yet so exciting, since it is neither physical nor social technology at stake but the so-called problem of social and cultural lag, on which there is almost no literature to mention—indeed, only a book or two. As far as the totalitarian countries are concerned, the only hope for their progress is that sufficiently many people from their intellectual and social elites, in part even their political elites, will be familiar with Western progress and impressed with it and willing to try and emulate it. The program for helping them try such ideas can be worked out now.

Yet the chief point is that all revolutions, or all rapid social processes, are small, ordinary actions that have some snowball effect. This simple fact is a very powerful technological tool, and I suppose the snowball must start rolling in the Western industrialized countries, with the democratization of education and of the quality of working life and such, with wide and bold reforms which will be quickly emulated whenever they happen to be successful.

THE COGNITIVE DIMENSION OF TECHNOLOGICAL CHANGE

Stanley R. Carpenter GEORGIA INSTITUTE OF TECHNOLOGY

INTRODUCTION

Though interdependence of science knowledge and technological action is currently extensive, this "marriage of science and technology" is hardly more than a century old.[1] The dimensions of the problem of understanding this merger include at the very least history, sociology, economics and political theory. One aspect, however, is fundamentally philosophical in nature. This has to do with the cognitive dimension of technological change. Progressive technical mastery has been fostered by an evolving rationality which in its most articulated form embodies science. I shall adopt a model which holds that rationality is a biological property, an evolved trait which up to the present has been species adaptive.[2] If rationality is an evolving property, rather than an ideal logical type, we deprive ourselves of the classical philosophical position which holds that this growth of rationality is eudaemonistic. On the other hand this position also makes it more difficult to adopt a fatalistic pessimism which places rational technique beyond human control.[3] Thus while the growth of rationality in technology leads to change, it does not guarantee progress, for change can result in domination as well as liberation.[4] What, then, is proposed is an explication of the phrase "the rationalization of technology" and an explanation of how science in method and substance contributes to this rationalization.

The rationalization of practical action is probably most commonly associated with systematization, routinization, and formalization according to rules. An example which comes immediately to mind is the rationalization of management as devised by Frederick W. Taylor.[5] Through time-motion studies Taylor systematized factory worker routines, instructed them in the precise manual behavior to follow and effected a demonstrable increase in productivity. "Taylor proved . . . that it was possible to manipulate human activity and to control it by logical procedures in much the same way that physical objects could be measured and controlled."[6] What needs to be observed, however, is that this process was more rational, not from the point of the worker, but from the perspective of management.[7] Consequently the concept of rational action must be refined to include mention of the agent. The sense of rationality which we shall adopt has to do with being the agent of the action, to act rather than being acted upon. As one author succinctly put it: "To become *more* rational, in this root sense, means to transform into ends things which previously were not ends. A man becomes more rational just insofar as he brings within the scope of his will some datum of experience which previously confronted him as independent of his will."[8] On this formulation Taylor's manager gained in rationality by establishing control over the uncoordinated behavior of the factory workers. Parenthetically, however, it becomes apparent that the process of rationalization contains within it the potential for domination whenever the object of control is another human being.

For action to be rational it must be brought under control of conscious ideas. Here cognition includes both an awareness of ends and the appropriate means to reach those ends.[9] It thus becomes possible to contrast rational action with those forms which are not rational at all or at least less rational. Coercion against one's will is a clear case. So also are instances of purely affective behavior where cognition and reflection are denied. Indeed, recent counterculture and anarchist writings grasp this clearly and are explicit in opting for spontaneous, erotic, and emotive alternatives to rationality.[10] Max Weber, whose well-known studies examine the many different forms of rationalization, would add tradition-bound behavior to the list.[11] Presumably an unreflective dependency on habitual cultural patterns substitutes for cognition in such cases.

What is the payoff of opting for the rational approach? By bringing action under the control of conscious ideas one escapes the bondage of sheer circumstance. Action becomes an object of decision.[12] One no longer simply reacts to the environment but rather acts to master and shape it according to conscious intention. With the progressive mastery that results there comes responsibility. What was before a matter of fact becomes a matter for policy deliberation. To the extent that the agent of the action adopts a policy that fails to consider the relevant means to the desired end or further fails to systematically relate immediate ends to longer range goals he is said to be acting irresponsibly or irrationally. However, to the extent that the freedom of agenthood results in bondage for others or ultimately to a condition which threatens their survival, or even his own if he so chooses, his actions may be rational though morally repugnant. Thus it is that rationality as an evolving trait though heretofore generally conducive to species survival can become maladaptive.[13]

RATIONALIZATION IN TECHNOLOGY

It is possible to trace the growth of rationality in evolving forms of technological action. I have argued elsewhere[14] for a continuum model of technology which begins with artisan skills at one extreme[15] and progresses through greater and greater systematization of technique to high technologies. The latter make such extensive use of science that they are frequently held to be nothing more than applied science, though this conclusion is surely wrong.[16] The continuum is understood as a progression from primarily prescriptive rules toward increasingly descriptive statements which are themselves ultimately systematized within a comprehensive scientific theory. At the level of artisan skills the process of rationalization is present, though minimally so. Manual skills are consolidated into routines while conceptualization is underdeveloped. Communication from master to apprentice is mimetic. What verbal behavior is used to guide the beginner takes the form of rules. Beyond the ad hoc rules with which the artisan supplements his performance, rationalization produces a distillation of technical maxims which are themselves explicitly prescriptive and incomplete in specification. They constitute a first level of generalization. A recipe book, for example, is a collection of technical maxims.

The formulation of a technical maxim is, however, a noteworthy milestone in the process of rationalization. Though mainly prescriptive, it reflects an implicit awareness of the limits of practical action. If the maxim prescribes routine *A* and warns against *B* it is at least implying that the operating context is such that *A* leads to success and *B* doesn't. The technical maxim begins to stretch out the limits of the possible.

Empirical generalizations which synthesize repeatable correlations among phenomena constitute the next level of rationalization. A gap is opening up between cognition and volition. It is a short step from learning that hickory burns slower than pine to "Burn hickory in the stove." It is a much greater one from observing that the sun is lower in winter than in summer to domestic window design that effectively exploits this fact. This gap is broadened as empirical generalization becomes a part of a simple explanation or is perhaps ultimately subsumed in a comprehensive scientific theory. The gap is only closed by the act of invention which, while abetted by cognition, seems itself paradoxically nonrational.

THE ROLE OF MODERN SCIENCE

1. Manipulation and Abstraction

The progressive rationalization of technology is accomplished by an increasing dependence upon discursive knowledge. Axiomatic to this approach is the claim that modern science embodies the best technique yet evolved for systematizing this knowledge. That is, modern science represents the zenith of human rationality, though this is no guarantee that it will remain an adaptive trait. What then are the ways that scientific knowledge contributes to rationalized technology?

A popular though facile answer holds that the impact of basic science on technology is virtually inevitable. But this "trickle-down theory," which relates today's new artifacts to yesterday's scientific insights, seems at times to be more a statement of faith than an accurate description of this complex process.[17] Furthermore, it has not escaped historians of technology that a science-based technology emerged as late as the nineteenth century, fully two hundred years after Newton revolutionized science.[18] However, certain fundamental mind shifts accompanied the new science and in a general way did affect the milieu of the practicing technologist. Furthermore, it is possible to pinpoint specific ways scientific knowledge currently aids technology.

The profound shift in ontological bearings which followed in the wake of Newton has been eloquently described.[19] Two features are especially relevant to subsequent technological practice. One concerns the manipulability of the physical environment as a new way to knowledge, the *vita activa*. The other involves the power of the new sciences to abstract from sensible experiences in ways that were frequently counterintuitive. Technology was an ultimate benefactor of each of these insights.

In his essay on the origins of modern technology Hans Jonas observes, "The very conception of reality that was fostered by the rise of modern science, i.e., the new concept of *nature,* contained manipulability at its theoretical core and, in the form of experiment, involved actual manipulation in the investigative process."[20] Though Bacon's insights were prescient in this direction, it was only with the geometrizing of nature and the mathematization of physics that controlled experimentation became a reliable road to knowledge. Galileo, of course, grasped this. And while, for the two hundred years which followed, the main purpose of experimentation was the generation of scientific knowledge, the circle was completed in the chemical and electrical industries at the turn of the century, at which point scientific knowledge became a guide to rational manipulation for practical purposes.[21]

Equally instrumental for the ultimate emergence of a science-based technology were the abstract assumptions which grounded modern science. If Copernicus' heliocentric model appeared to deny the data of the senses, the Galilean claim that uniform rectilinear motion continued without the constant push of an impressed physical force compounded the abstraction. Theory acquired a grounding in counterintuitive assumptions.

Much later, when technology was able to assimilate the substantive and methodological insights of the physical sciences, it was as a result able to proceed in totally surprising ways. This is most clearly seen in the emergence of electrical and electronic technologies. Arthur C. Clarke[22] illustrates this point with characteristic clarity. For hundreds of years before their invention men were able to imagine such artifacts as automobiles, flying machines, steam engines, submarines, space ships, telephones, and robots, to name a few. On the other hand, we are surrounded daily with other items which were totally unanticipated, such as X-rays, nuclear energy, radio, TV, electronics, photography, sound recording, transistors, lasers, superfluids, etc.

Thus while the emergence of the modern technological society clearly began with a mechanical stage, with new sources of motive power extending capacities of hands and backs, the sciences of chemistry and electromagnetism sent it charging off in unforeseen directions. As a result, novel commodities such as information have been created which in turn generate new requirements of management and control.

2. Substantive Contributions of Modern Science

The connection between science and technology can be specified in greater detail. This specification explains, in part at least, how science contributes to the rationalization of technological practice. Science currently influences technology in two more or less separable ways, substantively and methodologically. Substantively science provides a systematized data base and a deductive algorithm in the form of a theory to be used by the engineer. By restricting initial or boundary conditions to reflect the practical environment, the engineer gains explanatory and predictive utility. Methodologically the symbolization, quantification, and manipulative ability afforded by mathematical models provides the technologist with a powerful tool for simulating complex practical systems such as communication, transportation, or ecological networks. System simulation idealizes practice and facilitates the creation of "alternative futures" generated by large numbers of interactive variables. Let us examine these substantive and methodological modes further.

Technology bridges a gap between perceived need and what is known of the world. Invention constitutes this bridge, but some inventions are only partial successes while others are failures. An important way to minimize the chances of failure is to ascertain the outside limits for a particular practical application. To do so is clearly to bring within one's cognition some datum of experience that is relevant to rational behavior. Scientific knowledge functions in this regard in substantive ways. Applied science provides knowledge of environmental limits by means of properly bounded applications of theory.[23] Additionally, it furnishes constitutive knowledge of physical properties, such as tensile strengths or thermal properties, necessary for a rational selection of artificial materials. Science imposes principles of selectivity on the testing of inventions. In this case it is noteworthy that scientific experiment is undertaken not to test a scientific hypothesis but rather to evaluate a product or process.[24]

Bringing substantive scientific knowledge to bear on a practical problem can demonstrate the futility of some envisaged solutions. Action may be shown to be futile either because it demonstrably involves theoretical impossibility or because it can be shown to exceed the "state of the art" in some crucial way. Examples of impossibility range over the familiar failures of *perpetuum mobile* to Shannon's axioms about theoretical limits of information-carrying channels.[25] In candor, however, it must be observed that scientifically dictated statements of impossibility are far from infallible.[26] It is possible to be rational and wrong.[27]

Ex post facto science can provide an explanation of failure. It may demonstrate that the environment and artifact were incompatible, with the former driving the latter outside its operating range. It may expose theoretical violations of the type just mentioned. As a result scientific knowledge can offer a rational basis for improvement. In doing so it must frequently surmount the inherent conservatism of artisan technology which depends upon tradition and incompletely specified rules of thumb. For example, it took years and a theory of rigid bodies to force automobile-chassis makers to abandon the box frame of the horse-drawn carriage. Analysis of generated torques within the motorcar led to the A-frame as late as the 1930s.[28]

Substantive scientific theory thus rationalizes practice by providing knowledge of both the physical context *and* the operational characteristics and limits of the devised artifact or process. Although distinct from the inventive act, it provides reasonable assurance of compatibility between the context of need and the practical act while also offering insights into unanticipated directions that invention may profitably take.

3. Methodological Contributions of Modern Science

Scientific rationality provides an additional payoff for technology. Its methodology uniquely unites experience with deductive reasoning in the form of mathematical models. By means of symbolization and quantification, a tool of great precision and manipulability is made available for ordering dimensions of human social behavior itself. The rapid growth of systems engineering marks the successful mapping of mathematical models onto complex technological processes.[29] Aided by the electronic computer it is possible to apply dynamic programming, game theory, cybernetics, econometrics, operations research to practical objectives. The result is what Bunge has termed "operative technological theories" which employ not substantive scientific knowledge but the method of science.[30] In characterizing system behavior, considerations of the physical makeup are subsidiary. Whether electrical, mechanical, neural or economic, the focus of attention is elsewhere, viz., on how the system operates; i.e., on what patterns follow other patterns.[31]

The contrast between the technological fruits of substantive science and those of operative system theories is important for an understanding of the rationalization process. Historically the paradigm has been the physical sciences with their emphasis on analysis and reduction to elemental matter. Engineering has modeled itself after physics and chemistry with the result that practical solutions are all too frequently piecemeal. With characteristic overstatement Galbraith states it, "The real accomplishment of modern science and technology consists in taking ordinary men, informing them narrowly and deeply and then, through appropriate organization, arranging to have their knowledge combined with that of other specialized but equally organized men. This dispenses with the need for genius."[32]

On the other hand, the impact of systems engineering, dating no earlier than to the Second World War,[33] is only just beginning to be felt. The increased use of large-scale system models precisely illustrates the process of rationalization. Some complex phenomenon, say, worldwide resource use, is shown to contain regular patterns. Some of the causal determinants are discovered. Beyond this, ways are invented for influencing and possibly controlling the process. Here, just as in cases involving physical science, technology devises ways to interfere. What was before an independent datum of experience becomes for the first time an object of decision. Mankind acquires a new degree of freedom, at least in principle, to act rather than be acted upon.

4. Examples of Rationalization

To the extent that the object of decision is perceived as having a major impact on large numbers of people, this process of rationalization may be stoutly resisted. The explanation for this reaction is surely complex and beyond us here. In part it may hark back to some romantic ideal of simpler days, as the commune movement seems to do with its emphasis on spontaneity, eros, and primitiveness, and its explicitly antirational posturing.[34]

In other cases a growth in rationality may point up a new responsibility that was avoidable as long as the phenomenon in question was thought to be beyond control. Hardin[35] observes that industries refused to pay workmen's compensation as long as accidents were considered unpredictable. Only after the rationalization of the insurance industry, when quantitative statistical measures were made available, did the factory owners agree to assume compensatory responsibility and thereby acquire an additional "internal" cost.

A similar case, cited by Wolff,[36] involves macro-economic theory. Beyond the ongoing activities of production and exchange, economics uncovers apparently objective laws governing the relations of wages, profits, interest levels, etc. The Gross National Product is devised as an aggregate measure of this economic activity. Gradually it is discovered that regulation of the money supply may be used to "tune" the GNP. At this point, control of the economy becomes an object of decision. This possibility, however, is not embraced by all. Some prefer the "hidden hand" to an explicit exercise of control. Sensing the dimension of responsibility that accompanies mastery, they prefer prerational magic in which aggregate economic behavior is rendered inexplicable. To the extent, however, that econometric models have uncovered laws of macroeconomic behavior, and in addition have devised techniques for utilizing these regularities for control, a decision to do nothing becomes deliberate policy. To refuse to consider these latter developments in economic theory becomes thereby an instance of irrationality.

It is in this light, I believe, that a part, at least, of the vociferous opposition to the two Club of Rome studies, *Limits to Growth*[37] and *Mankind at the Turning Point*,[38] may well be viewed. While it is true, especially in the case of the Meadows study, that substantial methodological shortcomings have been pinpointed,[39] this does not explain the intensity of the criticism. Despite explicit disclaimers in both books that the findings do not amount to ironclad inevitabilities, they are labeled "models of doom." On the model of rationality presented herein they represent the first step in a truly global rationalization of economic processes.

Consider the Mesarovic and Pestel study, for example. An extensive data base on world resources such as metals, energy, arable land, etc., was collected. Worldwide demographic statistics were considered. Economic variables such as regional consumption patterns, investment, government expenditures were combined into gross regional product measures. Regional dietary needs were considered. These data, along with an ensemble of alternative economic policies, were combined with a computer-implemented hierarchical systems model. Ten world geographical regions were differentiated and resultant measures of population, petrochemical supplies, investment aid, and food supplies were extrapolated for fifty years hence.

Leaving aside the substantive refinements of this model which are even now being made,[40] the process is an exemplar of the growth of rationality to encompass matters of major social importance. The first step involves the uncovering of regularities in the interactions of food resources, land use, population, resource depletion. Secondly, system sensitivities to alternative social policies are discovered. Third, using an admittedly uncriticized utilitarianism, an optimum policy is recommended involving what the authors call "organic growth." By this procedure patterns of world economics have become explicit objects of decision.

We began by claiming that one task for the philosopher of technology is to analyze the role of knowledge as it shapes practical action. From its earliest forms the power of human reflection, structured by natural language and other symbolizations, transformed the brutish given-ness of existence into objects of human volition. Freedom to act was enhanced by the understanding of natural limits on that action. Rationalization has meant a progressive systematization of such understanding. This evolutionarily acquired capacity has realized its highest embodiment in scientific rationality itself. In ways which I have outlined, this knowledge, in concert with the enormous amplification of human strength afforded by nonhuman energy sources, has provided a means for interfering with natural processes on a truly worldwide scale.

5. Reactions to the Growth of Technological Mastery

A threshold has been crossed. Physical forces unleashed by human agency rival for the first time those encountered in nature.[41] The existence of life on the planet has become an object of human decision. Just as we are beginning to absorb this fact, we are faced with the novel possibility that our fundamental species characteristics may themselves become optional through genetic manipulation.

Having brought us to this point, the future of human rational development appears far from clear. Critics of rationality have always been quick to point out that rationality can be applied to repression, even genocide. Rationality leads not only to fitting means to ends but also weighs end states themselves. But what is to insure that those consciously selected states are acceptable to us?

Weber has highlighted the irony of rationality. In a fundamental sense the drive for rational action is powered by an emotional energy.[42] Values which orient and guide action are themselves nonrational. As a result, during periods such as the present when there is widespread disagreement over which values are worthwhile, reason, to a certain extent, becomes impotent. Even such a widely held norm as "the greatest happiness for the greatest number" becomes ambiguous when the limitation on world population becomes an object of decision.

Or again, note that human survival, the yardstick by which adaptability is itself measured, becomes problematic whenever biology enables us to fashion a race of mutants. In the process who—or perhaps what—it is that survives may be unrecognizable, even abhorrent to us.

Having admitted the possibility that rationality may itself have become ineffectual, or worse than that, maladaptive, it is tempting to respond with a passive fatalism. Faced, for the first time, with the stewardship of the entire planet, we retreat into the occult or the erotic rather than confronting the new responsibility.

Equally escapist and probably more potent a threat is the attitude John McDermott characterizes as *laissez innover*.[43] *Laissez-faire* economics, it is recalled, postulated a mysterious self-correcting principle, the "hidden hand," which a government might imperil through attempts at regulation. Nowadays, in analogous fashion, we are assured that technology itself, if left to its own devices, is capable of correcting its past excesses.[44] The rational alternative, however, is to make the technological process itself an object of decision, to direct it, even restrict its forward thrust in ways consonant with a representative societal voice.

The emergence of technology assessment methodologies constitutes an encouraging sign that this may be occurring. Especially noteworthy are quite recent experiments aimed at involving citizens alongside the technical experts in technology assessment efforts.[45] If this approach proves workable it will not only insure that technological design criteria reflect the widest possible consensus of societal values but will also represent significant progress in removing the communication barriers separating technocrat and citizen.[46]

Another noteworthy development is what has been variously called the "soft" or "appropriate" technology movement.[47] Its proponents explicitly reject the Luddite antipathy to technology characteristic of the counterculture. Instead, they wish to impose more stringent design criteria on the innovation process than has heretofore been the case. Acceptable technological solutions must be environmentally gentle and preserving of the finite resources and fragile regenerative process of the earth. The scale of production processes is to be kept small and the supportive forms of social organization kept as nonhierarchical as possible. Because this movement places a premium on scientific and technical expertise, it represents a rational alternative to present practice. Because it explicitly embraces the task of caring, and preserving the planet itself, it holds promise as a genuine advance in human rationality itself.

CONCLUSION

A philosophy of technology should include among its concerns an elaboration of the process by which practical action itself becomes an object of reflection and control. Greater understanding of this process helps to explain the relative successes of various forms of human practice. It has been argued above that progressive mastery occurs when artisanal skills and practical objectives are combined with human reason. This combination is by no means easy to achieve. Action must be brought under the control of conscious ideas which must themselves be systematized, refined, and criticized.

When action itself becomes an object of reflection it acquires for the first time an intelligibility. Experience is transformed from sheer immediacy to an object of decision. Through the exercise of his will man becomes an agent rather than a mere fact of change. Not only does he participate in the process, he controls its direction while simultaneously reflecting on the effectiveness of that control.

Scientific methodology represents the most highly refined form of human rationality yet devised. Its discovery and evolution has provided the human species with an adaptive edge. Its transmission is cultural and its efficacy contingent.

Within the past century, techniques have been discovered which effectively couple science, in substance as well as method, with technological practice. The result has been to bring within the scope of human control a vast range of physical phenomena which heretofore had held man captive. This process we have called the "rationalization of technology."

Specific ways that science has contributed to the rationalization of technology include the principle of manipulation, definitive of the experimental method. Though its immediate result was the generation of knowledge, its long-term effect has been to systematize and focus the search for solutions to practical problems. Additionally, the principle of abstraction from sense experience which led science into unanticipated worlds of the intellect eventuated in surprising technologies such as electronics, information control, and now genetic synthesis.

Science also rationalizes technology by providing knowledge of the milieu in which an artifact or process is to intrude. It furnishes constitutive knowledge of physical properties and sets theoretical limits the perception of which can minimize failures. Additionally, it explains past failures and offers a rational basis for improvement.

Recently the rationalization of practice has been augmented by complex mathematical theories which can be modeled on the electronic computer. The result has been a quantum increase in the scale of phenomena, economic, ecological, demographic, etc., which can be described. With such understanding comes a new potential for human control.

Reactions to these novel possibilities have been mainly negative. One approach is to seek to escape from reason itself by returning to the eternal present of intuition, eros, and enchantment. Another adopts a "business as usual" attitude toward technological practice while closing one's eyes to all but the most optimistic long-term extrapolations.

Both approaches abandon reason itself. While it may turn out that human rationality is unable to cope with the options it has created, the possibility will be assured if irrationality itself is accepted as the only rational option. We would do better to embrace the stewardship of the planet including the preservation and regeneration of the human species as an object of decision.

FOOTNOTES

1. Melvin Kranzberg, "The Unity of Science—Technology," *American Scientist*, 55, 1 (1967) 48–66. See also Hans Jonas's careful description of this merger in *Philosophical Essays: From Ancient Creed to Technological Man* (Englewood Cliffs, N.J.: Prentice-Hall, Inc., 1974), chapter 3.

2. This follows the position defended by Marx Wartofsky, "Is Science Rational," in Willis H. Truitt and T. W. Graham Solomons (eds.), *Science, Technology, and Freedom* (Boston: Houghton Mifflin Co., 1974), pp. 202–210.

3. The former would range from Plato and Aristotle to Kant and Hegel. The latter is boldly asserted by Jacques Ellul. Cf. esp., *The Technological Society*, Vintage Books, trans. by John Wilkinson (New York: Random House, 1967).

4. For a survey of the literature on technological domination, cf. Stephen Cotgrove, "Discussion Paper: Technology, Rationality and Domination," in *Social Studies of Science*, 5,1 (January, 1975) 55–78; cf. also Wartofsky, *op. cit.*, note 2.

5. Cf. Robert H. Guest, "The Rationalization of Management," in Melvin Kranzberg and Carroll W. Purcell, Jr. (eds.), *Technology in Western Civilization*, II (New York: Oxford University Press, 1967), pp. 52–64.

6. *Ibid.*, p. 53.

7. Of course if the worker received higher wages as a result, and thereby was more

satisfied than before, then one might argue for a subsidiary sense of "rational." There is, however, no necessary connection between this possibility and the rationalization of the process itself.

8. Robert Paul Wolff, *The Poverty of Liberalism* (Boston: Beacon Press, 1968), p. 90; note that excluded from the present discussion is the psychological meaning of "rationalize": "to ascribe [one's acts, opinions, etc.] to causes that superficially seem unrelated to the true, possibly unconscious causes." *The Random House Dictionary*, Unabridged Edition, 1966.

9. On the distinction between cognition and thinking especially as it relates to work, cf. Hannah Arendt, *The Human Condition* (Chicago: The University of Chicago Press, 1958), p. 171; concerning the rationalization of means and ends Weber makes it clear that his concept of instrumental rationality *(zweckrational)* is not to be limited to a rationalization of means to ends alone, but is also a rational ordering of the ends themselves: *Economy and Society* (New York: Bedminster Press, 1968), pp. 24–26. A helpful differentiation between "rationality," "rationalism" and "rationalization" as found in Weber's work is offered by Ann Swidler, "The Concept of Rationality in the Work of Max Weber," *Sociological Inquiry* 43, 1 (1973), 35–42.

10. For an anarchist alternative to the technological society, cf. Murray Bookchin, *Post-Scarcity Anarchism* (San Francisco: Ramparts Press, 1971), esp. pp. 9–30. Cotgrove examines the antiscience movement in "Anti-Science," *New Scientist* (July 12, 1973) pp. 82–84, cf. also Wartofsky, *op. cit.*, note 2, pp. 202, 203.

11. Weber *op. cit.* note 9, p. 25.

12. Wolff offers a convincing explication of the phrase "object of decision" in his chapter entitled "Power," *op. cit.*, note 8, pp. 86 ff.

13. Popper argues that the argumentative, critical and rational function of language, by providing the genesis of science, has created "what is perhaps the most powerful tool for biological adaptation which has emerged in the course of organic evolution" *Of Clouds and Clocks* (Washington, D.C.: George Washington University Press, 1965), p. 19; Wartofsky deals with the possibility that rationality may itself become maladaptive by transformation into "repressive reason" in "Is Science Rational?" *op. cit.*, note 2, pp. 202–210.

14. Stanley R. Carpenter, "Modes of Knowing and Technological Action," *Philosophy Today* (Summer 1974), 162–168.

15. Admittedly this ignores speculations that protohuman became *Homo faber* through fortuitous accidents such as the discovery that bones made good weapons or that burnt flesh tasted good.

16. Agassi has provided a convincing refutation in "The Confusion between Science and Technology in the Standard Philosophies of Science," *Technology and Culture,* 7, 3, (Summer 1966), 348–366.

17. For an examination of three familiar attempts to relate specific scientific insights to the process of innovation—Project Hindsight, TRACES (Technology in Retrospect and Critical Events in Science), and the Battelle study, "Interactions of Science and Technology in the Innovation Process"—cf., Patrick Kelly, Melvin Kranzberg, *et al., Technological Innovation: A Critical Review of Current Knowledge,* I, chapter 2; Alvin Weinberg has noted that justifications for societal support of basic science fall into two general categories, the "high culture" and the "overhead" theories. The former claims society should support basic research because its very existence serves to enrich the culture in a manner similar to the arts. The latter holds out the possibility that some abstruse scientific insight may turn out to have practical ramifications. Financial support is seen as an overhead cost of providing this possibility. Weinberg thinks the second theory springs more funds, *Reflections on Big Science* (Cambridge: M.I.T. Press, 1967), pp. 91–100. The "trickle-down" theory is sometimes called the "spin-off" theory.

18. For a hypothesis as to the reasons for this delay, cf. Jonas, *op. cit.* note 1, pp. 71-79.

19. E.g., Alfred North Whitehead, *Science and the Modern World* (New York: Mentor Books, 1925); John Herman Randall, *The Making of the Modern Mind* (Boston, 1926); E. A. Burtt, *The Metaphysical Foundations of Modern Science,* Anchor Books, (Garden City, N.Y.: Doubleday & Co., 1932); Herbert Butterfield, *The Origins of Modern Science* (New York: Free Press, 1965); Alexandre Koyré, *From the Closed World to the Infinite Universe,* Harper Torchbooks (New York: Harper & Brothers, 1957); Jonas's own presentation in *Philosophical Essays, op. cit.* note 1, ch. 3, relates these developments especially to technology.

20. *Ibid.,* p. 48.

21. Kranzberg, *op. cit.,* note 1, p. 52.

22. Arthur C. Clarke, *Profiles of the Future,* Revised Edition (New York: Harper & Row Publishers, 1973), p. 20.

23. Popper discusses the relationship between scientific theories—what he designates "universal laws"—and their explicit restriction for practical purposes. Such restrictions he designates "initial conditions": *The Logic of Scientific Discovery,* Harper Torchbooks (New York: Harper & Row Publishers, 1959), pp. 59-60. For an elaboration of this point, cf. my own analysis of Hooke's Law as it might be utilized in a practical invention: "The Structure of Technological Action," unpublished Ph.D. dissertation, Boston University, 1971, pp. 102, 103.

24. Cf. Marx Wartofsky, *Conceptual Foundations of Scientific Thought* (New York: Macmillan Company, 1968), pp. 200-204. Of course the desire to determine physical properties of a substance may require that the theory behind a promising application be developed in greater detail. Then "technically justified science" is called for. Cf. Michael Polanyi, *Pure and Applied Science and Their Appropriate Forms of Organization* (Oxford: Society for Freedom in Science, Occasional Pamphlet #14, December, 1953), p. 4. A frequently cited example involves the development of the transistor. Basic research was undertaken to verify the intuition that a controllable flow of electrons could be obtained in germanium and silicon. The practical application was already envisaged. Cf. Jack Morton, "From Research to Technology," *International Science and Technology,* (May, 1964), pp. 82-92.

25. Claude E. Shannon and Warren Weaver, *The Mathematical Theory of Communication* (Urbana: University of Illinois Press, 1964), esp. Theorem 10, pp. 68, 69.

26. Clarke gives a popular account of several "impossibility claims" made by noted scientists: Sir William Preece, "Subdivision of the electric light [as accomplished in Edison's power distribution scheme] is an absolute *ignis fatuus*"; Newcomb wrote a famous essay demonstrating the impossibility of man flying long distances through the air; a certain Professor Bickerton produced a quantitative proof of the impossibility of rocket flight; Lord Rutherford chided "sensation mongers" who predicted the liberation of energy from matter. Clarke humorously formulates the following law: "When a distinguished but elderly scientist states that something is possible, he is almost certainly right. When he states that something is impossible, he is probably wrong." *Op. cit.,* note 22, pp. 2-5, 14.

27. Michael Scriven, discussing this point, distinguished between "rational belief" and "the truth." The former is the best guide to the latter. Rational belief involves the best evidence that is available *at the time* but it will most certainly vary from time to time. *Primary Philosophy* (New York: McGraw-Hill Book Company, 1966), pp. 11-21.

28. Cf. Gordon L. Glegg, *The Design of Design* (Cambridge: Cambridge University Press, 1969), pp. 26, 27.

29. For an analysis of the epistemological and ontological status of theoretical models, cf. Wartofsky, *op. cit.,* note 24, pp. 283-287.

30. Operations research, for example, models complex logistical systems for such diverse

applications as fighting wars, stocking a nationwide chain of department stores, expediting vehicular traffic through toll booths, or planning commercial air traffic schedules.

32. John Kenneth Galbraith, *The New Industrial State* (New York: Signet Books, The New American Library, 1967), p. 73.

33. Cf. Philip M. Morse and George E. Kimball, *Methods of Operations Research* (Cambridge: M.I.T. Press, 1951). This seminal work first appeared in classified form just after World War II.

34. Cf. note 10.

35. Garrett Hardin, *Exploring New Ethics for Survival* (Baltimore: Penguin Books, Inc., 1972), p. 81. Hardin dates the internalization of the costs of industrial accidents at about 1875.

36. Wolff, *op. cit.* note 8, pp. 90–93.

37. Donella H. Meadows, Dennis L. Meadows, Jorgan Randers, Wm. W. Behrens, III, *The Limits to Growth* (New York: Universe Books, 1972).

38. Mihajlo Mesarovic and Edward Pestel, *Mankind at the Turning Point: The Second Report to the Club of Rome* (New York: E. P. Dutton, 1974).

39. Cf. H. S. D. Cole, *et al. Models of Doom: A Critique of the Limits to Growth* (New York: Universe Books, 1973).

40. In conversation with the writer Edward Pestel indicated that finer discriminations than the ten world areas are now being programmed.

41. For quantitative documentation of this point, cf. Gordon J. F. MacDonald, "The Modification of Planet Earth by Man," *Technology Review*, 72, 1 (October/November 1969), pp. 27–35.

42. From the viewpoint of instrumental rationality *(Zweckrational)* Weber observes, "Value-rationality *(Wertrational)* is always irrational. Indeed the more the value to which action is oriented is raised to the status of an absolute value, the more 'irrational' in this sense the corresponding action is." *Op. cit.* note 9, p. 26. Commenting on this distinction Swidler observes, "Bringing all action under control by conscious ideas requires active effort, and must be powered by concentrated emotional energy. It is this need for an irrational spur to rationality that gives the problem of rationality its particular poignancy." *Op. cit.* note 9, p. 41.

43. "Technology: The Opiate of the Intellectuals," *New York Review of Books* (July 31, 1969), 25–35.

44. Though he hedges his bet in the closing paragraphs, Alvin Weinberg comes close to this position in "Can Technology Replace Social Engineering?" *Bulletin of the Atomic Scientists*, 22 (December, 1966), 4–8.

45. Cf. James D. Carroll, "Participatory Technology," *Science* 171 (19 February, 1971), 647–653; for a description of an early attempt at public participation in a technology assessment project, cf. Sherry R. Arnstein and Alexander N. Christakis, *Perspectives on Technology Assessment* (Jerusalem: Science and Technology Publishers, 1975), pp. 165–179.

46. The technocrat/citizen dichotomy is an extension of Lord Snow's familiar essay, *The Two Cultures and the Scientific Revolution*, Rede Lecture, 1959 (Cambridge: Cambridge University Press, 1959). McDermott argues that the technological society has fostered a decline in popular literacy as decision makers retreat into a world of technical jargon; *op. cit.*, note 43, p. 31. Hardin sees malevolent intent in the choice of some terms. E.g., "pesticide"—something obviously good because it kills pests; or, "side effect"—something we may not want but which is really subsidiary to the main effect we are after; *op. cit.*, note 35, pp. 66–70. Edwin Newman catalogs the list of euphemisms selected by the Defense Department to anesthetize the average citizens against the full impact of the Vietnam War: *Strictly Speaking* (New York: Bobbs-Merrill, 1974), p. 63.

47. For a description of the movement, cf. John P. Milton, "Communities that Seek Peace with Nature," *The Futurist,* 8, 6 (December 1974), 265 ff.; see also Nicholas Wade, "The New Alchemy Institute," *Science,* 187 (February 28, 1975), 727–729; Joan Rothchild supports the compatibility of the soft technology approach and broader feminist concerns in "A Feminist Perspective on Society and the Future of Human Technology," paper delivered at the Second General Assembly of the World Future Society, Washington, D.C., June 2–5, 1975; E. F. Schumacher has coined the phrases "intermediate technology" and "appropriate technology" to describe alternatives to the "high" technology of the West. Though originally directed to Third World problems, it has subsequently gained favor among critics of technology here at home: *Small Is Beautiful,* Harper Torchbooks (New York: Harper and Row, 1973); cf. also Loren Eisley, "Alternatives to Technology," Aaron W. Warner, Dean Morse, and Thomas Cooney (eds.), *The Environment of Change* (New York: Columbia University Press, 1969).

TYPES OF TECHNOLOGY

Carl Mitcham, ST. CATHARINE COLLEGE

The word "technology" has, in current discourse, a narrow and a broad meaning—which roughly correspond to the ways it is used by two major professional groups, engineers and social scientists. It is important to recognize this at the outset, because tension between these two usages, which stretches out a spectrum of conceptual references, easily results in analytic confusion.

The engineering usage is more restricted. The word "engineer" itself, coming from the medieval Latin root *ingeniare*, meaning "to create," "to implant in," or "to produce" (our words "engender" and "ingenious" have similar origins), readily connotes making or producing.[1] Yet the engineer distinguishes between engineering and technology in his strict sense. For the engineer is not so much one who actually makes or constructs an artifact, as one who directs, plans, or designs—as is indicated by such metaphorical usages as, "The general engineered a coup," meaning he planned or organized it, thought it all out, not that he actually ever picked up a gun. Of course, "engineer" has its own restricted sense, referring to one who operates engines—as in the expression "railroad engineer."[2] But generally, engineering is identified with the systematic knowledge of how to design artifacts—a discipline which (as the standard engineering educational curriculum shows) includes some pure science and mathematics, the so-called "engineering sciences" (e.g., strength of materials, thermodynamics), and is actualized by some social need. But while engineering involves a relationship to these other elements, still it is design (and the technical ideal of efficiency which distinguishes engineering from, say, artistic design) that constitutes the essence of engineering; because it is design which orders or establishes the unique engineering framework to relate these other elements.[3] The term "technology" is reserved by engineers for the process of material construction.

For engineers, "technologist" and "technician" are closely related if not strictly synonymous.[4] Just as the adjective "technical" connotes a limited or restricted viewpoint, so the engineering technician works from a more limited standpoint than the engineer himself. The technician or technologist might, for instance, know how to perform a test, operate a machine, construct or mass-produce a device (and even be involved in directing others who have a less comprehensive view of some particular operation or construction project), but not necessarily how to conceive, design, or think out such a test or artifact. Consider, for example, the terms "lab technician," "medical technologist," "drafting technician," etc.[5] In each case the person referred to is designated proficient at performing some operation or construction, but not at fully organizing or understanding the activity with which he is involved. This distinction, it may be noted, is confirmed by the classical name for the engineer—Latin: *architectus;* Greek: $\mathring{\alpha}\rho\chi\acute{\iota}\tau\epsilon\kappa\tau\omega\nu$, from $\mathring{\alpha}\rho\chi\acute{\iota}$- (first, master or director) plus $\tau\acute{\epsilon}\kappa\tau\omega\nu$ (carpenter or builder). That is, the architect or engineer is one with a superior or more inclusive view of a material construction than the carpenter or technical assistant.

For social scientists, however, the term "technology" has a much broader meaning.[6] To begin with, it includes all of what the engineer calls technology, along with engineering itself. This use has some basis in engineering parlance, as when an engineering school is named an "institute of technology." Yet this continues to limit technology to those making activities and operations which have come under the influence of modern science. The art of potting, for instance, is not a conspicuous feature of the curriculum at MIT. Social science usage, stimulated by recognition of the social significance of these modern scientific making activities (*vide* the sociocultural reaction to, and now the sociology of, the Industrial Revolution), has desired to extend the term even further to refer to all making of material artifacts, the objects made, their use, together with their intellectual and social contexts. Even arts such as potting become technologies in this loose sense, (a) because there are certain modern technologies (e.g., industrial ceramics) which grew out of potting, and (b) because the ways potting affected premodern society are presumed to embody principles continuous with those exhibited by the impact which modern technology has exerted on the social fabric. Indeed, in the history of technology, which is the primary social science study of technology, technology has sometimes been defined so as to refer even to the making of nonmaterial things such as laws and languages—although in practice this definition has not been widely utilized.[7] Thus the tension between the narrow, engineering usage and the broad, social science usage given to the word "technology" seems to point, first, toward the conceptual primacy of the making of material artifacts then, second, toward a large number of elements and influences that go into and arise out of this primary process, determining its different forms.

My thesis, in Aristotelian language, is that "technology" is not a univocal term; it does not mean exactly the same thing in all contexts. It is often, and in significant ways, context-dependent—both in speech and in the world. But neither is it an equivocal like "date" (on a calendar and on a tree). There is a primacy of reference to the making of material artifacts, especially as this making has been modified and influenced by modern natural science, and from this is derived a loose, analogous set of other references. An initial need in the philosophy of technology is for clarification of this conceptual one and many, a conceptual one and many which evidently exists because it reflects a real diversity of technologies with various interrelationships and levels of unity.

The philosophical value of becoming more aware of this spectrum of conceptual references can be illustrated by two cases. First, in discussions of the social and ethical consequences of technology questions invariably arise about whether or not technology can be done away with. Partisans line up on both sides. But much of the disagreement results from a failure to clarify differences in presumed definitions. On the one hand, if what one means by technology is the making activity and the use of material artifacts in general, then obviously technology can never be abandoned, and is in fact coeval with if not prior to (since animals also make and use artifacts like bird's nests and spider webs) the emergence of human life. On the other hand, if what one means by technology is some particular form or social embodiment of this general human activity, then equally clearly technology is expendable; technologies have been abandoned over and over again throughout history, under both peaceful and violent circumstances. Indeed, the sociology of technology depends on this interpretation of the word when it analyzes cultures in terms of their own technological changes.

Second, in philosophical discussion a large number of apparently incompatible definitions have been offered for technology. Technology has been variously described as sensorimotor skills (Feibleman), applied science (Bunge), design (engineers themselves), efficiency (Bavink, Skolimowski), rational efficient action (Ellul), neutral means (Jaspers), means for economic purposes (Gottl-Ottlilienfeld and other economists), means for socially set purposes (Jarvie), control of the environment to meet human needs (Carpenter), pursuit of power (Mumford, Spengler), means for realization of the *Gestalt* of the worker (Jünger) or any supernatural self-concept (Ortega), human liberation (Mesthene, Macpherson), self-initiated salvation (Brinkmann), invention and the material realization of transcendent forms (Dessauer), a "provoking, setting up disclosure of nature" (Heidegger), etc.[8] Some descriptions evidently differ only in the matter of words. Yet even after this is taken into account there remains a wide variety of definitions, each of which—it is reasonable to assume—highlights some real aspect of technology in the general sense, but under the guidance of a tacitly employed restricted focus. Thus argument over the truth or falsity of such descriptions too often hinges on the exclusiveness of a limited perspective. The proper resolution of disagreement would be a structural and/or phenomenological analysis of technology, delineating its different types and their interrelationships. Only such an analysis would provide a foundation for assessing the relative truth and significance of each individual description.

Such reflection on usage of the word "technology" leads me to suggest that the term be stipulated to refer to the human making and using of material artifacts in all forms and aspects. Human making is only to be distinguished, in the Aristotelian sense, from human doing—e.g., political, moral, or religious actions.[9] I am aware that neither does this reflect the etymology of the word itself (which became current in the nineteenth century to refer to the industrial arts),[10] nor does it always accord with various feelings and intuitions that are well entrenched in the English language. Nevertheless, it does seem to me to demarcate what should be the full scope of a philosophical concern with technology, and to draw out what is unique about this study. Traditionally, the philosophy of human action has concentrated almost exclusively on doing (such, for instance, is the province of ethics, political philosophy, etc.), at the expense of making—the only exception being some limited discussion in aesthetics. Under the stress of contemporary problems and needs, however, man is called upon to reflect on his making in a more comprehensive and fundamental manner. Henceforth, then, unless otherwise specified, I will use the term in this general way, without wishing to presume or imply any distinctions or relationships. Differences and unities can only be disclosed through an analysis of the types of technology and their structural characteristics. Where I think analysis does warrant typical relations I would argue for denoting these either by qualifying adjectives attached to the general word (as with the expression "scientific technology") or by distinct words properly defined (as with "technique"). This, to my mind, sets forth the initial conceptual program for a philosophy of technology.

* * * * *

The full presentation of such a typology is, of course, beyond the scope of this paper.[11] Moreover, as I conceive this, it is beyond the scope of a merely conceptual analysis; if not limited to a bare explication of conceptual contents it will ultimately involve anthropological, epistemological, and metaphysical reflections. Let me nevertheless begin to treat this problem of the technological one and the technological many by putting forth a rough catalog, a kind of philosophical lexicon of distinctions which seem to be fundamental and in fact widely accepted—as a summary and restatement of the work and conclusions of others which points toward the synthetic resolution of some confusions.

Distinctions among technologies may be said to create a three-dimensional grid. First, there are obvious subject or material distinctions between chemical technology, electrical technology, etc. Second, there are functional or structural distinctions. Third, there are social or historical distinctions—although in principle these should be able to be restated in material or formal terms when they are philosophically significant.

Subject or material distinctions are the least interesting philosophically. In their common forms they are the basis of social science studies like Singer's massive five-volume *History of Technology* and the Kranzberg-Pursell *Technology in Western Civilization*.[12] For philosophers it is fundamental differences that matter most, at least initially, because what philosophy seeks is a real definition of the essence of technology that can be seen to underlie or be exemplified in its various modes and manifestations. (Philosophers should not, however, belittle historical studies, for they are one necessary test of the generality of their own theses.)

Functionally, technology can be divided into that which goes on internally in man, that which is part of his bodily activity and thus his social involvement, and that which becomes in a sense part of and interacts with the natural world by taking on a life of its own independent of his immediate bodily action. This corresponds to the distinctions between technology-as-knowledge, technology-as-process, and technology-as-product—or thoughts, activities, and objects. In the standard conceptual analysis, however, the anthropologically interior is restricted to an intellectual component, when actually the will is also involved in an element which might be called technology-as-volition. If this is taken into account, along with the recursive activity of use, we may summarize these basic distinctions with the following diagram (Fig. 1):

Figure 1. The Modes of Technology

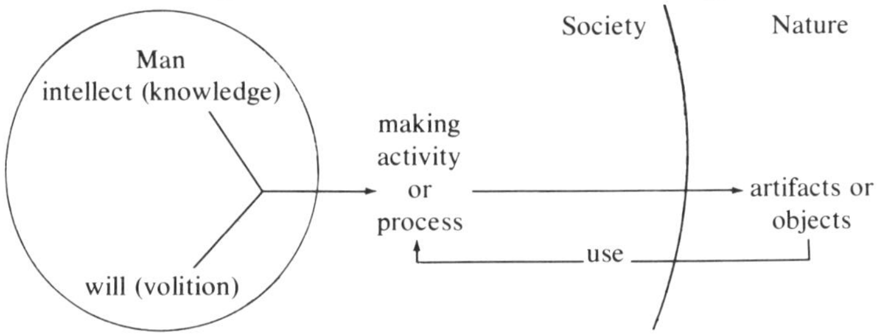

[Diagram: A circle on the left labeled with three sections — "Man intellect (knowledge)", "will (volition)" — connecting via "making activity or process" to "artifacts or objects" on the right, with a curved line separating "Society" and "Nature", and a "use" arrow returning from objects back to the process.]

I take this diagram and its distinctions to be more important than might first seem. For instance, since what philosophers are searching for in a definition of technology is an essence or whatness which lies behind and is manifested in these various modes, one good test of any definition would be to describe just how it is exemplified in each of these functional differentiations. Before suggesting other substantive implications, however, let me offer a catalog of the following further distinctions, and comments which seem appropriate to each category. In doing this I will work from the outside in, as it were; from objects to ideas and intentions; from the realm of the most obvious and well discussed to the least.

* * * * *

1. Technology-as-object: The class of technological objects has been divided (by Mumford) into utensils, apparatus, utilities, tools, and machines—without insisting that these are mutually exclusive or complete.[13] Slightly enlarged and elaborated, this list would include:
   (a) Utensils—e.g., baskets, pots; storage containers.
   (b) Apparatus—e.g., dye vats, brick-kilns; containers for some physical or chemical process.
   (c) Utilities—transformers; e.g., reservoirs, aqueducts, roads, buildings, lights. One subdivision would distinguish power utilities such as railroad tracks and electric power lines, which function only through the operation of power machinery.
   (d) Tools—instruments operated manually which act to move or transform the material world; implements which a worker uses to perform work. Perhaps, though, two kinds of tools should be distinguished: those used for making and those used for doing. A pencil is a tool which enables one to make letters, but the letters themselves are tools of communication, which can be (in song, for instance) a kind of doing.
   (e) Machines—tools which do not require human energy input because of an outside source of power (wind, water, steam, electricity, etc.), but do require human direction; a device which operates, under direction, to perform work. (Note the

equivocation on "work" here and in reference to tools. Tools perform work and produce works, whereas machines operate and produce products. When machines are spoken of as performing work, the definition of work becomes that of physics—force times displacement in the line of force.) One subdivision would distinguish machines powered by naturally given energies like wind and water and both of these from machines driven by more abstract forms of power such as electrical, chemical, or nuclear energy.

(f) Automatons or automated/cybernated machines—machines which require neither human energy input nor immediate human direction. Automated devices take part of their energy output and recycle it back into the instrument itself as a form of control. The most simple example is, of course, a thermostatically controlled heater, where part of the heat output is used to operate a thermocouple.

Comments:

A. All of these artifacts are meant in some way to be used, lived within or operated. Given the broad definition of technology there would seem to be at least two other types of technological object to be distinguished: that already mentioned under (d), tools of doing (letters, numbers, etc.) rather than making, and artifacts which are not meant to be used at all but only contemplated or worshiped, i.e., objects of art or religion. The problem with the distinction between tools of doing and those of making is that it appears to be highly context dependent and not always clearly discernable in the object itself. That is, sometimes numbers can be used for doing mathematics, at others for making money or even buildings. But is this latter use not really a doing of mathematics for purposes of making money or house building? Numbers themselves, as artifacts (it seems reasonable to argue) only realize their full potential or make the most sense when they are used for doing, even when it is possible—and, indeed, sometimes necessary—to subordinate this doing to a making, the doing becoming a means to some making. Language may be used to conduct business, yet it finds its full realization in poetry. This is so because a thing is defined, according to Aristotle, not by its innumerable possible uses but by what it can do only or best.[14] Further example: A carpenter's hammer is not in itself either an instrument for knocking out one's mother-in-law or digging weeds in the garden, although it could perform both functions. The problem is that one would fail to make sense of the claw, the other would fail to make sense of the mallet side of the head. Only when used as an instrument for fabricating with (driving or pulling) nails can all of its qualities be recognized as fitting together.[15]

B. Machines pose complex conceptual issues, partly because the term has shifted its meaning from the antique hand-operated instrument for working to the modern nonmanually operated instrument.[16] Thus historically "machine" can mean at least three different things:

1. It can refer to the simple machines of classical mechanics—lever,

wedge, wheel and axle (or winch), pulley (or block and tackle), screw, and inclined plane (to give the traditional list)—or some combination thereof.[17] Actually, since (as the science of mechanics has shown) the wedge = an adapted inclined plane; the pulley = a form of wheel and axle; and the screw = an inclined plane cut in a spiral, this traditional list of six can be reduced to three: wheel and axle, lever, and inclined plane.

2. "Machine" can refer to any implement or large-scale simple machine which requires more than one man to operate it because of its energy requirements. This is the definition found, for instance, in Vitruvius and applied to "projective engines and wine presses."[18]

3. Yet even in this form these "instruments for transmitting force or modifying its application" (to quote a common dictionary definition of machine) are not machines in the specifically modern sense, but special kinds of tools. Admittedly, as one author has observed, "the difference between the tool and the machine has never been clearly defined."[19] But the common notion is that a tool is a hand-operated machine, or at least the man-related element of a mechanical device; while "machine" denotes an instrument in its human independence, or at least that aspect of the device which is not dependent on man.[20] This corresponds to Hegel's definition of machine as a "self-reliant tool," tool being understood as any instrument of work,[21] and yields the third definition of machine as implement which does not depend on human energy—although it still requires some human monitoring or directing, "driving" in the sense that one "drives a car." (Note the two senses of "drive": in one sense the motor "drives the car," in another the "driver" does.) Of these one can readily distinguish three types: those which depend on animal power (horse-drawn plow), those which employ direct mechanical energy from nature (windmill, water wheel), and those which use some form of abstract energy (electricity). Of this last category there are also two types: those which generate or transform energy, and those which transmit power and perform work. The former (as in the electric dynamo) is the uniquely modern machine[22]; normally the latter will involve more intimate human direction.

Machine tools: Strictly speaking these are tools for metal cutting, tools used to make machines. But in common speech they are not always distinguished from power-driven hand tools such as the electric drill and saw or the air-driven jackhammer. A rotary electric handsaw is also quite different from the power-driven table or radial arm saw, i.e., stationary shop tools.

Finally, one should observe the resistance that is to be found in our uneasiness about calling an automatic device, which is constructed but neither energized nor directed by man, a machine. At most we seem willing to refer to it as an "automated machine." Nevertheless, this concept, as a natural extension of the previous conceptual development, does denote a fourth type of machine, the machine as cybernetic or self-regulating device.

C. The engineering analysis of machines: Classical mechanics is that branch of physics which deals with the motions of material bodies and the forces acting upon them.[23] So defined, mechanics is subdivided into statics (dealing with bodies at rest) and dynamics (dealing with bodies in motion). Prior to the development of vector analysis by Simon Stevin (1548–1620) mechanics consisted almost exclusively of formulas for the equilibrium of simple levers derived from Archimedes (third century B.C.). The work of Galileo (1564–1642) on falling bodies laid the foundation for the modern science of dynamics[24] with its two main branches—kinematics, dealing with the motion of rigid bodies without regard to the forces involved, and kinetics, dealing with the relations between forces and motions.

In modern engineering, machines are analyzed and described by means of the science of dynamics. Machines are, as it were, closed systems which can be analyzed in terms of motions (kinematics) and forces (kinetics). As such, a machine is commonly defined as "a combination of rigid or resistant bodies having definite motions and capable of performing useful work."[25] If a machine-like device fails to perform useful work in the technical sense, it is termed a mechanism. A clock or speedometer, for example, is a mechanism but not a machine.[26] To say the same thing in a different way: While the science of dynamics includes kinematics and kinetics, the primary function of a mechanical device can be either the modification of motion or the amplification of force. If the former, it is a mechanism; if the latter, a machine. Furthermore, the way the parts of a machine are interconnected to produce a required output motion from a given input motion, even when the purpose is force directed to useful work, is known as the mechanism of the machine. One can, for example, speak of the mechanism by which a steam engine, as a machine, works.[27]

This conception of a machine as "a closed kinematic chain" or "a combination of resistant bodies so arranged that by their means the mechanical forces of nature can be compelled to do work accompanied by certain determinate motions" goes back to Franz Reuleaux, *Kinematics of Machinery* (1875).[28] Reuleaux's definition was, however, formulated prior to the advent of electrical, chemical, and nuclear energies so that it does not strictly apply to, say, electronic "mechanisms" such as radios, car batteries, and computers—unless "kinematic chain" is interpreted very broadly to include the movement of electrons along a wire or within some chemical compound. Thus a question might arise concerning the possibility of reformulating Reuleaux's definition to make it suitably general, and within this genus to distinguish the various species of machines.

In fact, with the delimiting of the traditionally more general word "machine," the engineering term "device" has come to take its place as denoting any "mechanism, tool, or other piece of equipment designed for specific uses."[29] Device is not synonymous with artifact, for it denotes instrumentality and in most cases even dependence on some (internal or external) operation. A device accepts some input and modifies it in some unique way to produce a desired output. Thus devices are to be distinguished, for example, from structures; one would not normally speak of a house or a bridge as a device.[30] The concept of device includes tools, machines, and automatons, but it is questionable whether it could refer to utensils, apparatus, and utilities—not to mention tools of doing and objects of art—without becoming a metaphor. "Device" roughly corresponds to what Heidegger calls *Zeug,* gear or equipment.[31] As Heidegger argues, such an artifact is related in its essence to some functional totality which we might call a system. The interrelation of devices in a system as understood by engineering raises philosophical questions about systems as possible technological objects.[32] Conceived as systems, however, technological objects tend to collapse into processes.

D. On the phenomenology of tools and machines: The most common philosophical interpretation of machines (= early philosophy of technology) argues that tools and machines are extensions of man, organ projections. The idea is hinted at first by Aristotle; but it is also part of the typology of such late nineteenth- and early twentieth-century studies as those of Kapp and Lafitte.[33] Most recently it has been expanded to include even electronic media as extensions of the human nervous system (McLuhan[34]). But not to mention other questions, one can ask, "extensions" in what sense? There can be at least two kinds of extension: A hammer, for instance extends, by way of enlargement, the power of the arm muscle, while it extends by way of abstraction and magnification the hardness and formal properties of the fist. (The appeal here is to the distinction between the enlargement that goes on in a telescope, where light from a star is itself collected and concentrated, thus bringing one, as it were, nearer to it; and the magnification that goes on in a microscope which does not just bring one close to, but optically abstracts from, and transforms the visual properties of, an object.) A pile driver, however, not only magnifies form, it also magnifies power, by placing at man's disposal energies which he himself does not possess. If tools or machines in the first (classical) sense increased man's power, it was only by enlarging his own inherent energies. If machines in the second sense (as tools operated by more than one man) did the same, it was only by uniting the energies inherent in a group of men. Modern power machines achieve this effect in an entirely different way by making man the director of nonhuman energies. Thus, whereas tools are single-function instruments which break down, specify, distribute, or concentrate the total power resident in the hand, machines incorporate the hand as an instrumental link in man's multifaceted directing or governing of nonmanual energies.

Furthermore, "the technical advance that characterizes specifically the modern age is that from reciprocating motions to rotary motions."[35] Modern machines, unlike tools, typically achieve their effect by means of rotary rather than reciprocating motions. While this is most obvious in the rotary power saw as contrasted with the reciprocating handsaw, it is equally true of the pile driver, which develops its reciprocating power in a pneumatic operator which is dependent on the rotating shaft of some prime mover. As the historian, Lynn White, Jr., has argued in reference to the discovery of the crank,

> Continuous rotary motion is typical of inorganic matter, whereas reciprocating motion is the sole form of movement found in living things. The crank connects these two kinds of motion; therefore we who are organic find that crank motion does not come easily to us. The great physicist and philosopher Ernst Mach noticed that infants find crank motion hard to learn. Despite the rotary grindstone, even today razors are whetted rather than ground: we find rotary motion an impediment to the greatest sensitivity. The hurdy-gurdy soon went out of use as an instrument for serious music, leaving the reciprocating fiddle-bow . . . to become the foundation of modern European musical development. To use a crank, our tendons and muscles must relate themselves to the motion of galaxies and electrons. From this inhuman adventure our race long recoiled.[36]

Both kinetically and kinematically modern machines, as contrasted with traditional tools, involve a qualitatively distinct separation of man from his own body and its elemental awareness. Not all extensions are the same.[37]

This is confirmed by two separate connotations of the adjective "mechanical." In the traditional sense a "mechanical task" is one that has to be performed manually and is dependent on human energy. It is thus, in this special sense $\beta\alpha\nu\alpha\upsilon\sigma\iota\kappa\sigma\varsigma$, base or ignoble, for it focuses a man's attention on his own physical powers which are extremely limited.[38] It does not connect with higher, transhuman or spiritual powers but remains on the strictly natural plane. In modern usage, on the other hand, a "mechanical task" is one done without attention, repetitively, routinely, or even ritualistically (in the bad sense of that word). Modern machines, while connecting with nonliving physical energies, nevertheless are base or ignoble in a new sense. They alienate man from his own sensorimotor, mind-body complex; his body acts and does not need his mind. Consequently his attention is not focused at all, anywhere, and must be entertained by some extraneous sensations—music, colors, etc., as devised by industrial psychologists. This is why, from the contemporary perspective, a return to mechanical operations in the primitive sense can be seen as a desirable thing, reuniting mind and body; and this desirableness in turn is the source of the difficulty we experience in appreciating the ancient critique of manual work.

Along this line, Mumford (and others) have argued that "the skilled tool-user becomes more accurate and automatic, in short, more mechanical, as his originally voluntary motions settle down into reflexes."[39] Such a generalization lacks experience with tools. When an operation becomes mechanical, as in a machine, one loses control of it. To take a simple example, a power-driven saw cannot respond as sensitively to a piece of wood—a knot, say, or stringy grain that easily splinters and damages a particular work—as a handsaw. Admittedly, accuracy in one sense, the sense of following a superimposed, geometric line, becomes more fully realized; but only at the expense of a certain responsiveness to materials. As Sōetsu Yanagi summarizes the experience of craftsmen:

> No machine can compare with a man's hands. Machinery gives speed, power, complete uniformity, and precision, but it cannot give creativity, adaptability, freedom, heterogeneity. These the machine is incapable of, hence the superiority of the hand, which no amount of rationalism can negate.[40]

And as carpenters are well aware, a power saw easily gets "out of hand"; and a handsaw wound is usually a lot less serious than one from a power saw. It is no accident that finishing work or fine cabinetmaking continues to be done primarily by hand. Surely an artist does not become less sensitive or even necessarily more accurate in the geometric sense as his brush stroke techniques become internalized in sensorimotor reflexes. The loss of awareness at the level of technique increases the control of the artist at the level of his proper work. Thus there seem to be important differences between so-called automatic operations with tools and with machines, and these should not be too facilely identified. But such a phenomenology of use points away from technology-as-object and toward the category of technology-as-process.

E. Two final suggestions: Just as the relative proportions of these various types of technological objects available in some particular society will affect that society in various ways, so their presence can be one key to differences in types of technologies in general. For instance, machines run by abstract power (electricity, etc.) dominate modern society but are not found at all in traditional or primitive societies. This point is relevant to the contemporary discussion about so-called "intermediate" or "soft" technologies as opposed to "hard" technologies, and their social and ecological impacts in underdeveloped (and developed) countries.[42] But it is also important to remember that types will not always come in pure forms; usually there will be an admixture of various elements. The real question will be one of proportion and degree, not simple presence or absence—as in the case with machines driven by nonliving sources of power.

Moreover, these differences in the phenomenology of use are related to differences in the characters of the objects produced, although these differences are not easily conceptualized. Consider, for instance, the distinctions which the Mexican philosopher Octavio Paz draws among objects of fine art, industrial technology, and handcraft:

> The industrial object tends to disappear as a form and to become indistinguishable from its function. Its being is its meaning and its meaning is to be useful. It is the diametrical opposite of the work of art [in which the meaning is to be useless but beautiful]. Craftwork is a mediation between these two poles: its forms are not governed by the principle of efficiency but of pleasure, which is always wasteful, and for which no rules exist. The industrial object allows the superfluous no place; craftwork delights in decoration. Its predilection for ornamentation is a violation of the principle of efficiency. The decorative patterns of the handcrafted object generally have no function whatsoever; hence they are ruthlessly eliminated by the industrial designer. The persistence and the proliferation of purely decorative motifs in craftwork reveal to us an intermediate zone between usefulness and aesthetic contemplation. In the work of handcraftsmen there is a constant shifting back and forth between usefulness and beauty. This continual interchange has a name: pleasure. Things are pleasing because they are useful *and* beautiful. This copulative conjunction defines craftwork, just as the disjunctive conjunction defines art and technology: usefulness *or* beauty.[41]

But such differences in the phenomenology of production, even when they are reflected by the objects themselves, point away from technology-as-object and toward the category of technology-as-process.

Furthermore, there are what may be called material or imaginative distinctions among technological objects as opposed to the formal distinctions mentioned so far. To take an example that has been made the object of a well-known historical study, the stirrup (part of a tool for horse riding) is capable of an indefinite number of material embodiments while retaining its basic formal (or technical, functional) properties. Just as the concept of a triangle is capable of being imagined and drawn as red or green, isosceles or scalene, so is the idea of a stirrup as footrest attached to saddle capable of many different types of reification. The figures in the Appendix are designed to show this in a graphic way, since by its nature this distinction is easier to see than to conceptualize. Thus one could postulate an argument between Platonic and Aristotelian views of technology-as-object—the first maintaining that a stirrup is a stirrup on the basis of form or function alone, never mind its material embodiment; the other maintaining that a simplified English riding stirrup is substantially different from an ornate Spanish *vaquero* stirrup. The point is that in some abstract functional sense these objects are the same, but at the level of actual material realization there are significant class distinctions (not just individual distinctions) to be made on the basis of materials used and formal ornamentation which reveal differences in use contexts and cultural attitudes. Technological objects are subtly modified by specific use, even when there exist more universal functions. Once again, though, this points away from the object itself and toward technological processes.

* * * * *

2. Technology–as–process: The processes of making and using are commonly discussed in terms of four main kinds of human activity, although here there is less definiteness than in the distinguishing of artifacts. These are: (a) invention, (b) design, (c) making (in the sense of materially fabricating), and (d) using.

General comments:

A. Although these appear as the basic types of technology considered as a human activity, (a) and (b) are less essentially involved with bodily action than (c) and (d). Thus invention and design as activities already point back toward technology-as-knowledge or as-volition, leaving making and using as the root distinctions within technology-as-process. This priority is indicated both by the gerundival character of (c) and (d), and by the fact that in some sense invention and design are themselves aspects of making and using.

B. Actually, this basic typology is a simplification of the distinctions within engineering analyzed as a function.

i) Invention is sometimes broken down into research and development. Applied research = using scientific and mathematical knowledge plus experimentation to synthesize new materials or create new energy-generating or transforming processes. Development = utilizing these materials and energies to design and fabricate prototype products to solve particular problems or meet special needs. (Industrial research = development.[43])

ii) Design can be considered an activity of development or an activity in its own right ordered toward construction and production.

iii) Making can be subdivided into construction (of stationary structures) and production (of moveable devices). (Another general classification is fabrication; types of fabrication = artistic, craft, industrial, mass, etc.)

iv) Operation and management denote types of use, as do testing, service, maintenance, sales, etc.—although testing can also be construed as part of development and design.
v) The functions of planning, teaching, consulting, and systems engineering cut across these various distinctions.[44]

C. This typology of processes is also amenable to elaboration or reformulation in other directions, usually depending on some extrinsic relationship. Examples:

i) From an economic point of view the following technological processes are sometimes distinguished: labor saving (capital intensive) processes, capital saving (labor intensive) processes, neutral processes;[45] potential vs. realized technology; invention vs. innovation; and material vs. social technology.[46]

With regard to the invention-innovation distinction: in contemporary usage, innovation is the broader term, denoting "any thought, behavior, or thing that is new because it is qualitatively different from existing forms"[47] or the activity of engendering such thought, behavior, or thing. This readily includes invention or the creation of new material objects.

As terms of contrast, invention = creation of a new artifact; innovation = the economic development and exploitation of some artifact, new or existing, by means of a reorganization of goods, methods of production, sources of supply, industrial structures, marketing, etc. (Does this include political development and exploitation as in warfare or the U.S. space program?) Innovation is thus a kind of using, technological using. Hence potential technology = a technological invention awaiting economic (or political) exploitation by innovation or technological use. One further distinction along this line: invention vs. technological change. Invention commonly implies novelty; technological improvements of existing hardware based on what is already known do not qualify as inventions. For example, the original four-stroke-cycle internal combustion engine (the Otto Silent Engine, 1876) was an invention; combining two or more four-stroke-cycle cylinders into one engine was merely technological change. (Patent law, an instrument of political economy, usually protects the former but not the latter.)

ii) Aristotle hints at another distinction of technological processes, one between cultivation and construction.[48] Cultivation = helping nature to produce more perfectly or abundantly things which she could produce herself (e.g., medicine, teaching, and farming). Construction = forcing nature to produce things she would otherwise not be able to produce (e.g., carpentry). As Van Melsen characterizes this distinction: In farming, "although man performs all kinds of preparatory tasks, such as clearing, plowing, and sowing, nature itself has to do the rest. Once his preparatory task is done, man can only sit down and wait. It is the inner growing power of living nature which performs the work." In contrast, "the craftsman gives natural materials forms which would not naturally arise in them. The technical object is something which is not cultivated but constructed, i.e., its component parts are arranged in an artificial pattern. The fashioning of these parts forces them into forms and functions which are not naturally present in them." Thus "in the work of construction there is a far more direct intervention in the natural order than there is in the work of cultivation."[49] Note, too, that this difference could be stated in terms of intention or volition. Another distinction along the same line: those technological activities which are in some sense in harmony with nature and those which are not—with harmony being given biological, ecological, or other interpretations.

Specific comments:

A'. Invention, depending on what it is contrasted with, can take on one or more of the following meanings: As opposed to scientific discovery, technological invention refers to the creating or generating of something new (although there are problems with the concept "new") rather than the finding out of something already there but hidden. Bell invented the telephone; Newton discovered the laws of gravity. The telephone did not exist prior to Bell's work; gravity existed but was not conceptualized in the form of scientific law prior to Newton. To generalize: invention causes things to come into existence from ideas, makes world conform to thought; whereas science, by deriving ideas from observation, makes thought conform to existence. Indeed, it is precisely this bringing together of ideas and materials—this uniting of a transformation which takes place first at the level of conception or thought with experiential confirmation in practical fabrication and operation—which is the essence of invention. Sir George Cayle (1773-1857), founder of the science of aerodynamics, arrived at an accurate conception of the airplane, but its invention (dated from the first successful flight at Kitty Hawk in 1903) had to await both the development of a suitable power plant and the Wright brothers' technical skills (of fabricating and operating).[50] Did Leonardo invent the parachute merely by conceiving it, or did Lenormand by constructing and testing it? Invention may begin in some conceptualization, but it does not finally take place until an artifact is operationally tested and discovered to be able to perform its assigned task. It is this active, worldly engagement which keeps invention (despite strong ideational components) from becoming just an element of technology-as-knowledge—although the kinds of ideas involved certainly require epistemological analysis, i.e., some conceptual distinguishing from scientific ideas.[51]

This notion of invention as a conscious process originating in the mind and confirmed by the activity of some individual is, however, a distinctly modern notion.[52] As with scientific discovery, the process of invention can take place over a short period of time or be prepared for through gradual evolutionary development. Some observers especially emphasize this evolutionary character of invention as a counterbalance to popular notions.[53] With this second kind of invention, however, one can almost never speak of an individual inventor. Instead, one speaks of national groups or historical periods as having invented such objects as the astrolabe or compass. Indeed, the process or activity of invention in the second case is probably quite different from that in the first. It originates not so much in the mind seeking its practical satisfaction in material objects as in a haphazard alteration of the matter and form of artifacts over the course of time and the eventual recognition of something useful.[54] It is truly evolutionary in the sense of being a natural selection of chance mutations. As such it is almost wholly devoid of the act of designing, a conspicuous aspect of the modern inventive process.

As opposed to design, invention refers to a process which proceeds by nonrational, unconscious, intuitive, or even accidental means. Invention is, as it were, accidental design—and as such highlights the element of insight which plays an important role even in highly systemized design. Modern engineering design, as an attempt to rationalize and systemize the creative process, has been called the "invention of invention" (Whitehead).[55] Also, there is a sense in which invention connotes a singularity of creation, whereas design takes an invention and adapts it to circumstances of, say, mass production. Although some inventors have been engineers, generally speaking the engineer is not concerned with novelty; if existing materials and processes are able to meet his task he will design around them, only with such refinements as circumstances immediately suggest.

B'. The ambiguities of "design": A design specifies some material object in sufficient detail to enable it to be fabricated. In designing (from the Latin *designare*, "to mark out") one can thus be concerned primarily with the formal properties of the object to be fabricated (as these must needs be expressed in the design)—as is the case in art and architecture;[56] or one can approach a design as the end term of some specific process—as in engineering. In the latter case, design readily becomes shorthand for the activity of designing.

"Engineering design is the process of applying the various techniques and scientific principles for the purposes of defining a device, a process or a system in sufficient detail to permit its physical realization."[57] Or, alternatively, engineering design is "an iterative decision-making activity to produce the plans by which resources are converted, preferably optimally, into systems or devices to meet human needs."[58] So defined, design in this sense may be described as the attempt to solve in thought, on the basis of available knowledge, problems of fabrication that will save work (as materials and/or energy).

Consider, for example, the construction of a foundation for some specific structure. If a contractor builds it on the basis of his experience and intuitions alone he is almost certain to do one of two things: either he will make it too weak and the building it should support will eventually collapse and have to be rebuilt; or he will make it too strong, using more steel and pouring more concrete than is actually necessary. In either case more work will have been expended than need have been. Were an engineer to design this same foundation, he would attempt to calculate the weight of the building and other relevant stresses, then using the principles of physics plus engineering geology (i.e., geological knowledge interpreted in terms of what kinds of structures various earth formations can support), and a socially specified safety factor, he would describe a foundation that would be neither more nor less than what was needed to suit the particular situation. It is somewhat paradoxical, but the right construction (like Aristotle's golden mean) is difficult to attain; it takes effort. But when this effort is expended at the right time, in the long run it saves effort. Engineering design is thus an effort (primarily of a mental sort) to save effort (of a physical sort).[59]

This mental effort is, however, something distinct from knowing or coming to know in a scientific or theoretical sense, because it terminates not in an interior act experienced as inherently valuable, but in construction (if not production). The scientist often experiences a tension between his knowledge and what he can express; he makes a discovery and then has to push himself beyond what he feels is his proper sphere in order to write it up. But such tension is not a normal feature of design experience, because the construction of drawings or models (which are also means of communication) is intimately bound up with the design process.[60] Engineering drawings, with their unique language and system for abstraction and representation, are not just means for communicating results arrived at by interior activity; they are part of the process and the means by which the results themselves are reached.[61] To say the same thing in a different way: although the actual execution of a plan is not designing, except insofar as the plan may continue to develop in order to meet originally unanticipated situations, such continuing development through actual execution is in fact a normal situation; on large construction projects a draftsman will be continuously at work revising drawings in the light of exigencies and changed circumstances, thus seeking to anticipate their further consequences. Drawing is a kind of testing or interrelating of various factors[62] by miniature building. It is not thinking in the sense of conceptualizing or the relating of concepts; it is thinking as picturing or imagining, and the relating of specific materials and energies. The designer solves problems of relating parts the way an artist does, by seeing them in practice. It is action or process, only on a reduced sale, made as free of physical labor as possible, but nevertheless not entirely free. It is still an effort (of a miniature physical sort) to save effort (of a gross physical sort). (This particular miniaturization of construction is, however, intimately related to a kind of knowledge in a way which remains to be considered.).

Right in the basic intention of design, then, we have a desire or will toward efficiency. Indeed, engineers often describe their work as the pursuit of an equilibrium of forces (stress and support in the case of a foundation), which says the same thing in different words. This efficiency (conceived in terms of force equilibrium), which constitutes the technical ideal, further manifests itself in varied forms in the various engineering activities classified according to their subject matter. As Skolimowski has argued, efficiency in surveying is accuracy of measurement; in civil engineering it is durability of structures; in mechanical engineering it comes out as the mathematical ratio of physical energy output over physical energy input, with the mechanical engineer always striving for a value of one.[63] Electrical engineering (to extend Skolimowski's argument) uses the same principle of efficiency, only with a different form of energy; ditto for chemical engineering. Thus the principle of engineering design, the volition which, as it were, founds it or makes it a consistent action: Never use more material or energy to solve a particular problem than is absolutely necessary.

C'. The methodology of design, engineering vs. art: In its most general description, designing is a process which takes place as a result of human intentions interacting with the world as it is given (when that given does not immediately satisfy the intentions) in such a way as to produce an otherwise unavailable object or course of action. In contemporary theory of design, the structure of this interaction or process is thought to be describable as a method, differing from yet analogous to the scientific method of knowing—that is, as a method of practical action. As such it has been argued to underlie all practical activity not only in engineering but in business, education, law, politics, art, etc.—if not simply all human action. The method is one, the only differences are in goals pursued and, perhaps, materials used. One designs works of art (from poems to paintings) for beauty, production and sales ventures for financial profit, engineering products for technical efficiency. At the same time, it is primarily within the engineering field that the methodology of design has been most seriously investigated.[64]

Without going into detail regarding the discussion of the exact character of this method itself or its relationship to the method of science, perhaps I can make a few relevant observations by considering briefly the question of the relationship between engineering and art. Initially, one distinguishes engineering from artistic design on the basis of ideals or ends in view (sometimes psychologically stated as volitions). The ideal of artistic design, in contrast with engineering efficiency, is beauty. Beauty is not so much a question of materials and energy efficiency as one of form. About this the whole subject of aesthetics has more to say, whereas it is ethics or politics which would incorporate a philosophical evaluation of efficiency.

Yet the difference between these two types of design does not remain at the level of ideals; it penetrates to the design process itself. This is apparent once one notices that in fact the ideal of efficiency refers to a process, is the directing criterion of an activity or a functioning product, whereas beauty is the ideal of a product or object in itself. Does a potter aim at efficiency in creating a beautiful pot? No, his aim is a good work, one of proportion and harmony; efficiency in production, while not to be wholly ignored, is a distinctly secondary consideration. For the engineer, on the other hand, it is beauty which is of secondary importance; while not to be ignored, beauty is judged, in industrial design, in terms of its contribution to function or efficiency.[65] Even more clearly: the ends of artistic design must be formal whereas the final causes of engineering processes are readily susceptible to verbal articulation in terms of human needs, wants, desires, etc.

A further observation: Engineering design limits itself to material reality (speaking metaphysically, matter and energy are both matter as contrasted with form). This is to be grasped or approached, however, by means of a mathematical calculus of forces which is derived from classical physics (Galileo and Newton) and its specific mathematical abstraction. The picturing or imagining that goes on in engineering design is done, as it were, through the grid of this physics—the grid itself being articulated as the engineering science of mechanics. This view of matter and energy through the grid of classical physics gives to engineering design a rational character not to be found in art. Engineering images, unlike other images, are capable of mathematical analysis and judgment; this is their unique character and one which sometimes leads people to confuse them with thinking in a deeper sense. (It also accounts, perhaps, for the fact that design is ignored as a kind of knowing in all major epistemological studies.) For art also is concerned with design and imagining, but its images cannot be rationally analyzed, are not subject to any well-developed calculus. Thus art, in contrast to engineering, appears as both more intuitive and more dependent on the senses. Although the artist, too, is concerned to design artifacts,[66] he necessarily does so in drawings and models which remain close in being to the final product. (Compare, for instance, a Rembrandt sketch for a painting with an engineering drawing of a building.) Needless to say, though, such remarks are no more than speculative suggestions about a very complex subject.

D'. On the phenomenological difference between invention and design: Invention and design are sometimes contrasted by saying that an inventor creates whereas a designer plans or at most discovers. Often it is said that the designer remains within the familiar and systematic; he does not deal with the unknown but only orders the known along well-established methodological lines, so that, given a clearly specified problem, two equally competent engineers will arrive at approximately the same solution. Design, like science, is impersonal; it is capable of objective or intersubjective confirmation. Whereas the inventor is supposed to work with the unknown and to bring forth from it unique creations, acting alone.

Dessauer, however, has argued that invention or creation also involves the experience of discovery.[67] Indeed, the word "invention" itself, from the Latin *invenire,* means "to come upon, find, or discover." Moreover, invention seems to be capable of objective confirmation—as is evidenced by such things as the independent invention of the airplane by the Wright brothers in America and the Brazilian aviator Alberto Santos-Dumont (1873-1932) in Europe and, indeed, by the fact that these two airplanes can be judged by a common standard which rates the second as inferior to the first in realizing and applying certain basic principles.[68] So much is this element of discovery and objectivity present that Dessauer postulates the existence of a transcendental realm of pre-established solutions to technical problems to explain the phenomenon. The natural or external world explains or accounts for the objectivity of science and engineering; but since invention does not bear upon what already *is* in a material sense, there must be a transcendent *is*ness or being to account for its discoveries. Others, however, while agreeing with Dessauer on the moment of discovery in invention, account for this in less transcendental ways. David Pye, for instance, argues quite simply, that

> Invention is the process of discovering a principle. Design is the process of applying that principle. The inventor discovers a class of system—a generalization—and the designer prescribes a particular embodiment of it to suit the particular result, objects, and source of energy he is concerned with.
>
> The facts which inventors discover are facts about the nature of the world just as much as the fact that gold amalgamates with mercury. Every useful invention is a discovery about the way things and energy can behave. The inventor does not make them behave as they do.[69]

Developing Pye's suggestion, one could argue that the difference between invention and design lies not in the fact that the designer deals with what is and the inventor with what might be,[70] for this proposition is an equivocation on "deals." In the sense of what they work from, both work from a given material world and its relations; in the sense of what they work toward, both work toward the making of a new artifact, the establishment of a new set of relations. (In the proposition above, the designer is said to "deal" with what he works from, the inventor with what he works toward.) Instead, the difference is more one of systematization and analytical calculation of potentialities in the given. As mentioned earlier, engineering design can be described as "the invention of invention"—that is, organized invention. Whereas primitive invention relies on accident, fortitutous insight into possible relationships among elements in the given, design is the attempt to develop a calculus of such relationships that can be used to solve well-specified problems. The fact that such a calculus may still rely at crucial moments on a cultivated serendipity (brainstorming sessions, etc.) only reveals that irreducible essence of invention as creative insight which must so far remain as a structured aspect of design. And such insight is also discovery: not of previously unknown elements or laws (as in science), but of previously unknown ways of relating elements or reorganizing natural processes (otherwise described by laws)—usually to meet previously recognized specific social needs or fabrication problems.

To be even more explicit, perhaps the making process in the initial instance (it might be different with routine making) can be broken down into the following sequence of logical moments (Fig. 2). (Note, however, that this is a logical, not a phenomenological sequence; the various moments will obviously be existentially interrelated in considerably more complex fashion than can be schematically indicated.)

Figure 2. The initial making process.

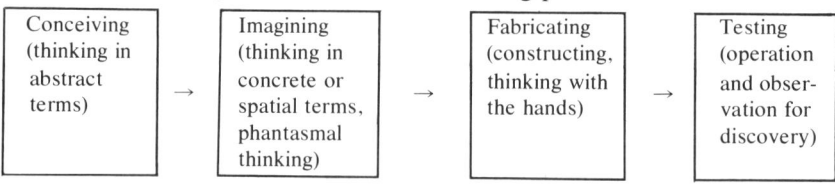

It is the second moment in this sequence that is designing in the proper sense. Invention is, as it were, a bipolar concept, referring both to conceiving and to the discovery manifested in testing; hence its ambiguity when it refers to only one of these elements in isolation from the other. In this sense, then, design will always be a moment linking these two aspects of invention; the only question would concern the degree of significance in particular cases. On the other hand, design itself, as phantasmal thinking or miniature construction, will invariably run up against certain barriers or problems which require new conceptions, a return to the conceptual moment. Thus design, too, will have a strong tendency to turn inside out into the invention sequence; the only question will concern how often and to what degree.

This articulation of the relationship between invention and design throws light on the characterization of design as the invention of invention. To invent (in the full sense) invention is to conceive and to put into operation the inventive process—or, to say the same thing in a different way, to consider the various circumstances or conditions under which invention readily takes place, to design (imaginatively plan) an institution which enhances precisely these factors, and to establish just such a working institution. Industrial research and development laboratories, applied (as opposed to basic) research institutions, are the result. At the same time, the invention of invention is capable of referring to any of these elements separately or in various combinations.

Finally, in stressing the moment of discovery in invention, questions are raised about the relationship between scientific and technological research. What is the difference between discovering a new form of life on Mars (basic research) and synthesizing a (perhaps identical) new type of DNA molecule in the laboratory (genetic engineering)? Other than intention, perhaps the fundamental difference is one of context. Technological discovery is dependent on a technological process—not just in the subjective sense of a human activity, but in the objective sense of an artificial process, i.e., one which is a function of artifacts and/or not part of the natural environment in which it goes on. Consideration of technological process in this sense, however, points back toward technology-as-object and the concept of system; further reflection on the sequence of moments in invention leads in the direction of fabricating and testing, making and using.

E'. Making and using: As already mentioned, these are the central categories of technology-as-process. The category of use by itself immediately calls to mind ethical analyses of the means-end relationship which it is not necessary to go into here.[71] But as human activities these two aspects are not necessarily so distinct. As Feibleman has argued, in each the key element seems to be a sensorimotor skill or technique.[72]

The word "technique" here raises another conceptual issue—the need to articulate the intuitive basis of a contrast between technique and technology. One limited distinction is that of the nineteenth century, in which "technology" means a systematic knowledge of the industrial arts, with "technique" being the means of practical application. Although this distinction continues to influence French (*technologie* vs. *technique*) and German (*Technologie* vs. *Technik*) usage, it has broken down in English for good reasons. It appreciates neither the inherently practical character of "technology" (as knowledge), nor the generality of "technique" (as skill, which can be of playing the piano or even reading a book).

Another proposed distinction argues that technological practice involves only interactions with artifacts, whereas technique can involve interaction with artifacts, natural objects, human beings, etc. There are techniques of swimming, wrestling, politics, computer programming, automobile construction and maintenance, etc.; but there are only technologies of computer programming, automobile construction and maintenance, etc.[73] In other words, there are techniques of both making and doing, but there are only technologies of making and using (when use involves artifacts)—and there could be no technology of making in the most primitive sense, making with the hands.

This way of distinguishing technology and technique has immediate commonsense appeal, yet by stressing a material differentia it glosses over a number of difficulties. First, it fails to explain the fact that sometimes we wish to speak not of the technology of, say, computer programming but of the technique. It is not just the presence or absence of artifacts in a human activity that determines whether or not it becomes a technology; the question is, instead, one of prominence or relationship to man. In drawing or writing or piano-playing, the development and training of the human psychomotor complex is much more central than any particular artifact, even though artifacts are undoubtedly employed. In the case of assembly-line production, on the other hand, the tools or machines are themselves more central. Thus tools or hand instruments tend to engender techniques, machines technologies—although even with machines, if one wishes to focus on the human manipulative processes, one might well speak of techniques. (In some cases, of course, the prominence of the tools is unclear and, as with glass blowing, processes can be spoken of both as techniques and technologies.) Second, "technique" and especially its adjective "technical" connotes singular making, whereas technology connotes multiplicity of production. An object may, for instance, be said to be technically feasible but not technologically feasible—meaning it can be made but not mass-produced. Finally, there is a sense in which technology, as opposed to technique, involves the greater use of rules, consciously articulated procedures and guidelines. As just suggested, technique is more involved with the training of the human body and mind (which is why one can speak of the "techniques of logic"—but not so easily of the "technology of logic"), whereas technology is concerned with exterior things and their rational manipulation. Techniques involve a large unrational or at least unrationalized (better still: unconscious) component. Techniques rely a lot on intuition, not so much on discursive thought. Technologies, on the other hand, are more tightly associated with the conscious articulation of rules and principles (which is why, in another sense, it is possible to speak of logic as a technology). Sometimes these rules are forced to remain at the level of heuristic principles. But at the core of technology there seems to be a desire to transform the heuristics of technique into algorithms of practice.

Types of Technology

When this is achieved, however, techniques become bound up with technology-as-knowledge as much as with processes.[74]

F'. Types of use: Philosophical analysis of the using of artifacts is an area readily susceptible to cross-fertilization with traditional ethics and politics as well as contemporary philosophy of action.[75] The following remarks are thus no more than propaedeutic.

Using is a more inclusive concept than making. Virtually all making involves the using of artifacts. But not all using results in fabrication. One type of using which does not directly produce artifacts, for instance, is living within. It may well be that there is a different type of use for each type of technological object—structures are used by being lived within, tools by being manipulated (in the strict sense, as used with the hand, handled), machines by being "driven," art objects by being viewed, systems by being managed, etc. Each of these types of using, however, can be practiced in, as it were, a private or a public (personal or social) manner. The shift from private to public use commonly goes under the name of innovation—which was described earlier (see 2, General Comments C, i) as the economic exploitation of invention. Innovation, it can be argued, is the paradigm of technological use. As such innovation, and what in sociological literature is termed technological transfer, deserves special philosophical consideration.

According to Toulmin's evolutionary analysis, innovation is part of a three-stage process: "(1) the phase of *mutation,* (2) that of *selection,* and (3) that of *diffusion* and eventual dominance."[76] The first is a conceptual or mental activity, the second involves practical testing, and the third is dependent on economic exploitation.

> The phase of *mutation* corresponds to the first half of the research and development operation, during which new techniques and processes are devised and prepared for testing and costing; the phase of *selection* is the one at which, within some specific area of application, the techniques or processes in question are shown to be feasible, both in technical and in economic terms; while the phase of *diffusion* and *dominance* . . . is that in which these skills spread into the general body of industrial and engineering techniques.[77]

While utilization in the broad sense may well have this sequential structure (with possible interrelationships[78]), questions of innovation are more often directly concerned with stage 3, that of diffusion. The exact character of this diffusion process itself can vary, however, from an unconscious long-term adaptation to a consciously stimulated acquisition. The complex changes in, say, traditional agriculture over long periods of time are quite different from the well-advertised exploitation of new products in a consumer-oriented society. Perhaps it may be suggested that this difference results in part from a difference in the way mutation (and hence testing) takes place.

Contrary to Toulmin, mutation can be either (a) material or (b) mental—that is, it can originate either in the artifacts themselves, as a result of wear, accidental variations in materials and fabrication techniques, etc., or it can take place consciously in the mind of an inventor. In the first case, it is possible that physical diffusion could even occur prior to recognition of utility in the object; in any case, it is a recognition (or discovery) of utility that will be primary. Yet if utilization is grounded in this recognition utility rather than in novelty of conceptualization, then utility will not be nearly as subject to conscious development. When mutation takes place as a result of creative conceptualization, however, utilization (testing and innovation) are likely to have to be planned. A mental framework at the beginning has implications for the mental structuring of diffusion. The conscious structuring of mutation, testing, and diffusion is, however, what under some circumstances is termed management.

G'. Management as a technological activity: The question of whether management is a technological activity is related, first, to the problem of relationships between economics and technology and, second, to the question of bureaucracy.

Historical background: classical economics identified three factors in the production of wealth: land, labor, capital (with technological objects, machines, etc. = fixed capital). The end of the nineteenth century saw the identification of a fourth (by Alfred Marshall): business organization or enterprise. From this fourth element has evolved the modern concept of management, or the organizing and directing of a business enterprise as a distinct profession if not a science. Schools of management theory (according to Koontz[79]):

i) Operational school—organized summary of public and private enterprise experience.
ii) Empirical school—case studies of successful managers with generalizations therefrom (similar to i).
iii) Human behavior school—emphasis on interpersonal relations as key factor in management; oriented toward psychology.
iv) Social system school—stress on cultural relationships; tends toward sociology.
v) Decision theory school—based on theories of decision-making as selection of optimal course of action from various possibilities; business game theory (tends toward iv).
vi) Mathematical school—proposes to quantify and analyze mathematically management and organization processes; includes operations research.

Now any of these schools of management theory is going to be part of technology-as-process because it is involved with making and using artifacts—although sometimes indirectly, through the organization of men. Technology-as-organization is part of the study of technology-as-process insofar as organization=structured process. But schools v and vi can be considered technologies in an even stronger sense, according to the contrast between technique and technology set forth above. For these management theories seek to make conscious a procedure for correctly using tools and machines, technological objects. In this they seek to be for use what design is for making; they are the engineering of use. (Note, also, how innovation=the managerial equivalent of invention; thus invention is to making as innovation is to use.)

As for the issue of bureaucracy, this would seem at first sight, in some governmental forms at least, to be not so much a kind of making as a doing. But given the dependence of bureaucracy on the modern technological infrastructure (typewriters, telephones, etc.), its rise in conjunction with attempts to control a highly technological civilization, and its tendency to be reduced to technocracy, it seems fair to identify this, too, as a facet of technology-as-process.

One more suggestion: perhaps the current program for technology assessment should be viewed as a type of management theory in a broad sense. For, clearly, it is an attempt to generate a calculus of use.

H'. Finally, as is the case with technological objects, the presence or dominance of these various aspects of technology-as-process in different social and historical settings points toward differences in technology in general. The most obvious distinction to be noticed in this respect is between the dominance of artistic design in the ancient world and engineering design in the modern. Prior to the development of modern mechanics and its calculus of forces architects and engineers (as well as artisans and other makers of artifacts) were forced to concentrate on formal, not to say aesthetic, properties in their constructions. With the development of the science of mechanics in the seventeenth and eighteenth centuries, attention seems to shift, especially by the nineteenth century, toward questions of materials and especially energy efficiency. Indeed, it is this which was a major contributor to the economic expansion which was part of the Industrial Revolution and might well have taken place to some extent independently of the development of new sources of power (steam engine, etc.); for the energy calculus also makes possible a precise economic assessment, once energy sources can be priced. In more ways than one, then, the scientific revolution of the modern period paved the way for the technological revolution. Another social consequence or indicator of the importance of this shift from artistic to engineering design is the fact that prior to, say, 1750 technological advances strengthened the artisan class; after that they undermined and eventually destroyed it.[80] Finally, as indicated above, the presence of bureaucracy as a specific structuring of technological processes is also a distinguishing feature of the modern period.

* * * * *

3. Technology-as-knowledge: It is this mode of the manifestation of technology which has so far received the most hard-core philosophical scrutiny. In this scrutiny philosophers have argued for the following distinctions—working from the least to the most conceptual:

(a) Unconscious sensorimotor awareness of how to make or use some artifact. Since these sensorimotor skills are unconscious they do not, of course, qualify as knowledge in the strict sense; as a further result they are acquired by apprenticeship to a master (someone who already possesses them) and intuitive training by example.

(b) Technical maxims (Carpenter[81]) or rules of thumb of prescientific work (Bunge[82]). These constitute the first attempt to articulate generalizations about the successful making or using skills. Example: "To cook rice, bring water to a boil, add one half volume of rice, and simmer for 20 minutes." Indeed, most cookbook recipes are composed of technical maxims.[83]

(c) Descriptive laws (Carpenter) or nomopragmatic statements (Bunge). These laws are of the form "If A then B," with concrete reference to experience. As Carpenter says, descriptive laws "are like scientific laws in being explicitly descriptive and only implicitly prescriptive of action, but they are not yet scientific in that the theoretical framework which could explain the law is not yet explicit."[84] Because they are usually generalizations derived directly from experience, such nomopragmatic formulae are called by engineers empirical laws. Example: Coulomb's empirical laws for constructing retaining walls, formulated not with the use of engineering geology and physics, but simply on the basis of the observations about which size and shape fortifications held up well in such and such conditions, etc. Note that there are also many descriptive laws of use, such as those developed by Taylor from his time and motion studies at the Watertown arsenal.

(d) Technological theories. Theories either systematically relate a number of laws or provide a broad conceptual framework to explain them. Technological theories, according to Bunge, are of two types: substantive and operative. "Substantive technological theories are essentially applications, to nearly real situations, of scientific theories."[85] Examples: aerodynamics or the theory of flight as an application of fluid dynamics; thermodynamics; electrical engineering, etc. Substantive theories, then, constitute the so-called engineering sciences and are applied science in the strict sense.[86] Substantive theory has also been called (by Polanyi) systematic technology, in constrast to the empirical technology of descriptive laws.[87] Operative technological theories "are from the start concerned with the operations of men and men-machine complexes in nearly real situations."[88] Examples: decision theory, operations research, etc. Substantive theory employs both the content and method of science; operative theory applies only the method of science to problems of action, to develop "scientific theories of action."[89] Thus the former are more tied up with making, the latter with use.

Comments:

A. According to Bunge, the central element in modern technology is the development of another type of technological knowledge in between technical maxims and technological theories. These are what he calls the grounded rules of applied science. They are really rules (i.e., like technical maxims they take the form "To get B do A"), but they are not attempts to generalize immediate experience; instead they are grounded in the nomopragmatic statement "If A then B," which in turn is warranted pragmatically (if not logically) by a scientific law which is part of a scientific theory. To Bunge's mind the attempt to understand "exactly what the foundation of rules consists in" constitutes "the core of the philosophy of technology."[90]

B. Once again, then, one can state a difference between ancient and modern technology: the former relied for guidance purely on sensorimotor skills, technical maxims, and descriptive laws, whereas the latter uses these plus technological rules and technological theories. It might be maintained, as well, that this presence of technological rules and theories undermines the importance of skills and maxims. There is a need, however, to explore the ways in which these technological rules and theories are made possible by modern science, and the ways in which they in turn make something like engineering design possible. There is need, that is, for a deeper epistemological and metaphysical analysis of this conceptual differentiation between two kinds of technology-as-knowledge.[91]

C. All types of technology-as-knowledge are generally to be distinguished from science. Science is based upon observation, and is the accumulation of information about the world; its basic element is a scientific law which describes the way the world is. Technology, on the other hand, is primarily thought about how to control the world; its basic element is a rule—if not a concrete invention. The full articulation of the conceptual structure and epistemological foundations of scientific laws, and the full articulation of the conceptual structure and epistemological foundation of technological rules are the subjects of philosophy of science and philosophy of technology, respectively. But the most common way to distinguish between science and technology is on the basis of ends or intentions: scientific knowledge is said to aim at knowing the world, technological knowledge aims at controlling or manipulating it.[92] This is the difference, for Bunge again, between scientific prediction (which is a means for confirmation of theory) and technological forecast (which, by suggesting how to influence circumstances, is a means to control). Such a difference in aims is also useful to explain the difference between scientific and technological experiments: the first is a test of the truth of some theory, the second a test for effectiveness. And these different aims, by their prolongation into action, produce different experimental structures.[93] Such an appeal immediately points, on the one hand, back toward technology-as-process (testing) and, on the other, forward toward technology-as-volition (intentions).

* * * * *

4. Technology-as-volition: If technology-as-knowledge is, so far, the best analyzed mode of technology, then technology-as-volition is the least. Partly this is because the nature of willing is itself so poorly understood by philosophy; will is the elusive Proteus of the philosophy of mind. And technology-as-volition perfectly illustrates the problem; technologies seem to be tied up with every imaginable will, motive, love, desire, need, intention, affection, choice. etc. Technology-as-volition has been described in terms of a) will to survive, or to satisfy some basic biological drive; b) will to power; c) will to freedom; d) will to help others, an altruistic will; e) will to make money, the economic will; f) will to be famous; and g) will to realize almost any self-concept. And each of these could arguably be expected to produce different types of technology.

Over and above the problem of a lack of philosophical consensus about the nature of willing, the difficulty with approaching technology from the perspective of volition or will[94] is threefold. First, volition is the most individualized and subjective of the four manifestations of technology. Thus there may well be a sense in which each person's motivation, being somewhat unique, becomes connected with making and using to give rise to a unique technology. But surely this is not only the least philosophically interesting thing to say, it is also the least practically meaningful. Such individuality never has social or public consequences until it is united with similar volitions of others to produce what might be called a social, public, or cultural act of willing.

Second, there is always the problem, in volition, of correspondence between subjective intentions and objective means. An act of willing, except in the case of oneself, cannot be directly known (some would even argue against its being known directly by oneself); it can only be inferred from action (including, of course, speech). But is the action or means chosen an adequate expression of the particular intention, so that one can legitimately infer from the character of one back to the character of the other? The same question arises, except in reverse, with knowing and ideas. Do one's ideas adequately correspond to what one knows in reality? In the case of knowledge we attempt to deal with this issue (at least heuristically) by clarifying our concepts so that individuals may judge for themselves, on the basis of their own experience of an object, which ideas are the best mental representatives of its character. In the case of technology, however, we are forced (on the same heuristic grounds) to clarify processes and objects in an attempt to discover and elucidate their objective tendencies and properties, again so that individuals may choose for themselves whether or not they adequately express their own volitions. Indeed, much of the present (at least popular) discussion about technology and values is vacuous precisely because it does not attempt to do this. Instead, it assumes that one can take a new value or volition, attach it to an existing process or object, and create a new technology. But is the existing process or object really commensurate with the new volition? Sometimes it is, sometimes it is not. The problem is obviously recognized on the macroscopic level, since people do not try to use guns as toothpicks. But they do say things like, "The problem with technology is just how man uses it"—thinking that all existing technology can be magically transformed by a change of volitions. To give a perhaps fatuous example: You cannot really use nuclear weapons for peaceful purposes— to dig canals and that kind of thing. You just cannot take a constructive will and harness to it a technological object with an intrinsic principle of massive destructiveness. The object resists; either it alters your volition, you fail in your project, or you abandon it for some more nearly adequate means. To avoid such mistakes what is needed is a clarification of the inherent nature of various technological processes and objects—which is what is coming about through contemporary historical, sociological and ecological studies of technology.

Third, there is the problem of self-understanding and levels of the will. According to Pfänder's phenomenology of willing and motivation,[95] willing in the general sense is only awareness of striving—a psychological phenomenon which impinges on the ego, but does not involve its center or core. In its lowest form striving would be experienced as a biological urge or instinct, although it might also be felt as a peripheral wish, hope, longing, desire, fear, etc. Striving is simply characterized by an awareness of something absent which attracts—and can be composed of numerous even conflicting impulses. What converts this striving into an act of willing is its being taken into the center of the ego. Willing in the strict sense is constituted only after one comes to believe that the goal of a striving can be realized through one's own actions, and when (as sometimes happens) the ego spontaneously sides or identifies with such striving. As Pfänder says, "Thus willing, but not striving, includes the immediate consciousness of self." In other words, "The act of willing is . . . a *practical act of proposing filled with a certain intent of the will* which issues from the ego-center and, penetrating to the ego itself, induces in it a certain future behavior. It is an act of self-determination in the sense that the ego is both the subject and the object of the act."[96] But since it is of "the very essence of the performance of an act of willing" that the ego "appears as the agent,"[97] willing is dependent on the self-concept possessed by the ego. Only if one sees oneself in a certain way can one identify with some particular striving. The question for technology, then, is what self-concept enables or urges one to identify with certain strivings the proper means to the realization of which are particular types of technology? The question becomes one of the understanding, or better self-understanding, of man.

Why is it necessary, though, to raise such questions in precisely this form with regard to technology? Clearly all three difficulties immediately take us out of the present limited conceptual typology. The second returns us to questions of technology-as-process and as-object—at once physical and metaphysical. The first and third point toward the anthropological foundations of technology, to the question of the relationship between technology and man. The answer is that whereas knowing is, as it were, its own volition, making and using are not. Knowing needs no independent act of the will to be set in motion in the psyche; this is one thing it means to say man is a rational animal. In knowing, activity (insofar as activity is involved) is a means to something which is already experienced as an end in itself; knowing is so intimately involved with man's nature he simply does it without questions. (In fact, he only questions not doing it.) The same does not hold, however, for making and using. Here the questions are always not "Why not?" but "Why?"[98] Strivings are always initially experienced as eccentric, outside one's core self. Only after investigation and questioning in terms of oneself is it possible for them to be affirmed or identified with. They are only undertaken as self-projects if the ego can, through its self-understanding, side with them. To say the same thing in a different way: Kierkegaard defines the soul as a relation which relates itself to itself.[99] In so doing it is always questioning its other relationships. In Kierkegaard's anthropology man is always trying to identify himself with various projects—striving for pleasure, striving for fame, etc.—but in each case the development of this self-concept undermines the identification. What is it about the ego, then, that allows such a "siding" with technology to take place in the stabilized form that history evidently presents? Do some self-understandings encourage and others discourage particular sidings? Again, such questions point directly toward philosophical anthropology and the origins of technology in the nature of man.

* * * * *

Let me now summarize the typology I have developed and briefly indicate two conclusions. Technology, to my mind, should be differentiated modally and generically. Its four modes are technology-as-object, as-process, as-knowledge, and as-volition. Each of these modes is composed of certain more or less definite elements: utilities, utensils, tools, machines, devices, etc., for technology-as-object, and so on. Any generic differences in types of technology should be able to be stated in terms of the presence and organization of elements under each of these four modes—although this, by itself, will not give a definition of the essence of a technology. The essence is what stands behind and manifests itself differentially (but with equal strength) in each mode. The one clear generic difference I have found distinguishes between ancient and modern technology. Each species, however, can be further subdivided by the matter and imagination in and through which it is embodied into an indefinite number of subspecies—although in the case of modern technology, I would suggest that one of its characteristic features, grounded in the mathematization of its possible modalities, is a tendency to remain highly unified. Such a morphology can be conveniently diagrammed in the following manner (Figure 3):

Figure 3. Morphology of Technology

TECHNOLOGY

(all making and using of artifacts)

Ancient

Modal characteristics:

as-knowledge: skills, maxims, with some descriptive laws

as-process: artistic design

as-object: utensils, apparatus, utilities, tools and machines (but only powered by naturally given human, animal, or physical energies)

Indefinite material and cultural differentiation

Modern

Modal characteristics:

as-knowledge: technological rules and theories

as-process: engineering design, bureaucracy

as-object: machines, especially those powered by abstract energies, plus automated devices

Indefinite material and cultural differentiation. (The diagramic difference here is meant to represent the greater unity obtaining among modern types of technology.)

Underlying this morphological distinction there is, as it were, an anatomy and physiology. Generic differences are clearly evident only in pure types; but the structural identification of pure types is problematical. Technology is practiced as often in impure as in pure forms. Thus it is necessary (to extend the biological metaphor) to indicate the structural possibilities for natural variations from the ideal type. To say the same thing in a different way: The problem of grasping the essence of technology within a species is complicated by the fact that each of its modes can, at any one point in time or social history, exist independently of others and thus manifest the essence its own way. Through this temporary independence a complex set of aggregates and subspecies, exaggerations of varying stabilities, readily comes into being. There is, then, need for the recognition of what may be called the *concept of pure technology* and its relationships to technology in its various impure or incomplete forms. The requisite anatomical analysis can be indicated by means of a modified Venn diagram (Figure 4):

Figure 4. Anatomy of Technology[100]

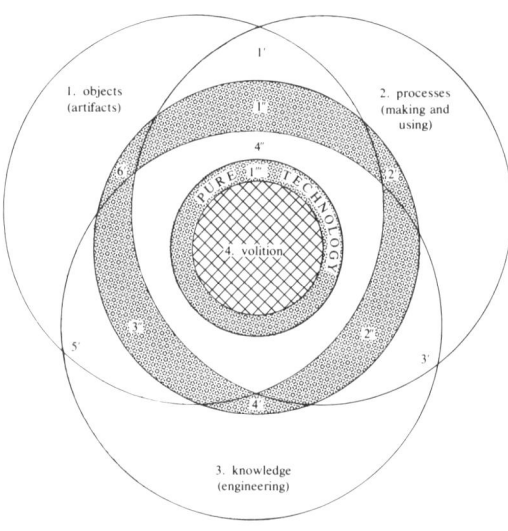

Circle 1, exclusive of the cross-hatched inner circle, represents technology-as-object; circle 2, exclusive of the cross-hatched inner circle, technology-as-process; circle 3, exclusive of the cross-hatched inner circle, technology-as-knowledge. The shaded ring and the two concentric circles at the center represent technology-as-volition.

1'	objects + processes (without knowledge or volition)
2'	processes + volition (without objects or knowledge)
3'	processes + knowledge (without objects or volition)
4'	knowledge + volition (without objects or processes)
5'	knowledge + objects (without processes or volition)
6'	objects + volition (without processes or knowledge)
1"	objects + processes + volition (without knowledge)
2"	processes + knowledge + volition (without objects)
3"	knowledge + volition + objects (without processes)
4"	objects + processes + knowledge (without volition)
1'''	objects + processes + knowledge + volition

Types of Technology

To flesh out this anatomical skeleton, consider the following possible varieties of technological man:

1 = man the car owner
2 = man the worker on an automobile-assembly line and/or the operator of a car
3 = man the automotive engineer
4 = man the consumer of cars, one who strongly desires to own a car and sees his self-realization bound up with owning a car
1' = one who both owns and operates a car in a playful or detached manner
2' = skilled mechanic, especially one whose self identity is tied up with proficient operation of a car
3' = engineer who designs fantasy cars as a hobby
4' = automotive inventor or research engineer
5' = car collector
6' = car owner who views his automobile as a magical object, a fetish
1" = assembly-line technician
2" = automotive design engineer
3" = automotive testing engineer
4" = manager of an automobile plant in an underdeveloped country
1''' = pure technologist: automotive engineer who designed and manufactured his own car and defines himself in terms of that ability and ownership

Transforming this anthropology into a more general conceptual map, one can analyze modern technology (since this is more familiar to us) in the following terms:

1 = technological products or objects
2 = activities of making and using
3 = engineering
4 = technological volition, the will to power
1' = play (see the tradition of making and using imaginative gadgets and machines for their own sake[101])
2' = technical skills
3' = operations research, decision theory, etc.
4' = invention (meaning especially research and development engineering)
5' = automata theory
6' = magic (what Bunge calls pseudo-technology[102])
1" = mass-production facilities or organizations (bureaucracies)
2" = design
3" = testing or experimentation
4" = technology in an underdeveloped country (technological civilization without technological culture[103])

1‴ = pure modern technology (modern technological objects, processes, and knowledge embedded in a modern technological culture)

Ancient technology would have its own corresponding conceptual anatomy.

It is important to recognize that this scheme is not necessarily exhaustive; nor do all aspects denote equally real possibilities. Some probably denote no more than abstract fictions. Furthermore, in modern technology objects and knowledge tend to collapse into processes in ways that make process a much more dominant category than in ancient technology. Nevertheless, such a two-dimensional typology provides, I think, a basis for practical reflection on our polytechnical environment and capabilities, at the same time that it points toward philosophical resolutions to many of the conflicting descriptions of the nature and meaning of technology.[104] However, as has already been stated, this conceptual framework is merely a speculative presentation lacking full development and detail. The anatomical description is put forth with special tentativeness. And morphologically, no suggestion has been made for a definition of either species of technology, while my notes and comments have stressed the need for deeper epistemological and metaphysical studies. But the anatomical analysis nevertheless hints that the essential nature of technology, ancient or modern, will be found most clearly revealed at the point of its greatest density—in the confluence of objects, processes, knowledge, and volitions. One way of reading this paper is as a prolegomenon to a critique of pure technology. And as my conspicuous omission of a description of specific differences in terms of technology-as-volition is meant to indicate, it is through a philosophical analysis of the relationship between technology and this aspect of man that such a critique ought now to be developed. Yet that attempt lies beyond the scope of the present paper, and must be left for another occasion.

FOOTNOTES

1. The term engineer (Latin *ingeniator*) was first used in the Middle Ages to designate builders (and sometimes operators) of battering rams, catapults, and other "engines" of war. Before this, and even still at this time, the general name for one who designs and directs the construction of large-scale artifacts was architect. Thus Vitruvius' *De architectura*, a work in ten books published in Rome in the first century A.D., deals with urban planning, building materials, aesthetic principles, construction techniques, hydraulics, geometry, clocks, mechanics, etc. John Smeaton (1724–1792), British architect of the Eddystone Lighthouse and other structures, was apparently the first person to call himself a civil engineer—a term which came to refer to one who designs roads, bridges, water-supply and sanitation systems, railroads, etc., that is, architectural projects which are approached more from the point of view of utility and efficiency than aesthetic form or life functions. In the

Types of Technology

nineteenth century, as an offshoot of the inventions of James Watt (1736–1819) and the textile machinists of the Industrial Revolution there arose the field of mechanical engineering—perhaps the key event for the emergence of engineering in the modern sense. With the development of a profession separate from either artisans (and inventors) or scientists—that is, engineers, as men with scientific-mathematical training and technological involvement—the 1800s realized the Enlightenment vision of a union between science and the practical arts, in which science would provide a method for solving practical problems, and thus a foundation for systematic progress. Since then engineering has been divided into other branches, according to the kinds of materials, energies, or products being dealt with by this method—i.e., electrical engineering, chemical engineering, electronic engineering, aeronautical engineering, nuclear engineering, computer engineering, etc. And these branches have been further subdivided by fields—e.g., some of the mechanical engineering fields are machines, heat, fluids, automotive, refrigeration, gas turbines, etc.

2. Engineering can be divided not only into various branches, according to what the engineer works *with*, but also into functions, determined by *how* the engineer works, the various functional elements of engineering method. Engineering functions range from research and development through design, production, and construction, to operation, sales, service, and management. The relationship between these functions can be schematized in a flow diagram of the following sort:

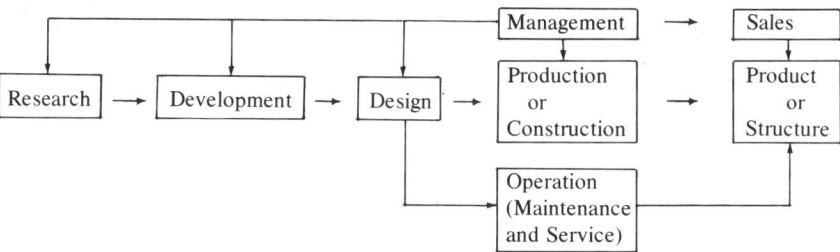

As with the branches of engineering, however, there is no one universally agreed-upon list of these functions; there is simply a spectrum of activities which can be divided and subdivided in numerous ways. A flow diagram has its own limitations and oversimplifications as well. The point is that there are not just electrical engineers, but electrical research engineers (doing applied research on electrical energy), electrical design engineers (designing either a specific electrical device for factory production or an electrical system for on site construction), electrical service engineers (maintaining and servicing some electrical product or system), etc. For a good elementary discussion of these and other aspects of engineering from the viewpoint of the engineer, see Ralph J. Smith, *Engineering as a Career,* 3rd edition (New York: McGraw-Hill, 1969). Supplementary reference: A. W. Futrell, Jr., *Orientation to Engineering* (Columbus, Ohio: Charles E. Merrill, 1961), esp. ch. 16, "Functions of Engineering."

3. *Webster's Third International Dictionary* (1961) and the *McGraw-Hill Dictionary of Scientific and Technical Terms* (1974) define engineering as "the science by which properties of matter and the sources of energy (Webster's)/power (McGraw-Hill) in nature are made useful to man in structures, machines, and products." For a sample of definitions running from that of the British architect and civil engineer Thomas Tredgold (1788–1829) to that of the Engineers Council for Professional Development (in 1963), see Smith, *op. cit.,* pp. 8–9. Smith's own definition is that "engineering is the art of applying science to the optimum conversion of natural resources to the benefit of man" (p. 10). Commenting on this definition he observes that "the conception and design of a structure or device or system to meet

specific conditions in an optimum manner is engineering" (pp. 11 and 13—p. 12 is a picture). Furthermore, "it is the desire for efficiency and economy that differentiates ceramic engineering from the work of the potter, textile engineering from weaving, and agricultural engineering from farming" (p. 13). "In a broad sense," he writes later, "the essence of engineering is *design*, planning in the mind a device or process or system that will efficiently solve a problem or meet a need" (p. 160). Thomas T. Woodson, in *Introduction to Engineering Design* (New York: McGraw-Hill, 1966), while noting that not all engineering functions involve design in the technical sense, also identifies "engineering design" as "the essential activity of professional engineering" (p. 8). Variations on this theme can be found in all the texts on design cited in note 64 below. Cf. also Joseph Edward Shigley, *Machine Design* (New York: McGraw-Hill, 1956), which characterizes design "as an intermediary between those areas of study which are basically scientific and analytic in content and the more practical engineering courses which are best described as teaching the art of engineering" (p. vii); and Robert E. Parr, *Principles of Mechanical Design* (New York: McGraw-Hill, 1970), which describes design as *"the creative part of engineering"* (p. 1). For some historical discussion of the "ability to design" as the crucial requirement for membership in the engineering profession, see Edwin T. Layton, Jr., *The Revolt of the Engineers: Social Responsibility and the American Engineering Profession* (Cleveland: Press of Case Western Reserve Univ., 1971), pp. 26–27, 30, 39, 49 (note 33), 51 (note 60), 80, 88–89; and William McClellan, "A Suggestion for the Engineering Profession," *Transactions of the American Institute of Electrical Engineers* 32, pt. 2 (May–December 1913), p. 1272. For further discussion and references see the commentary on design, B' under section 2, technology-as-process.

4. Some differences in connotation and use will be discussed below under 2, E'. Engineers themselves, reflecting and influencing the culture at large, tend to take the two cognate chains technic-technical-technician (abstract noun-adjective-practitioner) and technology-technological-technologist and amalgamate them to form the grammatical hybrid technology-technical-technician. This explains how the terms "technical" and "technician" can be in greater currency when describing the concrete art of making, in which one is more likely to need an adjective or name for the practitioner, while this same process can be referred to abstractly as "technology." As illustrative of this latter use, consider Douglas M. Considine, ed., *Chemical and Process Technology* (New York: McGraw-Hill, 1974). According to the editor's Preface, this handbook "is *not* a compilation of generalities, but rather is packed with detailed information" (p. xxii) concerning "the traditional spheres of interest in industrial chemistry and chemical technology as reflected by the petroleum, petrochemical, chemical, paper, textile, and other long-established process industries" as well as "more recent applications of an advancing and broadening chemical technology, including, as examples, the materials and processes now required by the electronics, optics, and aerospace industries" (p. xxi). The following diagram from p. xxviii gives a conceptual overview of the kind of empirical data contained in such a volume:

DIAGRAM

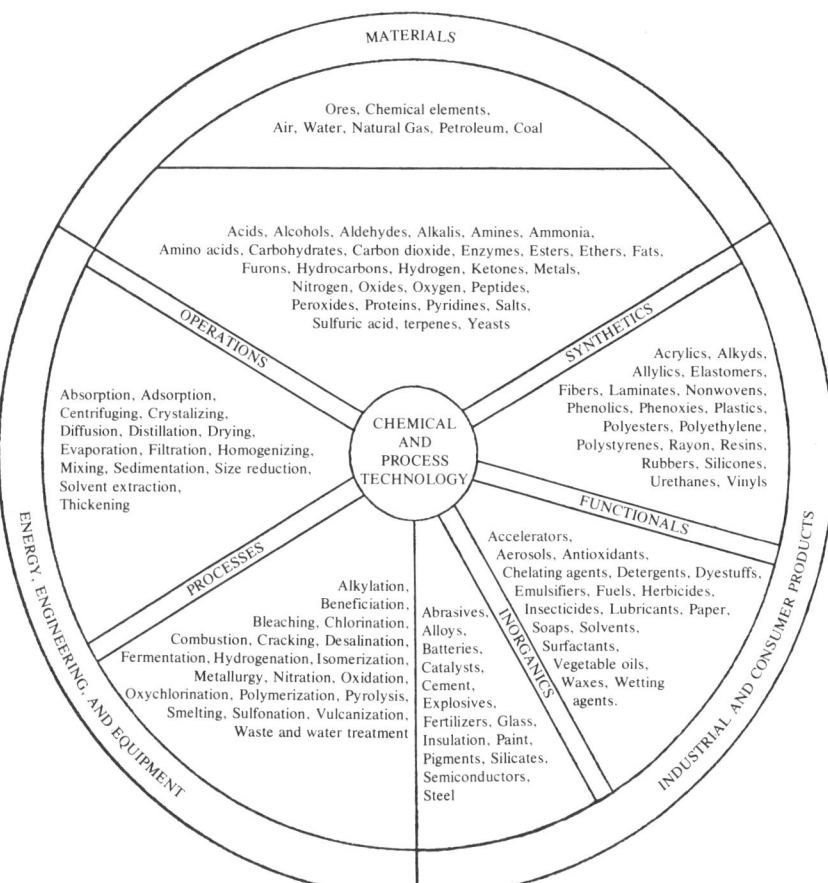

From Douglas M. Considine, ed., *Chemical and Process Technology,* p. xxviii.

Similar examples of this use of the term "technology" to refer not to theory but empirical data concerning the raw materials of industrial processes and their products can be found in, e.g., Alexander S. Craig, *Dictionary of Rubber Technology* (New York: Philosophical Library, 1969), and Alan Gilpin, *Dictionary of Fuel Technology* (New York: Philosophical Library, 1970).

This "materialist" or practical usage is also the foundation of the term "technological sciences" (= not just making, but systematic *knowledge* of making, or sciences of the industrial arts), which is meant to include traditional military and civil engineering, agricultural engineering, and the new disciplines related to space, computers, and automation. For more on this term, see Carroll W. Pursell, Jr., "Technological Sciences," *New Encyclopaedia Britannica* (1974), Macropaedia, Vol. 18, pp. 19–21. It is this term, in the condensed form of just "technology," which is defined in the *McGraw-Hill Dictionary, op. cit.*, as "systematic knowledge of and its application to industrial processes; closely related to engineering and science," or in T. C. Collocott, ed., *Dictionary of Science and Technology* (New York: Barnes and Noble, 1971), as "the practice, description, and terminology of any or all of the applied sciences which have practical value and/or industrial use."

For one representative statement of the relation between technician, scientist, engineer, and craftsman from the engineering perspective see Ralph J. Smith, *op. cit.*, pp. 210–211:

> The engineer is a man of ideas and a man of action. . . . He develops mental skills but seldom has the opportunity to develop manual skills. In concentrating on the application of science he can obtain only a limited knowledge of science itself. . . . The primary objective of the *scientist* is "to know," to discover new facts, develop new theories, and learn new truths about the *natural* world without concern for the practical application of the new knowledge. . . . The engineer is concerned with the *man-made* world. He has primary responsibility for designing and planning research programs, development projects, industrial plants, production procedures, construction methods, sales programs, operation and maintenance procedures and structures, machines, circuits, and processes. . . . The *technician* usually specializes in one aspect of engineering, becoming a draftsman, a cost estimator, a time-study specialist, an equipment salesman, a trouble shooter on industrial controls, an inspector on technical apparatus, or an operator of complex test equipment . . . [The] technician occupies a position intermediate between the engineer and the skilled *craftsman*. The craftsman, such as the electrician, machinist, welder, patternmaker, instrument-maker, and modelmaker, uses his hands more than his head, tools more than instruments, and mathematics and science rarely.

These basic distinctions are confirmed by the description of "Technology" in the *McGraw-Hill Encyclopedia of Science and Technology* (1971), vol. 13, pp. 428–429. See also the description of the Bachelor of Engineering Technology (B.E.T.) degree in John Dustin Kemper, *The Engineer and His Profession* 2nd ed. (New York: Holt, Rinehart and Winston, 1975), pp. 263–265: "The intention of the creators of the B.E.T. degree is that its holders would occupy a middle ground between the craftsman and the engineer." "In the words of the Advisory Committee for Engineering Technology Education Study: 'The technologist should be a worker of detail, the engineer, of the total system. . . . The development of methods or new applications is the work of the engineer. Effective use of established methods is the work of the technologist.'" James K. Feibleman, "Pure Science, Applied Science, and Technology: An Attempt at Definitions," in Carl Mitcham and Robert Mackey, eds., *Philosophy and Technology* (New York: Free Press, 1972), pp. 33–41, offers an introductory philosophical articulation of such distinctions, as does C. David Gruender, "On Distinguishing Science and Technology," *Technology and Culture* 12:3 (July 1971),

456–463. For Gruender the chief distinction between applied science and technology "is in the scope or generality of the problem assigned. Those of broader scope we are inclined to think of as problems of 'applied' science [= engineering?]; those that are closer to being specific and particular we think of as 'technology'."

For a moderately different perspective, see Philip Sporn, *Foundations of Engineering* (New York: Pergamon Press, 1964), pp. 18–19, where technician, technologist, and engineer are distinguished on the basis of the comprehensiveness of their abilities; technicians make particular devices (motors), technologists have mastered some whole field (electric-power production), and engineers are concerned with a whole system, including the socio-economic context (electric power systems). This distinction corresponds to that in some engineering curricula which offer a two-year degree in engineering technology.

5. Note, once more, the complexity of usage being so summarily appealed to here. Consider how the terms "medical technologist" and "medical technician" easily refer to one and the same individual who is proficient in medical technology (the use of technical medical hardware without even the nurse's limited comprehension of how it contributes to a patient's health), whereas one who is skilled in the manipulation of educational technology (audio-visual devices, etc.) is not called an educational technologist and only seldom a technician. Too often, alas, such an individual is called a teacher.

6. See, e.g., Tom Burns, "Technology," in Julius Gould and William L. Kolb, eds., *A Dictionary of the Social Sciences* (New York: Free Press, 1964), pp. 716–717, where the term is defined, first, in regard to primitive societies as denoting "the body of *knowledge* available for the fashioning of implements and artifacts of all kinds," and second, in regard to industrial societies as denoting "the body of *knowledge* about (a) scientific principles and discoveries and (b) existing and previous industrial processes, resources of power and materials, and methods of transmission and communication, which are thought to be relevant to the production or improvement of goods and services." Yet Burns's further commentary leaves something to be desired. He seems to misconstrue the meaning of the Greek word τεχνολογεῖν; he fails to differentiate *Technik* and *Technologie* (German), and *technique* and *technologie* (French); he overemphasizes a divergence between archeological and historical-sociological usage; he does not acknowledge contemporary economic definitions. For this last, see John Kenneth Galbraith, *The New Industrial State*, 2nd rev. ed. (Boston: Houghton Mifflin, 1971), p. 12: "Technology means the systematic application of scientific or other organized knowledge to practical tasks." Both Burns and Galbraith, however, probably stress too heavily the knowledge factor in "technology" as a social science concept.

> Technology in its broad meaning connotes the practical arts. These arts range from hunting, fishing, gathering, agriculture, animal husbandry, and mining through manufacturing, construction, transportation, provision of food, power, heat, light, etc., to means of communication, medicine, and military technology. Technologies are bodies of skills, knowledge, and procedures for making, using, and doing useful things. They are techniques, means for accomplishing recognized purposes.—Robert S. Merrill, "The Study of Technology," *International Encyclopedia of the Social Sciences* (New York: Macmillan and Free Press, 1968), pp. 576–577.

But cf. D. S. L. Cardwell's careful use of the terms "technics" and "technology" in his article on "Technology" in Philip P. Wiener, ed., *Dictionary of the History of Ideas* (New York: Scribner's, 1973). See also Peter F. Drucker's argument that the subject matter of technology is not so much "how things are done or made" as "how man does or makes" in "Work and Tools," *Technology and Culture* 1:1 (Winter 1959), 28–37. This important article has been reprinted in Drucker's *Technology, Management and Society* (New York:

Harper and Row, 1970); and in Melvin Kranzberg and William H. Davenport, eds., *Technology and Culture: An Anthology* (New York: Schocken Books, 1972). (But for the blindness of the senior editor, it should have been included in Mitcham and Mackey, eds., *Philosophy and Technology, op. cit.*)

7. For a brief critique of some historical studies and definitions of technology see Carl Mitcham, "Philosophy and the History of Technology," forthcoming in George Bugliarello and Dean B. Doner, eds., *The History and Philosophy of Technology*, Proceedings of an International Symposium on the History and Philosophy of Technology held at the University of Illinois at Chicago Circle, May 14–16, 1973 (Urbana: University of Illinois Press).

8. References for these definitions are as follows: James K. Feibleman, "Technology as Skills," *Technology and Culture* 7:3 (Summer 1966), 318–328; Mario Bunge, "Toward a Philosophy of Technology," in Carl Mitcham and Robert Mackey, eds., *Philosophy and Technology* (New York: Free Press, 1972), pp. 62–76; Bernard Bavink, "Philosophy of Technology," *The Natural Sciences* (New York: Century, 1932), pp. 562–574; Henryk Skolimowski, "The Structure of Thinking in Technology," in Mitcham and Mackey, eds., *op. cit.*, pp. 42–49; Jacques Ellul, *The Technological Society*, trans. John Wilkinson (New York: Knopf, 1964); Karl Jaspers, "The Nature of Technology," in *The Origin and Goal of History* (New Haven: Yale University Press, 1953), pp. 100–102; Friedrich von Gottl-Ottlilienfeld, *Wirtschaft und Technik*, 2nd ed. (Tübingen: Mohr, 1923); I. C. Jarvie, "The Social Character of Technological Problems" and "Technology and the Structure of Knowledge," in Mitcham and Mackey, eds., *op. cit.*, pp. 50–61; Stanley R. Carpenter, "Modes of Knowing and Technological Action," *Philosophy Today* 18:2 (Summer 1974), 162–168; Lewis Mumford, "Technics and the Nature of Man," in Mitcham and Mackey, eds., *op. cit.*, pp. 77–85; Oswald Spengler, *Man and Technics* (New York: Knopf, 1932); Ernst Jünger, "Technology as the Mobilization of the World Through the *Gestalt* of the Worker," in Mitcham and Mackey, eds., *op. cit.*, pp. 269–289; Ortega y Gasset, "Thoughts on Technology," in Mitcham and Mackey, eds., *op. cit.*, pp. 290–313; Emmanuel G. Mesthene, "Technology and Wisdom," in Mitcham and Mackey, eds., *op. cit.*, pp. 109–115; C. B. Macpherson, "Democratic Theory: Ontology and Technology," in Mitcham and Mackey, eds., *op. cit.*, pp. 161–170; Donald Brinkmann, "Technology as Philosophic Problem," *Philosophy Today* 15:2 (Summer 1971), 122–128; Friedrich Dessauer, "Technology in Its Proper Sphere," in Mitcham and Mackey, eds., *op. cit.*, pp. 317–334; Martin Heidegger, "Die Frage nach der Technik," in *Vorträge und Aufsätze* (Pfullinge: Neske, 1954). Friedrich Dessauer, *Streit um die Technik* (Frankfurt a/M: J. Knecht, 1956), pp. 230–235, has another collection of definitions of technology. For still further references, see Carl Mitcham and Robert Mackey, *Bibliography of the Philosophy of Technology* (Chicago: University of Chicago Press, 1973); and Paul T. Durbin, "Philosophy of Technology," *New Catholic Encyclopedia*, vol. XVI, Supplement 1967–1974, pp. 446–448. For a good Marxist analytic attempt to define technology, see L. Tondl, "On the Concepts of 'Technology' and 'Technological Sciences'," in Friedrich Rapp, ed., *Contributions to a Philosophy of Technology: Studies in the Structure of Thinking in the Technological Sciences* (Boston: D. Reidel, 1974), pp. 1–18.

9. Aristotle, *Nicomachean Ethics* VI, 4 (1140a 1–7).

10. For a fuller etymology of the word "technology" see Mitcham, "Philosophy and the History of Technology," *op. cit.*, the beginning of section 2, "*Techne* and Technology."

11. Numerous other typologies have been proposed. In fact, any serious attempt to analyze technology is forced, by the sheer massiveness of the subject, to develop some kind of classification scheme. But so far these typologies have been mainly:
   a) for technical purposes—i.e., the engineering branches and functions mentioned above in notes 1 and 2.

Types of Technology 271

    b) for historical purposes—i.e., the various subject matters of technology in their typical historical periodization (Greek, Roman, Medieval, seventeenth century, etc.); but see also Mumford's distinctions between the eotechnic, paleotechnic, and neotechnic phases of technical activity in *Technics and Civilization* (New York: Harcourt, Brace and World, 1934), and the critical symposium on "The Historiography of Technology," *Technology and Culture* 15:1 (January 1974), especially the papers by Robert P. Multhauf and Eugene S. Ferguson;

    c) for educational or heuristic purposes—e.g., those of Hugh of St. Victor in his *Didascalicon* (c.1125); of Daniel Callahan, "Modes and Manifestations of Technology," chapter 3 in *The Tyranny of Survival* (New York: Macmillan, 1973); and of John G. Burke, described by H. J. Eisenman, "Technology, Society, and Values in 20th-century America: The UCLA 1973 Summer Seminar," *Technology and Culture* 16:2 (April 1975), 185–186.

Group a) would, of course, include the typologies of engineering education; group c) is more specially concerned with understanding technology in a liberal or humanistic framework, and thus tends to rely on various external perspectives or relationships—see e.g., Donald W. Shriver, Jr., "Man and His Machines: Four Angles of Vision," *Technology and Culture* 13:4 (October 1972), 531–555, plus many of the economic classifications of technology. In contrast, the present paper attempts to develop a typology based on the internal structure of technology itself analyzed philosophically. Along this line, cf. the typological distinctions in Mario Bunge, "Philosophical Inputs and Outputs of Technology," forthcoming in *The History and Philosophy of Technology, op. cit.*, note 7, and even more important ones in E. Schuurman, *Techniek en Toekomst* (Assen: Van Gorcum, 1972), which will be mentioned again below. D. Teichman, "On the Classification of the Technological Sciences," in F. Rapp, ed., *Contributions to a Philosophy of Technology, op. cit.*, note 8, considers five different bases for the "'internal' classification of the technological sciences"—historical development, types of science or natural law utilized, kinds of production supported, functional place in the general productive process, and structural characteristics of the objects produced—and concludes that "the only meaningful classification of the technological sciences will be one which takes into account both the objective structure of technology, the classification of the natural sciences and the teaching structures" (p. 138). Teichmann fails, however, to offer any classification to meet these criteria. (As is also clear from his discussion, his last four types of classification merely consider different aspects of my a) above.)

    12. Charles Singer, E. J. Holmyard, A. R. Hall, Trevor I. Williams, *et al., A History of Technology,* 5 volumes (New York: Oxford University Press, 1955–1958), Melvin Kranzberg and Carroll W. Pursell, Jr., eds., *Technology in Western Civilization,* 2 volumes (New York: Oxford University Press, 1967).

    13. Lewis Mumford, *Technics and Civilization* (New York: Harcourt, Brace and World, 1962—first edition 1934), pp. 9–12. See also Mumford's article on "Machines," *Encyclopedia Americana,* International Edition (1972), vol. 18, pp. 57–62.

    14. Cf. Aristotle, *Nicomachean Ethics* I, 7 (1097b24–1098a17).

    15. As demonstrated by the fact that tools do tend to impose certain forms of operating on their environment, even when their user might desire otherwise. "If the only tool you have is a hammer, you tend to treat everything as if it were a nail."—Abraham Maslow, *Psychology of Science* (New York: Harper and Row, 1966), pp. 15–16.

    16. The noun "machine" and hence the adjective "mechanical" come from the Greek μηχανή (Latin *machina*), meaning "instrument for lifting heavy weights," with the cognate verb meaning "to make by art, to construct, to contrive by skill or cunning." Going back even further, the word appears related to the hypothetical Indo-European roots *mogh-* and *megh-*, and thus the German root *mazan* (from which come *Macht* and *machen*), all meaning

"to have power," hence the English "may." Since all mechanical (physical) power was traditionally from the hand—even when cunningly structured so that it could lift heavy weights—the adjective came in later Latin and in English up until the seventeenth century to have strong associations with manual work. Thus John Donne in his sermons could speak of writing, carving, and acting—indeed, anything that "belongs to the hand"—as "mechanical offices" (*LXXX Sermons* [1640], no. 37, 253–254). And only such an understanding of mechanical as manual or bodily activity explains the unity of Hugh of St. Victor's scheme of the seven mechanical arts of weaving, weapons forging, navigation, agriculture, hunting, medicine, and acting. Yet with the development of nonmanual sources of power in the modern period the noun changed its meaning, and so did the adjective. Note, e.g., how "manuscript" used to mean what was written by hand, whereas now to ask for the MS of a paper often is to ask for a typescript.

17. The simple machines of antiquity are sometimes numbered as five, others as six. When five, either the inclined plane or wedge is omitted as an application of the other. In modern mechanics the gear drive and hydraulic press are often considered to be simple machines.

18. Vitruvius, *De architectura* X, 1, 3; quoted from Frank Granger trans. in the Loeb Classical Library series, Vitruvius, vol. II, p. 277.

19. F. M. Feldhaus, "Machines and Tools: Ancient, Medieval and Early Modern," *Encyclopaedia of the Social Sciences* (New York: Macmillan, 1933), vol. 10, p. 14.

20. In this sense all modern "mechanical" devices (if not electronic, chemical, etc., ones) are made up of simple machines as elements, some of which will invariably be tools—e.g., the on-off switch.

Thus, too, by utilizing appropriately shaped natural objects, man used tools before he made them—but not machines. With machines he had to make them (using tools) before he could use them. "Technology . . . is characterized by the fact that it itself is the product of technological means. Technology was not created by man with his bare hands and capacity; it is one technology that produces another."—Nathan Rotenstreich, "Technology and Politics," in Mitcham and Mackey, eds., *Philosophy and Technology, op. cit.,* p. 152.

Perhaps it is also necessary to distinguish between "instrument" in the strict sense, as a measuring or recording or observing device—hand-operated or otherwise—and "tool" as a manually operated machine "for transmitting force or modifying its application." There is a sense in which measuring instruments can be spoken of not as tools of man, but as tools man gives to nature to use in speaking to himself. Cf. Vincent Edward Smith, "Toward a Philosophy of Physical Instruments," *The Thomist* 10:3 (July 1947), 307–333; and F. V. Lazarev and M. K. Trifonova, "The Role of Apparatus in Cognition and Its Classification," in F. Rapp, ed., *Contributions to a Philosophy of Technology, op. cit.,* pp. 197–209.

21. For Hegel's most fully developed discussion of tools and machines and their relationship to work, see the two sets of lecture notes from his period at Jena: *Jenenser Realphilosophie I*, edited by J. Hoffmeister (Leipzig: 1932)—notes from 1803/4; and *Jenenser Realphilosophie II*, edited by J. Hoffmeister (Leipzig: 1931)—notes from 1805/6. *Jenenser Realphilosophie II* has been republished as *Jenaer Realphilosophie* (Hamburg: Felix Meiner, 1967). See especially *Realphilosophie I*, pp. 236ff; and *Realphilosophie II*, pp. 197ff. Other relevant comments can be found in Hegel's *Encyclopaedia,* section 526; and in the *Philosophy of Right,* section 198, and the addition to section 203. (It is worth noting in passing, however, that in all Hegel's system—which includes virtually every other type of consciousness—there is no clear place for technology or engineering as such.) See also Marx's discussion of tools and machines in *Capital* (1867), vol. I, ch. 15, "Machinery and Modern Industry," section 1, "The Development of Machinery." Cf. also Aristotle's discussion of the possibility of tools which would perform their own work and his definition of a

Types of Technology 273

slave as a living tool in *Politics* I, 4 (1253b23–1254a18). Finally, Edmond Barbotin, *The Humanity of Man* (Maryknoll, N.Y.: Orbis Books, 1975), pp. 197–200, gives a brief phenomenological description of the tool-machine relationship which parallels the present analysis.

22. The *locus classicus* for this argument is, of course, Henry Adams (1838–1918), "The Dynamo and the Virgin," ch. 25 in *The Education of Henry Adams* (Boston: Houghton Mifflin, 1918). But cf. Lynn White, Jr., "Virgin and Dynamo Reconsidered," in *Machina Ex Deo* (Cambridge, Mass.: MIT Press, 1968); and Harvey Cox, "The Virgin and the Dynamo Revisited: An Essay on the Symbolism of Technology," *Soundings* 44:2 (Summer 1971), 125–146.

23. Classical (or Newtonian) mechanics, which deals with material bodies of a size appreciable by the unaided human senses moving at speeds small in comparison with the speed of light, is to be distinguished from quantum mechanics, which describes the behavior of atoms and subatomic particles, and from relativity mechanics, which deals with speeds approaching the speed of light. Celestial mechanics is an astronomical application of classical mechanics; fluid mechanics applies classical mechanics to nonrigid bodies.

24. "This dynamics is entirely a modern science. The mechanical speculations of the ancients, particularly of the Greeks, related wholly to statics. Only in mostly unsuccessful paths does their thinking extend into dynamics."—Ernst Mach, *The Science of Mechanics: A Critical and Historical Account of Its Development* (La Salle, Illinois: Open Court, 1960 [first published 1883]), p. 151.

25. *McGraw-Hill Dictionary, op. cit.,* and *McGraw-Hill Encyclopedia, op. cit.* Cf.: Alexander Cowie, "Machines and Machine Components," *New Encyclopaedia Britannica* (1974), Macropaedia, vol. 11, p. 231: a machine is "a device consisting of two or more resistant, relatively constrained parts that may serve to transmit and modify force and motion in order to do work"; and George H. Martin, *Kinematics and Dynamics of Machines* (New York: McGraw-Hill, 1969), p. 3: "A machine is a device for transforming or transferring energy." See also Mumford, *Technics and Civilization, op. cit.,* p. 9.

26. Mechanism = "that part of a machine which contains two or more pieces so arranged that the motion of one compels the motion of the others."—*McGraw-Hill Dictionary, op. cit.;* see also *McGraw-Hill Encyclopedia, op. cit.,* under "Machine." Joseph Edward Shigley, *Theory of Machines* (New York: McGraw-Hill, 1961), p. 3: "A machine is a device which uses power to accomplish a physical effect. But a mechanism is a combination of machine elements to achieve a certain motion."

27. A slightly different conceptualization of the machine-mechanism distinction is contained in Anthony Esposito, *Kinematics for Technology* (Columbus, Ohio: Charles Merrill, 1973): "A mechanism is a device, consisting of two or more members, which accepts an input and modifies it in some way to produce a desired output. Inputs and outputs are generally considered to be either forces or motions" (p. 3). "A machine consists of one or more mechanisms and, as such, usually performs various functions" (p. 6). Esposito uses "mechanism" to denote any component of a machine complex.

According to classical mechanics, the components (which themselves can be machines) of some machine complex or system are the prime mover, generator, motor, and operator. A machine system typically receives an energy input from some natural source (air or water in motion, coal, gasoline, uranium, etc.) and transforms it into mechanical energy, often in the form of a rotating shaft, in what is called the prime mover. This mechanical energy is then transformed by some generator into electrical, hydraulic, or pneumatic power, which is in turn (no pun intended) used to drive a motor, which then drives what is called an operator. An operator can have a variety of outputs such as materials processing, packaging, conveying, sewing, washing, etc. This sequence can be represented schematically as follows:

On occasion, a prime mover can directly drive an operator or motor, as indicated by the dotted line. Operators also include direct manually operated implements such as typewriters and calculating machines, as indicated by the double dotted line. This conception of an operator is obviously closely related to that of tool.

28. Franz Reuleaux, *The Kinematics of Machinery; Outlines of a Theory of Machines,* trans. Alex B. W. Kennedy (London: Macmillan, 1876), pp. 502 and 503.

29. *McGraw-Hill Dictionary, op. cit.* As this scientific and technical dictionary indicates, too, the term "device" is peculiarly common in electrical engineering. As conceived there, device = "an electronic element that cannot be divided without destroying its stated function; commonly applied to active elements such as transistors and transducers." Thus circuits are said to guide energy within devices which are combined into systems; or devices perform some function (generation, amplification, modulation, etc.), and systems incorporate circuits and devices to accomplish a desired result.

The *Oxford English Dictionary,* having noted its other meanings of "wish or will," "speech or discourse," "division or separation," defines device in the present sense as "the result of contriving; something devised or framed by art or inventive power; an invention, contrivance; *esp.* a mechanical contrivance (usually of a simple character) for some particular purpose."

30. In engineering parlance, structure (as opposed to machine) is usually thought of as not having moving parts.

31. Martin Heidegger, *Being and Time,* trans. John Macquarrie and Edward Robinson (New York: Harper and Row, 1962), pp. 96ff. See also Heidegger's discussion of *Bestand,* technological objects which are "in stock" or "supply," in his "Die Frage nach der Technik," in *Vorträge und Aufsätze* (Pfullingen: Neske, 1954). Cf. too Albert Borgmann, "Orientation in Technology," *Philosophy Today* 16:2 (Summer 1972), 135–147.

32. Cf. Norris W. Clarke, "System: A New Category of Being," *Proceedings of the Twenty-Third Annual Convention of the Jesuit Philosophical Association* (Woodstock, Md.: Woodstock College Press, 1961), pp. 5–17. See also H. Rombach's monumental *Substanz System Struktur: Die Ontologie des Funktionalismus und der Philosophische Hintergrund der Modernen Wissenschaften,* 2 vols. (Freiburg-Munich: Alber, 1966).

33. Aristotle, *De Anima* III, 8 (431a 1-3), *Eudemian Ethics* VII, 9 (1241b17–24); Ernst Kapp, *Grundlinien einer Philosophie der Technik* (Braunschweig: Westerman, 1877); Jacques Lafitte, *Reflexions sur la science des machines* (Paris: Bloud and Gay, 1932).

34. Marshall McLuhan, *Understanding Media: The Extensions of Man* (New York: McGraw-Hill, 1964).

35. Mumford, *Technics and Civilization, op. cit.,* p. 80, himself citing Reuleaux.

36. Lynn White, Jr., *Medieval Technology and Social Change* (New York: Oxford University Press, 1962), p. 115. Cf. also Siegfried Giedion's remarks on the natural movements of the hand in his *Mechanization Takes Command* (New York: W. W. Norton, 1969 [first published 1948]), pp. 46–47.

37. Although the analysis here has focused on the term "extension," one could also analyze the notion of projection, making use of its various psychological meanings. What really needs to be done, though, is through a careful phenomenological description of the interactions between hand, tool, and machine to relate these suggestions to Heidegger's analysis of gear as *Zuhanden* (*Being and Time, op. cit.,* circa p. 98), as differentiations of readiness-to-hand.

38. Cf., e.g., Xenophon, *Oeconomicus* IV, 2–3.
39. Mumford, *Technics and Civilization*, op. cit., p. 10.
40. Sōetsu Yanagi, *The Unknown Craftsman: A Japanese Insight into Beauty*, adapted by Bernard Leach (New York: Kodansha International, 1972).
This point is confirmed by psychological studies of the modularization of skills and the occasions for positive and negative transfer of skill modules. Cf., e.g., Jerome S. Bruner, *Beyond the Information Given* (New York: Norton, 1973). When an individual develops a tendency to respond to some general type of task in a predetermined way, he is said to have acquired a *set*. Although obviously useful, set also readily becomes what is called *functional fixedness*—that is, an inability to respond to a variation in the task in an appropriate manner. Power tools, it may perhaps be suggested, suffer from reified functional fixedness.
41. Octavio Paz, "Use and Contemplation," in *In Praise of Hands* (Greenwich, Conn.: New York Graphic Society, 1974), p. 21. Cf. also: "It was a long room of agreeable shape. The thick clay walls had been finished on the inside by the deft palms of Indian women, and had that irregular and intimate quality of things made entirely by the human hand. There was a reassuring solidity and depth about those walls, rounded at door-sills and window-sills, rounded in wide wings about the corner fire-place." — Willa Cather, *Death Comes for the Archbishop* (New York: Knopf, 1927), pp. 32–33.
With reference to this whole discussion of a phenomenology of tools and machines, see also ch. VIII, "Ancient Crafts and Modern Industry," in René Guénon, *The Reign of Quantity and the Signs of the Times* (Baltimore: Penguin, 1972 [first published in French, 1945]). Another reference with a lot of good historical information on this topic is Gillian Naylor, *The Arts and Crafts Movement* (Cambridge, Mass.: MIT Press, 1971).
42. For an impressionistic account of this distinction see the section on "Soft Technology" in Stewart Brand, ed., *Whole Earth Epilog* (Baltimore: Point/Penguin Books, 1974). But see also E. F. Schumacher, *Small Is Beautiful* (New York: Harper and Row, 1973); David Dickson, *Alternative Technology and the Politics of Technical Change* (New York: Universe Books, 1975), and R. Clark, "The Pressing Need for Alternative Technology," *Impact of Science on Society* 23:4 (October–December 1973). Another set of terms sometimes used to indicate this distinction: high versus low technology. Some remarks below under 2, general Comments C', are also relevant.
43. "It is the function of the engineering and development department of a modern corporation to take equipment which it has been decided by management to manufacture, and to do the detailed study of the design and manufacturing process which is necessary if the device is to be produced cheaply and in volume, and if it is to be free from minor defects under field operation."—Francis Russell Bichowsky, *Industrial Research* (New York: Arno Press, 1972 [first published 1942]), p. 26. Bichowsky also points out that such design "is never a fixed thing" because of changes in demand, experience, the availability of materials, etc.
44. The suggested relationships between various engineering functions and the basic categories of technology-as-process can be summarized in the following diagram:

45. Paul A. Samuelson, *Economics*, 9th ed. (New York: McGraw-Hill, 1973), pp. 748–749.

46. Material technology = the efficient production of goods (with efficiency judged in terms of matter and energy); social technology = the efficient organization of society (with efficiency judged either in terms of technological productivity or psychological stress). Social technology is closely related to B. F. Skinner's conception of a "technology of behavior"—not to mention his technology of teaching. See B. F. Skinner, *Beyond Freedom and Dignity* (New York: Knopf, 1971).

47. H. G. Barnett, *Innovation* (New York: McGraw-Hill, 1953), p. 7. But cf. W. F. Ogburn, *On Culture and Social Change: Selected Papers,* ed. Otis Dudley Duncan (Chicago: Univ. of Chicago Press, 1964), p. 23: "Invention is defined as a combination of existing and known elements of culture, material and/or nonmaterial, or a modification of one to form a new one." (Quotation comes from a paper first published 1950.) Note, too, that this discussion is limited to social science usage; for the engineer, innovation is simply small scale or minor invention.

48. Compare Aristotle, *Physics* II, 1 (193a12–17); *Politics* VII, 17 (1337a2); and *Oeconomica* I, 1 (1343a26–1343b2). Until the eighteenth century the term "art" covered all forms of human skill and was categorized into two main types, servile and liberal, depending on whether the work produced was primarily material or mental. (The arts of cooking and painting are both servile, those of logic and rhetoric both liberal. Thus this is clearly not the same as the modern distinction between the useful and the fine arts, which is grounded in the subjective differentiation between utilitarian and aesthetic needs or senses.) But Aristotle also seems to distinguish arts on the basis of process employed as well as object made, and it is this which results in the construction-cultivation distinction.

49. Andrew G. Van Melsen, *Science and Technology* (Pittsburgh: Duquesne Univ. Press, 1961), pp. 235–236.

50. A second example along this line: In 1862 Alphonse Beau de Rochas wrote a pamphlet on improving the efficiency of locomotives in which he clearly and in detail conceived the four-stroke-cycle internal combustion engine and gave a correct theoretical explanation of its working principles. But it was not until fourteen years later that Nicolaus August Otto invented the four-stroke-cycle engine—on the basis of an incorrect theory. Notice how these examples point up the weakness of, e.g., R. J. Forbes's definition of invention as "a mental process in which various discoveries and observations are combined and guided by experience into some new tool or operation"—"The Beginnings of Technology and Man," in Kranzberg-Pursell, eds., *Technology in Western Civilization, op. cit.,* vol. 1, p. 14.

51. So far the only serious epistemological analysis of invention is that to be found in the work of Dessauer. Yet Dessauer's highly idealistic conclusions must be supplemented by a more strictly phenomenological approach.

52. Francis Bacon (1561–1626) is the first historical figure to argue explicitly and at length for the need to cultivate inventions. In so doing he distinguishes between those inventions which have been made on the basis of an appropriate understanding of nature and those which have been virtually independent of scientific knowledge—and, we would add, method. The former are what today would be called science-based inventions; the latter are more traditional evolutionary inventions. See, e.g., Francis Bacon, "Thoughts and Conclusions," pp. 90 ff., trans. from the Latin in Benjamin Farrington, *The Philosophy of Francis Bacon* (Chicago: Univ. of Chicago Press, 1966).

53. See, e.g., S. C. Gilfillan, *The Sociology of Invention* (Cambridge: MIT Press, 1963 [first published 1935]). Two other primary sources for philosophical reflection on the nature and meaning of invention: John Jewkes, David Sawers, and Richard Stillerman, *The Sources of Invention,* 2nd ed. (New York: Norton, 1969); and H. Stafford Hatfield, *The Inventor and His World* (New York: Dutton, 1933).

54. It is this failure to appreciate ancient invention as a process *sui generis* that vitiates much historical speculation on the origins of certain technological objects. See, e.g., Lynn

White, Jr., "The Act of Invention: Causes, Contexts, Continuities, and Consequences," *Technology and Culture* 3:4, 486–500, esp. the first numbered example, where the author seems mystified by the Chumash Indian "invention" of plank boats.

55. "The greatest invention of the nineteenth century was the invention of the method of invention. A new method entered into life. In order to understand our epoch, we can neglect all the details of change, such as railways, telegraphs, radios, spinning machines, synthetic dyes. We must concentrate on the method itself; that is the real novelty, which has broken up the foundations of the old civilization."—Alfred North Whitehead, *Science and the Modern World* (New York: Free Press, 1967 [first published 1925]), p. 96. Thomas Edison (1847–1931) was perhaps the key figure in this development with his establishment in the 1870s of an "invention factory" in Menlo Park, New Jersey, designed to make "inventions to order." For Edison, "discovery is not invention. . . . A discovery is more or less in the nature of an accident." Invention is the product of organized purpose. Although invention had become well recognized and culturally prized by the early 1800s, and thus pursued and encouraged in various ways, it remained an activity based largely on individual initiative and intuition, and divorced from direct large-scale financial backing. It was Edison who, especially with his massive, methodically directed trial-and-error search for a suitable filament for the incandescent light in conjunction with the systematic development of related elements necessary to its commercial exploitation (vacuum bulbs, parallel circuits, dynamos, voltage regulators, metering devices, etc.), first created the industrial research organization tied to capitalistic economic structures. Cf. also John B. Rae, "The Invention of Invention," in Kranzberg-Pursell, eds., *Technology in Western Civilization, op. cit.,* vol. 1, pp. 325–336; and Daniel J. Boorstin, *The Americans: The Democratic Experience* (New York: Random House 1973), ch. 56, "The Social Inventor: Inventing the Market" and ch. 57, "Communities of Inventors: Solutions in Search of Problems."

56. Art and architecture books on the subject of design invariably concentrate on issues of form. A book on roof design, for example, provides an inventory of various ways to build roofs—not ways as processes, but ways as forms, patterns, shapes; one on lighting design contains pictures and drawings of various formal solutions to lighting design problems. A work whose subtitle aptly illustrates this approach: *Designing Architectural Facades: An Ideas File for Architects,* by Kurt Hoffmann, Helga Griese, and Walter Meyer-Bohe (New York: Whitney Library of Design, 1975). For three comprehensive architectural discussions of design that approach the philosophical, see Paul J. Grillo, *What Is Design?* (Chicago: P. Theobald, 1960); Christopher Alexander, *Notes on the Synthesis of Form* (Cambridge: Harvard University Press, 1964); and David Pye, *The Nature of Design* (New York: Reinhold, 1964).

57. "Report on Engineering Design" (MIT Committee on Engineering Design), *Journal of Engineering Education* 51:8 (April 1961), p. 647.

58. Thomas T. Woodson, *Introduction to Engineering Design, op. cit.,* p. 3. Cf. *McGraw-Hill Dictionary, op. cit.,* design = "the act of conceiving and planning the structure and parameter values of a system, device, process, or work of art." But see also the short article on "Design" in *Encyclopaedia Britannica* (1969).

59. Ortega y Gasset, in Mitcham and Mackey, eds., *Philosophy and Technology, op. cit.,* pp. 295 ff.

60. Cf. Joseph Edward Shigley, *Theory of Machines, op. cit.:* "The use of the drawing board in kinematics instruction is very desirable and usually necessary" (p. v). "The most direct method of attacking a kinematics or dynamics problem is the graphical one" (p. vi). The importance of drawing is also hinted at in E. F. O'Doherty's remarks about the difference between sensori-motor skills, phantasmal capacity, and conceptual capacity. "The *phantasma,* or sensory representation at whatever level of complexity . . . is what we are concerned with."—"Psychological Aspects of the Creative Act," in J. Christopher Jones

and D. G. Thornley, *Conference on Design Methods* (New York: Macmillan, 1963), pp. 197–203. Drawing is, however, only one way of performing the more general engineering activity of modeling.

One source of confusion in thinking about design is the tendency to identify design with one of its languages, drawing. . . . Design, like musical composition, is done essentially in the mind, and the making of drawings or writing of notes is a recording process. The designer, however, uses drawing for self-communication just as everyone uses words for thinking. This use of drawing as an extension of the mind, a sort of external (and reliable) memory can be a very important part of the design process. Drawing should be taught not primarily to give the student facility in the use of tools—pencil, triangle, tee square, and most important, the eraser—but to give him practice in pictorial extension of the mind. It is not to be expected that all students are equally endowed with the ability to think pictorially any more than to think mathematically. Somehow educators tend to look upon mathematical ability as a more desirable quality than the ability to think in terms of spacial relations. Before dismissing the latter as something of lesser merit, it may be well to reflect that one of the greatest engineers of all time, Leonardo da Vinci, was essentially a draftsman, not a mathematician.

Pictorial language is especially well adapted to expressing particular physical form and physical space relationships. Functional relationships are often better expressed by a symbolic language. Such languages have particular facility in expressing generalizations without specifying detail. The chemical engineer uses the flow sheet, the electrical engineer the circuit diagram, and all kinds of engineers use the block diagram as important tools in the conceptual process.

The designer often uses the symbolic languages of mathematics but usually in connection with the analysis of a design rather than directly in the conceptual process.— "Report on Engineering Design," *Journal of Engineering Education, op. cit.,* note 57, 647–648.

Yet by pointing out the essential nature of drawing, this discussion merely confirms the argument of the text. For an empirical account of the high correlation between drawing and general engineering abilities, see Steve M. Slaby and Arthur L. Bigelow, "Engineering Graphics—A Predictor for Academic Performance in Engineering," *Journal of Engineering Education* 51:7 (March 1961), 581–587. At the conclusion of their statistical presentation, the authors suggest that "graphics is part of the thinking of an engineer" (p. 586). For "the ability to 'think' space is a necessary condition if we are to define engineering correctly" (p. 587).

For an elementary discussion of engineering modeling which unintentionally brings out its inherent character as miniature construction see *The Man-Made World: Engineering Concepts Curriculum Project* (New York: McGraw-Hill, 1971), ch. 4, pp. 139–178. "Models are used, not only to describe a set of ideas, but also to evaluate and to predict the behavior of systems before they are built. This procedure can save enormous amounts of time and money. It can avoid expensive failures and permit the best design to be found without the need for construction of many versions of the real thing. Models evolve, and it is customary to go through a process of making successive refinements to find a more suitable model" (*Ibid.,* p. 177).

For a historian's analysis of the centrality of design which parallels my own, see Edwin T. Layton, Jr., "Technology as Knowledge," *Technology and Culture* 15:1 (January 1974), 31–41. After noting how for artists thinking means something different than for philosophers, and how "technologists display a plastic, geometrical, and to some extent nonverbal mode of thought that has more in common with that of artists than that of philosophers" (p. 36), Layton describes design in the following terms:

The first stages of design involve a conception in a person's mind which, by degrees, is translated into a detailed plan or design. But it is only in the last stages, in drafting the blueprints, that design can be reduced to technique. And it is still later that design is manifested in tools and things made. Design involves a structure or pattern, a particular combination of details or component parts, and it is precisely the gestalt or pattern that is of the essence for the designer.

We may view technology as a spectrum, with ideas at one end and techniques and things at the other, with design as a middle term. Technological ideas must be translated into designs and tools to produce things (pp. 37–38.)

My only criticism would be to argue that drafting of one form or another is something that goes on at all stages of design, not just the final one of making blueprints. It is also well to remember (as Layton notes one paragraph later) that

The designs for the final products of technology do not exist in isolation. They are intimately associated with production and management, which, as Frederick W. Taylor insisted, also require design (p. 38).

Finally, for a related discussion of phantasmal thinking see L. R. Rogers, "Sculptural Thinking—I" and "Sculptural Thinking—III" (part II is a commentary by Donal Brook), in Harold Osborne, ed., *Aesthetics in the Modern World* (New York: Weybright and Talley, 1968).

61. Contrary to some modern philosophies of mathematics, Aristotle recognizes how this is true even in the case of geometry. See *Metaphysics* IX, 9 (1051a22–34).

62. These factors are usually enumerated as four: materials, interrelation of parts, methods of construction, and effect of the whole upon those who will become involved with it. It is worth noting, however, that only the first three can actually be analyzed in the drawing itself; the fourth denotes a social context which is not amenable to quantification and is, in fact, a stumbling block and source of frustration to many engineers. For example, technically speaking a large bridge can be constructed which is completely safe but which, if it is flat, because of the curvature of the earth, will appear to an approaching driver to be sagging; as a result the public will be afraid to use it. This compels the engineer to arch the bridge in a way which is not required by any of the first three factors. A second example: floors in concrete buildings have to have almost twice as much concrete in them as they really need to support a designated load in order to keep them from vibrating in ways which pose no structural dangers but do make the occupants nervous.

63. Skolimowski, "The Structure of Thinking in Technology," in Mitcham and Mackey, eds., *Philosophy and Technology op. cit.*, pp. 42–49. I. C. Jarvie, in "The Social Character of Technological Problems: Comments on Skolimowski's Paper" (*ibid.*, 50–53) objects that what the engineer strives for is really determined by the social definition of the problem. For instance, in warfare, there are times when civil engineers are called upon to design a bridge for speed of construction not durability. But Skolimowski's point is that within such historically and socially set parameters as materials, cost, time, etc., a civil engineer *qua* civil engineer will always strive for as much durability as possible. In fact, Jarvie's own example tells against him. For would it not be a military (or transportation) engineer rather than a civil engineer who would be called upon to design a pontoon bridge for maximum military efficacy (or mobility)?

64. Some representative engineering texts are: Morris Asimov, *Introduction to Design* (Englewood Cliffs, N.J.: Prentice-Hall, 1962); W. Gosling, *The Design of Engineering Systems* (New York: John Wiley, 1962); John R. M. Alger and Carl V. Hays, *Creative Synthesis in Design* (Englewood Cliffs, N.J.: Prentice-Hall, 1964); E. V. Krick, *An Introduction to Engineering and Engineering Design* (New York: John Wiley, 1965); R. Dixon, *Design Engineering: Inventiveness, Analysis, and Decision Making* (New York: McGraw-Hill, 1966); L. Harrisberger, *Engineersmanship, A Philosophy of Design* (Belmont, Calif.: Brooks-Cole, 1966); Thomas T. Woodson, *Introduction to Engineering Design* (New York:

McGraw-Hill, 1966); William H. Middendorf, *Engineering Design* (Boston: Allyn and Bacon, 1969); Myron Tribus, *Rational Descriptions, Decisions and Designs* (New York: Pergamon, 1969); Joseph P. Vidosic, *Elements of Design Engineering* (New York: Ronald, 1969); Percy H. Hill, *The Science of Engineering Design* (New York: Holt, Rinehart and Winston, 1970); George C. Beakley and Ernest G. Chilton, *Introduction to Engineering Design and Graphics* (New York: Macmillan, 1973).

But for general discussion of the questions of design methodology see: J. Christopher Jones and D. G. Thornley, eds., *Conference on Design Methods* (New York: Macmillan, 1963); Gerald Nadler, "An Investigation of Design Methodology," *Management Science* 13:10 (June 1967), B-642 to B-655; Bohdan Walentynowicz, "On Methodology of Engineering Design," *Proceedings of the XIVth International Congress of Philosophy*, Vienna, Sept. 2–9, 1968 (Vienna: Herder, 1968), vol. 2, pp. 586–590; Herbert A. Simon, *The Sciences of the Artificial* (Cambridge: MIT Press, 1969), esp. ch. 3, "The Science of Design: Creating the Artificial," pp. 55–83; Lord Hinton of Bankside. *Engineers and Engineering* (New York: Oxford Univ. Press. 1970), esp. ch. 6, "The Engineer and Design," pp. 28–35; S. A. Gregory, ed., *Creativity and Innovation in Engineering* (London: Butterworths, 1972). Simon's argument is the most comprehensive and sophisticated. As he points out, his analysis of design is closely allied with theories of business administration, decision theory, computer programming, and psychological studies of creativity and problem solving—especially those involving the computer simulation of cognitive processes. On the latter topic of the psychology of problem solving see, e.g., Allen Newell and H. A. Simon, *Human Problem Solving* (Englewood Cliffs, N.J.: Prentice-Hall, 1972; Donald M. Johnson, *Systematic Introduction to the Psychology of Thinking* (New York: Harper and Row, 1972); and Zbigniew Pietrasinski, *The Psychology of Efficient Thinking*, trans. Boguslaw Jankowski (New York: Pergamon Press, 1969), esp. the section on "The Psychology of Technical Invention," pp. 122–135. F. Rapp, ed., *Contributions to a Philosophy of Technology, op. cit.*, also reprints two articles on design methodology by M. Asimov and R. J. McCrory. Jacques Hadamard's *An Essay on the Psychology of Invention in the Mathematical Field* (Princeton, N.J.: Princeton Univ. Press, 1945) is of unexpected interest to this area.

65. Industrial design, especially the Bauhaus school of industrial design, is an attempt to bridge this gap between art and engineering, and either to include aesthetic formal properties in the design process, or to find aesthetic value in purely functional designs. In fact, however, the attempt has led to the triumph of engineering efficiency as influenced by economic pressures. For a brief overview, see the article on "Industrial Design," *New Encyclopaedia Britannica* (1974), Macropaedia, vol. 9, pp. 512–520.

66. Although the artificial character of its products is most apparent in the plastic arts (sculpture, painting, etc.), this remains true in some sense even in poetry and music.

67. Friedrich Dessauer, "Technology in Its Proper Sphere," in Mitcham and Mackey, eds., *op. cit.*, pp. 317–334; see also the commentary on Dessauer in Mitcham and Mackey, pp. 22–25.

68. Some better known examples of double invention: Henry Bessemer (1813–1898) and William Kelly (1811–1888) invented the same process of steel production; C. M. Hall (1863–1914) and Paul L. T. Héroult (1863–1914) both devised the electrolytic process for refining aluminum.

69. David Pye, *The Nature of Design* (New York: Reinhold, 1964), p. 19.

This question of the relation between invention and design recalls another about the relation between art (as creation) and science (as discovery). Artists, too, often describe their creative efforts as culminating in discovery. T. S. Eliot has said that for the poet there is no such thing as "free verse"; William Butler Yeats once wrote that, unlike prose which has "no fixed laws," a poem "comes right with a click like closing a box"; and André Gide maintained that he did not know in advance what the characters in his novels were going to do, he had to discover it as he wrote. On the other side, scientists (who are supposed to discover the laws of nature rather than create them) say that their experience involves a high degree of creativity—see, for instance, the argument of J. Bronowski in *Science and Human Values,* rev. ed. (New York: Harper & Row, 1965), on the common characteristics of science and the humanities. Further interesting comment on this topic is available in *Impact of Science on Society* 24:1 (January–March 1974), on the theme "Art and Science," with articles by Piet Hein, Frank J. Malina, Francesco d'Arcais, Jasia Reichardt, Robert Preusser, David Dickson, and Rolf-Dieter Hermann.

70. See, e.g., Skolimowski, "Problems of Truth in Technology," *Ingenor* (University of Michigan, College of Engineering) 8 (Winter 1970/71), p. 42.

71. But see, e.g., in Mitcham and Mackey, eds. *Philosophy and Technology: op. cit.* Yves R. Simon, "Pursuit of Happiness and Lust for Power in Technological Society," esp. pp. 173–175; Hans Jonas, "The Practical Uses of Theory," esp. pp. 339–341.

72. Feibleman, "Technology as Skills," *op. cit.* note 4. Cf. also Michael Polanyi, *Personal Knowledge* (London: Routledge and Kegan Paul, 1958), chap. 4, "Skills."

73. Although widely appealed to, I have found this distinction most clearly stated in an unpublished course outline by Prof. Robert Nash, Department of Electrical Engineering, Vanderbilt University. But see also Max Weber's comments on the universality of technique in human activity in *Economy and Society,* trans. Guenther Roth and Claus Wittich (New York: Bedminster Press, 1968 [first published 1922]), vol. 1, p. 65; and the quotation from Nathan Rotenstreich in note 20 above.

74. This tension between heuristics (i.e., a problem-solving strategy that indicates a solution without testing all possible operations) and algorithms (a problem-solving method that exhausts all possible operations and thus guarantees a solution if one is possible) is intimately involved with the ultimate nature of matter and energy, its knowability, and the dream of a complete technology. Laplace's scientific dictum that if given a complete description of matter and motion at some point in time he could deduce the remainder of the world may be restated technologically as: given a complete description of matter and energy at some point in time, man can intervene to produce whatever he desires. Heisenberg's uncertainty principle and various other aspects of quantum mechanics, not to mention contemporary ecological problems, raise fundamental questions about the foundations of Laplace's postulate and its technological correlate.

75. Insofar as they transcend sociology, all studies of science, technology, and society are forced into ethical-political analyses of technological using. As for the relationship between philosophy of action and philosophy of technology, consider Richard J. Bernstein, *Praxis and Action* (Philadelphia: Univ. of Pennsylvania Press, 1971), which attempts to relate analytical philosophy and recent developments in Marxism, existentialism, and pragmatism focusing on the subject of human activity. It is noteworthy that, despite its omnipresence, the only discussion of technology occurs in the chapter on Marx.

76. Stephen Toulmin, "Innovation and the Problem of Utilization," in William H. Gruber and Donald G. Marquis, eds., *Factors in the Transfer of Technology* (Cambridge: MIT Press, 1969), p. 25. See also Stephen Toulmin, *Human Understanding,* vol. 1 (Princeton: Princeton Univ. Press, 1972), pp. 364–378.

Bernard Lonergan's discussion of what he calls common sense and its merging with science to produce technology is also relevant to this discussion, although Toulmin's focus is more limited than Lonergan's. See Lonergan, *Insight* (New York: Philosophical Library, 1956), pp. 207 ff. Lonergan is developing a point first made by Alexandre Koyré in a review of Mumford's *Technics and Civilization* entitled "Dur Monde de l'à peu près à l'Univers de la Precision," *Critique* 4, no. 28 (September 1948), pp. 806–823.

77. *Ibid.,* p. 27.

78. A similar biological model of the utilization-innovation process—but one which em-

phasizes more the interrelationships of a feedback-based ecological system—has been developed by the Georgia Tech Innovation Project. See the executive summary, *Technological Innovation: A Critical Review of Current Knowledge* (1975), available from the Advanced Technology and Sciences Studies Group, Georgia Tech, Atlanta, Georgia 30332.

79. Harold Koontz and Cyril O'Donnell, *Principles of Management,* 4th edition (New York: McGraw-Hill, 1968).

80. On this topic, see E. P. Thompson's masterful *The Making of the English Working Class* (New York: Pantheon, 1963).

81. Carpenter, "Modes of Knowing and Technological Action," *op. cit.,* note 4.

82. Bunge, "Toward a Philosophy of Technology," *op. cit.,* note 4.

83. It is possible, too, that the heuristic strategies of problem-solving could be interpreted as technical maxims.

84. Carpenter, p. 165.

85. Bunge, pp. 62–63.

86. Actually the engineering sciences, as defined in the authoritative James H. Potter, ed., *Handbook of the Engineering Sciences* (Princeton: Van Nostrand, 1967), include what are called the basic engineering sciences (mathematics, physics, chemistry, graphics, statistics, theory of experiments, and mechanics) and the applied engineering sciences (thermal phenomena, heat and mass transfer, chemical energy conversion, turbomachinery, nuclear reactor engineering, aeronautics and astronautics, field theory, electromechanical energy conversion, physical electronics, electronic circuits, system dynamics, materials science, machine elements, control systems, operations research, information retrieval, preparation of reports, computers).

87. Polanyi, *Personal Knowledge, op. cit.,* p. 179.

88. Bunge, p. 63.

89. *Ibid.*

90. *Ibid.,* p. 68. Some historians of technology, while admitting it as an element of modern technology, would disagree with the stress Bunge places on this theoretical grounding of technological rules. Indeed, because this grounding usually lacks direct practical efficacy, it is questionable whether it is a pursuit of science (knowledge) or technology (making and using). For a concise historical account of the character of modern engineering knowledge, see Edwin Layton, "Mirror-Image Twins: The Communities of Science and Technology in 19th Century America," *Technology and Culture* 12:4 (October 1971), 562–580. According to Layton there are two trends in the origins of modern technology: one would "build directly on the foundations of science"; the other would "borrow the methods of science to found new sciences built on existing craft practices" (p. 566). And contrary to what might be expected, the movement is often from the first to the second. Layton cites the example of strength of materials:

As the strength of materials moved from the community of science to that of technology, it went through an important transformation. Its ties with physics were weakened, and it developed in ways uncharacteristic of the basic sciences. At the same time, its range of technological usefulness was gradually expanded. Scientists tended to explain their findings by reference to the most fundamental entities, such as atoms, ether, and forces. But these entities cannot always be observed directly. To be useful to a designer, however, a formulation must deal with measurable entities, particularly those of importance to the practical man. These need not be fundamental in the scientific sense. The scientists who had done so much to found a science of the strength of materials . . . strove to found this study on the same ontological basis as classical mechanics—that is, they sought to explain their results in terms of molecules and the forces between them. Although not without interest, these efforts were not wholly successful. They were also needless complications from the technological point of view. A few engineers pioneering in this field . . . continued the quest, but in the end the attempt was abandoned. Instead, engineers were content with a simple macroscopic model—for example, viewing a beam as a bundle of fibers (pp. 568–569).

This paper by Layton also has some good material on the relationship between technology and graphics, a subject which was mentioned earlier (2, Specific Comments B'). Indeed, questions also need to be raised about the epistemological status of design theory.

91. Along this line, one speculative aside cannot be resisted. Robert E. Ornstein in *The Psychology of Consciousness* (San Francisco: W. H. Freeman, 1972) distinguishes between the types of thinking which take place in the right and left hemispheres of the brain.

The left hemisphere (connected to the right side of the body) is predominantly involved with analytic, logical thinking, especially in verbal and mathematical functions. Its mode of operations is primarily linear. This hemisphere seems to process information sequentially. . . .

If the left hemisphere is specialized for analysis, the right hemisphere (again, remember, connected to the left side of the body) seems specialized for holistic mentation. Its language ability is quite limited. This hemisphere is primarily responsible for our orientation in space, artistic endeavor, crafts, body image, recognition of faces. It processes information more diffusely than does the left hemisphere, and its responsibilities demand a ready integration of many inputs at once (pp. 52f).

Perhaps the shift from sensorimotor skills and technical maxims to descriptive laws and technological theories involves a transfer of responsibility for the making activity and thought from one side of the brain to the other. If so, in conjunction with Ornstein's general thesis about the relationship between spiritual disciplines and the two sides of the brain, this would have implications for the relationship between modern technology and religion.

92. See, e.g., Henryk Skolimowski, "Problems of Truth in Technology," *Ingenor* (Univ. of Michigan, College of Engineering) 8 (Winter 1970/71), 5–7 and 41–46; Michael Polanyi, *Pure and Applied Science and Their Appropriate Forms of Organization,* Occasional Pamphlet no. 14 (Oxford: Society for Freedom in Science, December 1953); and James K. Feibleman, "Pure Science, Applied Science, and Technology: An Attempt at Definitions," in Mitcham and Mackey, eds., *op. cit.,* pp. 33–41.

93. For more discussion of the distinction between scientific and technological experiments see Simon Moser, "Toward a Metaphysics of Technology," *Philosophy Today* 15:2 (Summer 1971), esp. 137–138. Cf. also Hilbert Schenck, Jr., *Theories of Engineering Experimentation,* 2nd ed., (New York: McGraw-Hill, 1968).

94. My use of term "will" here and following is not meant to imply any particular faculty of the will. Even followers of Gilbert Ryle, who rejects the notion of a faculty of will as a "ghost in the machine" (see Ryle, *The Concept of Mind* [New York: Barnes and Noble, 1949]), still admit the reality of acts of willing. Thus for "will" simply read "acts of willing," which is all it is necessary to admit for purposes of the present discussion.

Contemporary analytic discussion of willing has its roots, of course, in Wittgenstein's question about the difference between "raising my arm" and "my arm going up" (*Philosophical Investigations* [1953] I, 621). And if some philosophers have argued that the difference is in following rules, still others have questioned the adequacy of this theory of action. For some analytic discussions relevant to the concept of willing and will, see: R. S. Peters, *The Concept of Motivation* (London: Routledge and Kegan Paul, 1958); A. I. Melden, *Free Action* (London: Routledge & Kegan Paul, 1961); Anthony Kenny, *Action, Emotion and Will* (London: Routledge & Kegan Paul, 1963); D. F. Pears, ed., *Freedom and the Will* (New York: St. Martin's Press, 1963). Again, it is significant that none of these discussions of human action so much as mentions technology.

95. Alexander Pfänder, *Phenomenology of Willing and Motivation,* trans. Herbert Spiegelberg (Chicago: Northwestern University Press, 1967). I adopt here Pfänder's highly

condensed description of the act of willing because it is the most philosophically neutral, and leaves open the question of philosophical interpretation of the foundations of this activity. Not to mention other resources, a full account of volition would include consideration of William James's analysis of "Will," ch. 26 of *The Principles of Psychology* (1891); Austin Farrer's *The Freedom of the Will* (London: Adam & Charles Black, 1958); Paul Ricoeur's comprehensive *La Philosophie de la volonte* (vol. 1, 1950; vol. 2, parts I and II, 1960; vol. 2, part III and vol. 3 apparently in progress); and Roberto Assagioli's *The Act of the Will* (Baltimore: Penguin, 1973). For a good outline of various philosophical theories of the will see Vernon J. Bourke, *Will in Western Thought: An Historico-Critical Survey* (New York: Sheed and Ward, 1964); James N. Lapsley, ed., *The Concept of Willing* (Nashville: Abington, 1967), is also important.

96. Pfänder, pp. 22–23.

97. *Ibid.*, p. 26.

98. In response to the objection that there are times when ones does not want to know, is even afraid to know ("Don't tell me, I don't want to know"), it can be replied that the phenomenology of this experience only proves the point at issue. In the act of turning away from knowing one immediately feels guilty, feels as if this needs to be justified. Sometimes the appeal is simply to weakness ("I cannot stand it"); at other times it is fear that the knowledge will not really be true or will overshadow in its limited perspective larger principles and truths. Partial knowing is rejected in the name of more comprehensive knowing. Yet no matter what the case, the rejection of knowing is something which needs to be explained in a way which does not occur with the rejection of making and using—although contemporary culture, in its commitment to busyness, goes a long way toward overriding this natural bias. Needless to add, however, the argument here is no more than a suggestion in need of significant elaboration.

99. Sören Kierkegaard, *Fear and Trembling and Sickness Unto Death,* trans. Walter Lowrie (Princeton, N.J.: Princeton Univ. Press, 1954), p. 146: "The self is a relation which relates itself to its own self, or it is that in the relation [which accounts for it] that the relation relates itself to its own self; the self is not the relation but [consists in the fact] that the relation relates itself to its own self." (Material in brackets supplied by translator.) This is, admittedly, one of the most paradoxical and obscure passages in the Kierkegaard corpus. It is also not without its irony.

100. Proof that this diagram includes all possible nonordered combinations, C, can be had by applying the standard formula $C = \dfrac{n!}{r!(n-r)!}$ for n number of things taken r at a time:

C_1, for 4 things taken 1 at a time $= \dfrac{4!}{1!(4-1)!} = 4 \rightarrow \{1,2,3,4\}$

C_2, for 4 things taken 2 at a time $= \dfrac{4!}{2!(4-2)!} = 6 \rightarrow \{1',2',3',4',5',6'\}$

C_3, for 4 things taken 3 at a time $= \dfrac{4!}{3!(4-3)!} = 4 \rightarrow \{1'',2'',3'',4''\}$

C_4, for 4 things taken 4 at a time $= \dfrac{4!}{4!(4-4)!} = 1 \rightarrow \{1'''\}$

101. "Daedalus" (a pseudonym), "Pure Technology," *Technology Review* 72:8 (June 1970), 39–45, contains a good description of this tradition which, in his mind, includes things

as diverse as Hero's mechanical toys and *Scientific American*'s "Great International Paper Airplane Competition." See also the argument of Lynn White, Jr., in "The Flavor of Early Renaissance Technology," in Bernard S. Levy, ed., *Developments in the Early Renaissance* (Albany, N.Y.: State Univ. of New York Press, 1972), and other examples in Stacy V. Jones, *Inventions Necessity is Not the Mother Of: Patents Ridiculous and Sublime* (New York: Quadrangle, 1973).

102. Mario Bunge, "Philosophical Inputs and Outputs of Technology," forthcoming in *The History and Philosophy of Technology, op. cit.,* note, 7, above.

103. As terms of contrast, culture = general expressions of life such as religion, art, literature, and ultimate moral ends; civilization = systems of social organization and techniques of material production. "Thus one can have modern industry [civilization] without modern culture [beliefs, values, scientific theories] provided one imports technological know-how—and does not expect great technological innovations. And one can have scraps of modern culture [beliefs, values, scientific theories] without modern industry [civilization]—provided one is willing to put up with a one-sided and rickety culture. But no creative [or stable?] technology is possible outside modern civilization—which includes modern industry—and modern culture, which of course includes modern technology [as a system of beliefs, values, and knowledge]."—Mario Bunge, "Philosophical Inputs and Outputs of Technology," *Ibid.*

104. By way of brief commentary along this line: Consider how once the four modal aspects of technology are clearly distinguished, it becomes apparent that, say, Bunge and Ellul are not in any fundamental disagreement when they define technology as applied science and rational efficient action, respectively. They are simply choosing to emphasize two different elements (knowledge and process) of a complex whole. Furthermore, a view of the whole enables Ellul's argument to the effect that technology is not to be identified with machines (objects) to be seen as no doubt true but somewhat less than comprehensive. In each of these three cases, while technology is really being pointed out, the effect is nevertheless to divert attention from that point of greatest substantiality.

Three thinkers who have attempted to group technology in terms of more than one of its modal manifestations are Dessauer, Heidegger, and Schuurman. Dessauer's most considered statement of a definition of technology occurs in *Streit um Technik* (Frankfurt: J. Knecht, 1956), pp. 234–235. Following a catalogue of other definitions, Dessauer offers his own "Proposal for an Essential Definition of Technology":

> Technology is real being from ideas
> through purposeful designing and moulding
> from resources given in nature.

The first line gives the *ontological* (in the mode of being) definition: real being proceeding from "ideas" in the sense of human creative conceptual images, which anticipate a spatial or temporal configuration (instrument or process) in its nature (its being-such) in the imagination; hence it is true that: *essentia praecedit existentiam* (essence [nature] precedes existence).

The second line gives the mode of actualization: human consciousness, taking aim, forms tools and building elements mentally and by hand in such a way that, in their interconnection (totality), they fulfill as their *purpose* what was consciously aimed at.

The third line gives the cause that makes technology possible: matter, energy, laws of nature are storeroom and limitation of technological production.

Contrary to the Socratic understanding of *techne,* this definition does not rest on the personal skill (of the fisherman, flute player, horseman, etc.) which, being subjectively acquired and habitually exercised, passes away with the agent. When we speak of technology as a world-force, as fate, as the basis of civilization, we mean that which is objective, transferrable, carrying the purpose or end in itself, which detaches itself from its creator and entering into the stream of time continues to be effective there. (N. B. All parentheses and brackets are from Dessauer's own text.)

This definition contains at least passing reference to all four modes of technological manifestation: objects as instruments and products (spatial and temporal configurations), processes of designing and fabricating (moulding), knowledge (ideas), and volition (aims or purposes). And all are related in such a way as to explain or account for technology as an objective or impersonal activity. Yet as his own commentary indicates, this is primarily a definition of modern rather than ancient technology. And in his development of this definition, which first appeared in basically this same form as early as *Philosophie der Technik* (1927), one can easily cite numerous shortcomings: weaknesses in the discrimination of technological objects; a highly idealistic interpretation of the processes of inventing and designing; lacunae in the analysis of technological ideas (i.e., inventions concepts), especially in their relationship to scientific conceptualizations of the "resources given in nature"; and the various aims and purposes of technological activity almost always construed in either immediate economic terms or dubious religious ones. Nevertheless, it is this definition which first recognizes and brings together in a clearly conceived way all four modes of the manifestation of technology—thus marking the beginning of the serious philosophical understanding of technology.

In a commentary on Dessauer's definition of technology, Simon Moser ("Toward a Metaphysics of Technology," *Philosophy Today* 15:2 [Summer 1972], in a supplement available from the translation editor) has noted how Dessauer actually accounts for modern technology by describing it in terms of Aristotle's four causes: material, formal, efficient, and final. Material cause = resources given in a nature; formal cause = ideas; efficient cause = designing and moulding; final cause = purposes or aims. In what is undoubtedly the most complicated, penetrating, and original attempt to understand technology, Heidegger—having first rejected as inadequate the ideas of technology as either object (tools, appliances, etc.) or process (human activity, means)—begins by scrutinizing the traditional four causes and their interrelationships. He then asks what it is that unites them, makes them able to function together to bring forth or produce some entity or object. How is it that the four causes together can cause, can make unconcealed or present what was previously concealed or not present? In this I take it that Heidegger is pointing precisely in the direction of what I have called pure technology—the conjunction of objects, processes, knowledge, and volition—and trying to name the character that this pure technology has in its modern form. For only this will be able to account for the character of modern technological objects, processes, etc., when taken in isolation. According to Heidegger any unconcealing of the concealed is a form of truth. Thus technology is a kind of truth. The framework that is both source and product of this truth Heidegger calls *Ge-stell*. "*Ge-stell* is called the way of unconcealing which rules in the essence of modern technology and is not itself anything technological"; this way is a "provoking, setting-up disclosure of nature." Now this is not the place to venture an explication of Heidegger's meaning. Suffice it to say that, if I understand him correctly, my own meager effort to approach the essence of technology is not out of harmony with Heidegger's intentions.

According to Schuurman (cited at the end of note 11 above) modern technology should be analyzed into three basic structures: tools (technological objects), execution (technological form-giving), and preparation (technological designs). Once again these categories illustrate what I have attempted to describe as the basic modal features of technology. Some isolated points: Schuurman analyzes modern tools as "technological operators" designed to produce a "technological event." His "execution" corresponds, of course, to my "fabrication" (Dessauer's "moulding," which is also called "form-giving" or "processing"). Preparation or design in Schuurman's analysis is expanded to include engineering knowledge. In Schuurman's view "the decisive characteristic of modern technology is that *preparation* (the designing) has been separated from the actual execution."

Finally, for two more modest studies which tend to confirm my basic analysis, see David P. Billington, "Structures and Machines: The Two Sides of Technology," in conjunction with Todd R. LaPorte, "Beyond Machines and Structures: Bases for the Political Criticism of Technology," *Soundings* 57:3 (Fall 1974), 275–304; and F. Rapp, "Technology and Natural Science—A Methodological Investigation," in F. Rapp, ed., *Contributions to a Philosophy of Technology, op. cit.,* pp. 93–114. On pp. 102–103 of his paper, for instance, Rapp breaks technology down into four basic elements: skills, the academic disciplines of the engineering sciences, production processes, and objects and their uses.

* * * * *

Apologia pro tractatu meo. One need in the philosophy of technology is for a broad view of the relationship between technology and the traditional branches of philosophy; another is for a more intimate acquaintance on the part of students of philosophy with engineering and technology. In an attempt to respond to both needs this paper has been weighted with more references and notes than might otherwise have seemed appropriate. If my pedagogical, or bibliographical, impulses have gone too far in this direction I beg the reader's indulgence and pardon.

Acknowledgment for help in working my way through more than one conceptual quagmire is also due to Robert Mackey, a friend whose mind and work too often get lost in a more than conceptual bog. Contrary to the standard disclaimer, for failure to contribute more criticism than he did, he must share responsibility for many of the weaknesses that remain.

Types of Stirrups

Illustrated Appendix
(See pp. 241–242 in text.)

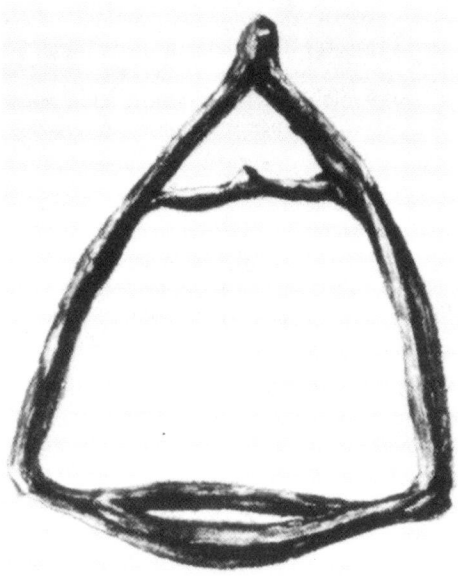

Figure I.   Iron stirrup used in Eastern Europe in 11th and 12th centuries.

Types of Technology

Figure II. Arab wooden platform stirrups from the 14th century. These are forerunners of the later Spanish *vaquero* stirrups (see figure VI).

Figure III. Modern English stirrups, starkly functional and free of all ornamentation.

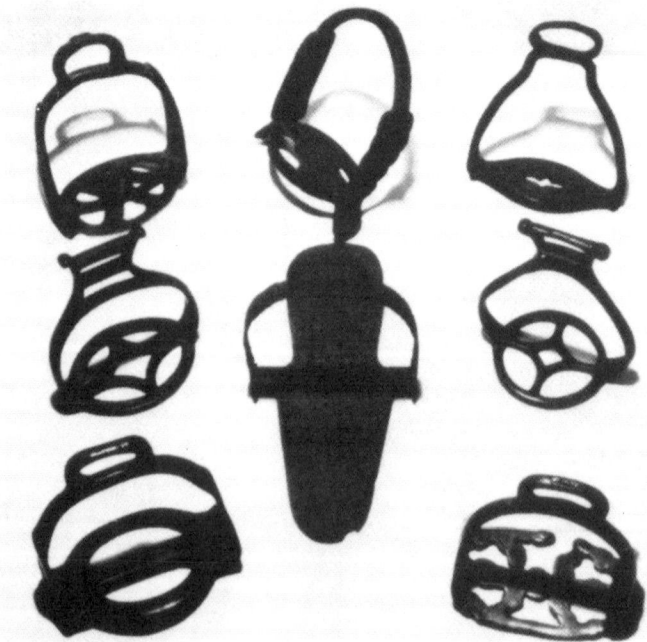

Figure IV. Spanish and French stirrups from 17th century. Note grace of design and center stirrup, which was designed to give exceptionally comfortable support.

Types of Technology

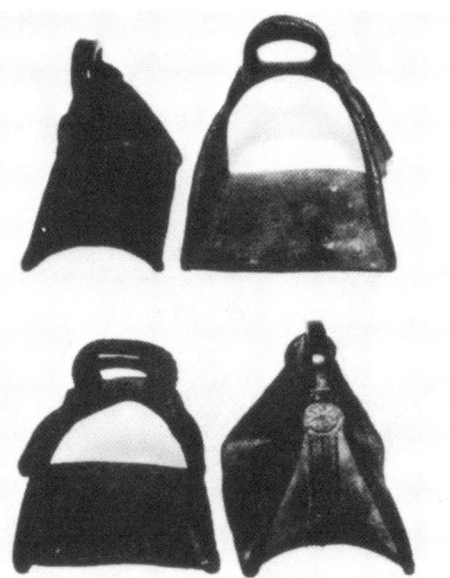

Figure V. A metal Spanish *vaquero* stirrup (lower right) shown with similar styles from Peru. Peruvian show stirrup at upper left; remaining two are work stirrups.

Figure VI. Round wooden stirrup, used by California *vaqueros* in 18th century.

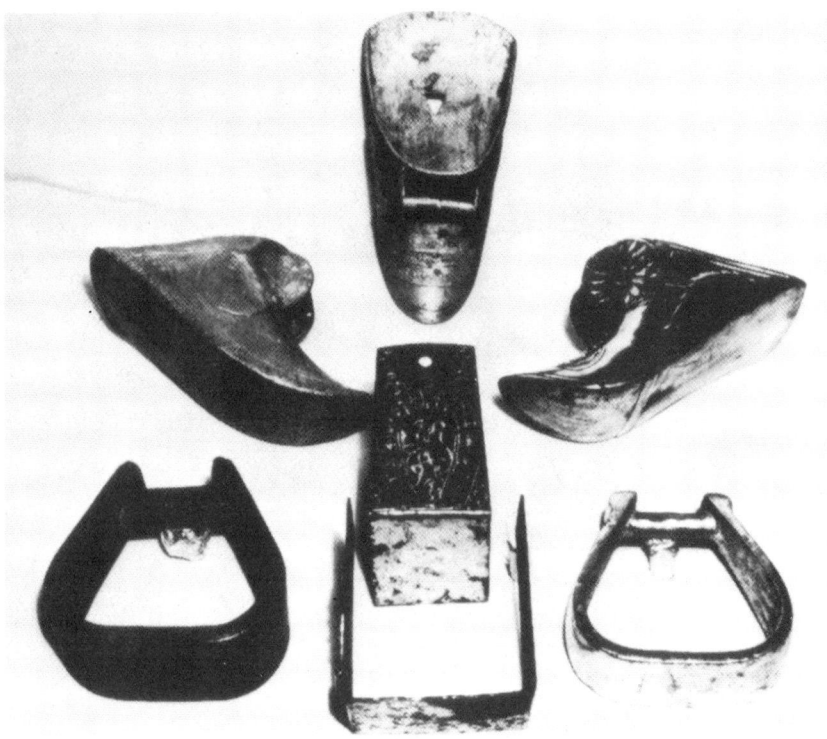

Figure VII. Contemporary stirrups from North and South America. Top: Aluminum copies of old Spanish stirrups, which are still in use in South America. Center: Mexican stirrups. Bottom left and right: Contemporary North American Western stirrups. Note that Mexican stirrups, although similar to U.S. types, are much more ornate.

Types of Technology 293

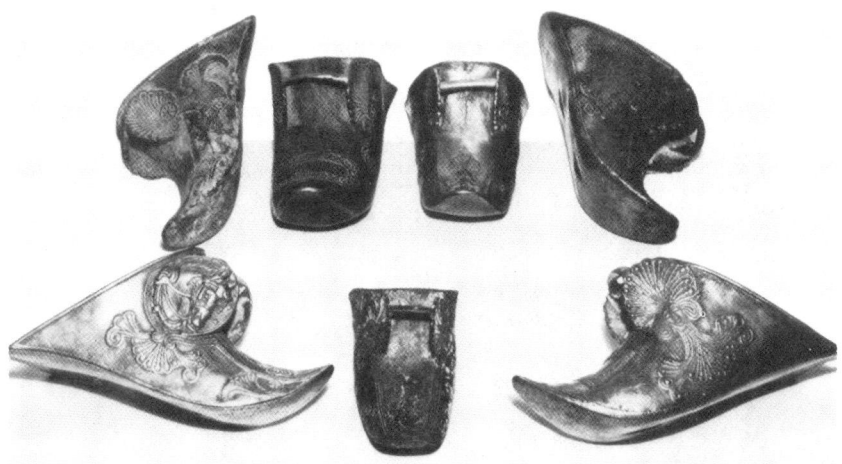

Figure VIII. Spanish stirrups from 16th and 17th centuries illustrating high-quality metal art work. Note that while some of the designs are of Spanish origin, others are derived from American Indians.

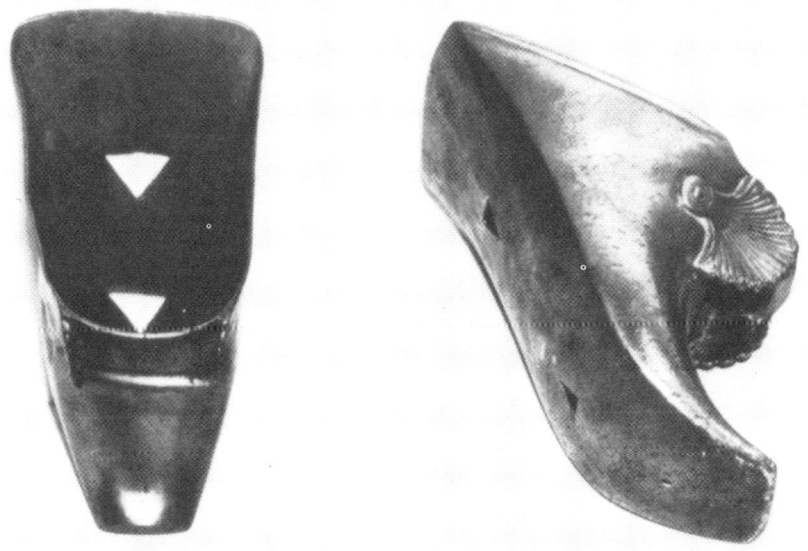

Figure IX. Close-up of Spanish stirrups of a square-toe design. Materials used are primarily brass and bronze. Holes in bottom are to drain water.

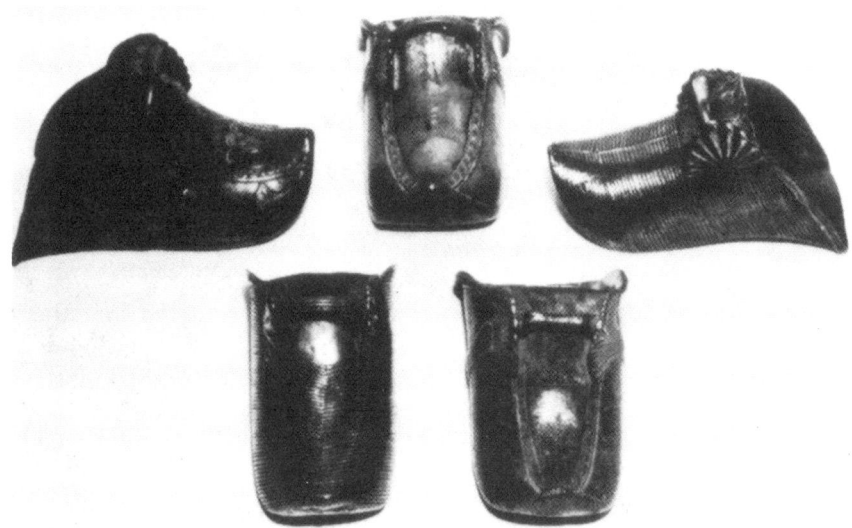

Figure X. Spanish stirrups of a "Dutch shoe" design. No two are exactly alike, because the mold had to be broken to remove the finished product. The best examples of this type bear maker's signature and date on the bottom.

Adapted from Verne R. Albright (photos by Carlos Sarmiento), "Ages of Stirrups," *Western Horseman* 35:9 (September 1970), pp. 108–109; and Doreen Bush, "Those Amazing Stirrups," *Horseman* 18:9 (April 1974), pp. 59-60.

PART III
REVIEW AND BIBLIOGRAPHY

REVIEW AND BIBLIOGRAPHICAL SURVEY

By Carl Mitcham,

With the Assistance of Jim Grote

Introductory Note: As mentioned in the general introduction, the addition of brief reviews and bibliographical surveys on various authors and issues seemed a desirable addition to the current bibliography. This year provides no more than a sample of the kind of material that, it is hoped, will be included in the future. If they are interested, authors are especially encouraged to submit bibliographical surveys on specialized topics such as cybernetics, systems theory, technology and values, etc. There are also plans to include one or more bibliographical surveys of European thought on the philosophy of technology in the second and third numbers of this series.

*Note: A name index planned for this survey unfortunately could not be included. It is being published separately in the *Philosophy & Technology Newsletter* and is available from the editors.

TECHNOLOGY, NIHILISM, AND ETHICS

Jonas, Hans. *Philosophical Essays: From Ancient Creed to Technological Man*. Englewood Cliffs, N.J.: Prentice-Hall, 1974. Pp. xviii, 349.

Jonas is perhaps the single most important U.S. philosopher currently writing on the philosophy (especially the ethics) of technology. This volume collects his essays from 1965 to 1973 on three main themes: Part One, "Science, Technology, and Ethics"; Part Two, "Organism, Mind, and History"; Part Three, "Religious Thought of the First Christian Centuries." The book includes a brief intellectual autobiography that explains the movement in Jonas's life and thought from his studies of gnosticism and the philosophy of biology to questions of technology. Presently Jonas is at work on what he suspects will be his last major philosophical effort, a book on technology and ethics. It is the eight essays of Part One which are of primary interest to philosophers of technology.

The first essay is "Technology and Responsibility: Reflections on the New Tasks of Ethics." Since ethics is concerned with action, and modern technology has changed the very nature of human action, Jonas argues for the necessity of a change in ethics. Traditional ethics was anthropocentric since it assumed that Nature was immutable and could take care of herself. Now, because of the vulnerability of Nature to man's technological intervention, a new ethics is needed which includes Nature herself among the objects of human responsibility.

Two basic problems confront the ethical task in the technological age. The practical problem concerns the overwhelming and unforseeable consequences which modern technological action initiates. Ethics today requires not only good will and ordinary intelligence as in the past, but also, more crucially, an immense predictive knowledge. On the theoretical level, a deeper dilemma presents itself. Modern scientific knowledge, while giving humanity great power, undermines all notions of "objective value and truth" which are necessary for regulating such power. More technology cannot solve but only aggravate this problem. "Now we shiver in the nakedness of a nihilism in which near-omnipotence is paired with near-emptiness, greatest capacity with knowing least what for" (p. 19). Jonas considers this ignorance to be reason enough for an ethics of restraint.

"Jewish and Christian Elements in Philosophy: Their Share in the Emergence of the Modern Mind," the second essay, is the most thought-provoking article in this collection. This essay examines the tradition of voluntarism from its roots in the biblical doctrine of creation *ex nihilo*, through Duns Scotus and the nominalists of the late Middle Ages, to its secularized apex in Nietzsche's "will to power." Jonas argues that it was the intrusion of Hebrew revelation into Greek philosophy that undermined the "essentialist" or "intellectualist" tradition and thus made possible modern nihilism. "To put it as briefly as possible, the biblical doctrine pitted contingency against necessity, particularity against universality, will against intellect" (p. 29). Without the world of intelligible essences, faith alone could hold up the notions of law and truth. As faith vanishes today (at least publicly) in the technological society, so must wisdom.

The third essay, "Seventeenth Century and After: The Meaning of the Scientific and Technological Revolution," is a discussion of the "revolution of thought" that comprised the scientific revolution of the seventeenth century and set the stage for the technological revolution of the nineteenth century. In contrasting the views of Copernicus, Galileo, and Newton with those of Aristotle and the ancients, Jonas provides an excellent analytical summary of the key concepts involved in the new cosmology and physics (e.g., the homogeneity of nature and the mathematization of physics). Jonas implies from this analysis that the modern scientific neutralization of nature (i.e., the rejection of final causality) and of man himself has put us in a peculiar predicament. This neutralization, "while giving us the license to do as we wish, at the same time denies us the guidance for knowing what to wish" (p. 79).

In the fourth essay, "Socio-Economic Knowledge and Ignorance of Goals," Jonas contends that economics is not quite as "value free" as the positivists assume. He argues that the biological fact of reproduction gives rise to the social goal of responsibility for future life. This responsibility, "indigenous to economics, becomes the source of categorical, normative judgment" (p. 97). Given the ecological crisis of today, the economist has the professional responsibility of making goal-decisions that will steer humanity away from disaster.

The fifth essay, "Philosophical Reflections on Experimenting with Human Subjects," lacks a central argument, as the title implies. Two themes, however, stand out from the rest: the "social contract" theory, and the problem of the choice of subjects. Given the primacy of the individual in the Western philosophic tradition, the "social contract" theory does not oblige an individual to risk his life for the common good except in an emergency (i.e., war) when the survival of society is in question. In light of this theory, human experimentation for modern medical research is often difficult to justify since much of this research deals with improving society rather than actually saving it.

In regard to the choice of human subjects, Jonas argues for the rule of the "descending order of permissibility." Permissibility descends according to education and financial resources. The call for volunteers should begin with the research community itself since their identification with and understanding of the experiment renders their decision the freest possible. Following the principle that "utter helplessness demands utter protection," experiments on the sick, except in relation to their own disease, are absolutely impermissible.

The next two essays, "Against the Stream: Comments on the Definition and Redefinition of Death" and "Biological Engineering—A Preview," are the only previously unpublished essays in Part One. "Against the Stream" is a postscript to the fifth essay and deals with the definition of death as "irreversible coma" (see "A Definition of Irreversible Coma," Report of the *Ad Hoc* Committee of the Harvard Medical School to Examine the Definition of Brain Death, *Journal of the American Medical Association* 205, no. 6 [August 5, 1968]: 337-340). Here Jonas expounds his suspicion that "removal of controversy on obtaining organs for transplantation" is the primary motivation behind the committee's definition. "Biological Engineering—A Preview" is a description of various modes of engineering (e.g., controlled mating, fetal screening, and cloning), along with a discussion of the practical and philosophical consequences of biological engineering.

In "Contemporary Problems in Ethics from a Jewish Perspective," the final essay, Jonas begins by contrasting the modern and the biblical views of nature and man. On nature, the Bible states that God deemed his creation to be good, while modern science regards nature as indifferent to good or evil. On man, the Bible tells us that man was created in the image of God. This image is destroyed, Jonas regrets, by Darwinism, historicism, and especially modern psychology—three movements which have shaped the modern view of man. Beginning with Nietzsche's "genealogy of morals," psychology has sought to expose man's higher aspirations as a disguise or sublimation of his "essentially base drives." Biological engineering is the final devaluation of man as the "image of God." In what image will man remake himself? Jonas argues that in this technological age Judaism must counsel a reverence for creation and man, along with a humility that engenders caution. Because of the paradoxical situation of our "excess of power to do" combined with our lack of guidance in what to do, Jonas claims that we need "a morality of don'ts, a criterion of rejection," like the Ten Commandments.

The paradoxical tension between power to act and ignorance about the ends of acting provides the impetus for Jonas's thinking. Over and over again, in various and sometimes overlapping ways, Jonas traces this paradox back to its origins in the revolutionary thought of such men as Bacon, Copernicus, Descartes, and Spinoza. The idea that modernity is a rejection of and not a dialectical progression from antiquity permeates all of Jonas's writings. The modern notion that doing, and not knowing, is the end of theory, comprises an absolute break with the past. In fact, Jonas argues, praxis-oriented theory is not really theory at all (in the Platonic sense). Without a transcendent theory, notions of limit, measure, and justice are impossible; in one form or another nihilism reigns. Yet at the same time, the power of practical thinking makes such limits all the more necessary. Thus the crisis in ethical thought is how to argue for restraint and limit in a world which worships power and excess. In such an impasse Jonas inevitably takes a "negative" approach, focusing on the catastrophe inherent in modern practice. It is hoped that in his future work on technology and ethics Jonas will systematically develop and more felicitously articulate his analysis of the ethical dilemma created by modern technology.

Six of the above articles have been previously published as follows:

"Technology and Responsibility: Reflections on the New Tasks of Ethics." Originally published in *Religion and the Humanizing of Man,* ed. James M. Robinson (New York: Council on the Study of Religion, 1972). Subsequently published in *Social Reserach* 40, no. 1 (Spring 1973): 31–54.

"Jewish and Christian Elements in Philosophy: Their Share in the Emergence of the Modern Mind." Originally published in *Commentary* 44 (November 1967) with the title "Jewish and Christian Elements in the Western Tradition." A revised version appears in *Creation: The Impact of an Idea,* ed. Francis Oakley and Daniel O'Connor (New York: Scribner's, 1969), pp. 241–258. The present annotated text is the definitive version.

"Seventeenth Century and After: The Meaning of the Scientific and Technological Revolution." An earlier version of this essay was published in *Philosophy Today* 15, no. 2 (Summer 1971): 76–101, with the title "The Scientific and Technological Revolutions."

"Socio-Economic Knowledge and Ignorance of Goals." An earlier version of this essay was published in *Economic Means and Social Ends,* ed. Robert L. Heilbroner (Englewood Cliffs, N.J.: Prentice-Hall, 1969), with the title "Socio-Economic Knowledge and the Critique of Goals."

"Philosophical Reflections on Experimenting with Human Subjects." Originally published in *Daedalus* 98, no. 2 (Spring 1969): 219–247, and in its present revised version, with a comment by Arthur J. Dyck, in *Experimentation with Human Subjects,* cd. Paul A. Freund (New York: Braziller, 1970).

"Contemporary Problems in Ethics from a Jewish Perspective." Reprinted, with minor modifications, from the *Central Conference American Rabbis Journal* (January 1968).

Jim Grote
University of Louisville

ANALYTIC PHILOSOPHY OF TECHNOLOGY, EAST AND WEST

Rapp, Friedrich, editor. *Contributions to a Philosophy of Technology; Studies in the Structure of Thinking in the Technological Sciences.* Dordrecht, Holland, and Boston, Mass.: D. Reidel, 1974. Pp. xiii, 228.

This is a collection of articles by engineers and philosophers from Czechoslovakia, Canada, Israel, the United States, the Federal Republic of Germany, the German Democratic Republic, Poland, England, and the Soviet Union (to list them simply in the order of their appearance) on what the editor properly calls "analytic philosophy of technology." The analytic perspective concentrates on methodological and epistemological studies of the structure and procedures of modern technology at the expense of what the editor less accurately calls "technological philosophy," which would deal with the human element in the making and using of technical objects and thus with how society affects and is affected by technological change. The signifcance of this volume lies not simply in the fact that it is the first book in English to focus consistently on analytic themes, but with the collaborative dialogue it reveals and engenders across disciplinary and ideological lines. Although one may well question the editor's view of logic and methodology as "basically neutral to any *Weltanschauung,*" interdisciplinary and interideological discussion of a subject so prone to the rhetoric of special interests cannot but be welcomed. (The only drawback to this work is its exorbitant price of $28.00; the whole book could be Xeroxed for less than one quarter of that cost.)

Friedrich Rapp, the West German editor, comes to philosophy from a prior background in physics, which no doubt contributes to his own analytic interests. However, he is clear that philosophy of technology, even in its strictly analytic form, is something quite different from the analytic philosophy of science. The centrality of the concept of design, and the recurring emphasis on design methodology and productive processes in the essays of this volume, bear out this point.

Following Rapp's short introduction, the first paper is by the Czech philosopher Ladislav Tondl, entitled "On the Concepts of 'Technology' and 'Technological Sciences'." This is an English version of Tondl's "Slovo o filosofii techniky," *Filosofický časopis* 12, no. 3 (1964): 281–293, which has also appeared in Russian as "O ponjatijax 'texnika' i 'texničeskie nauki'," *Organon* 3 (1966): 111–125. (The English seems to have been made from the Russian rather than the Czech text.) Here Tondl first examines various definitions of technology, by taking issue with the Russian historian Alexsandr A. Zvorikine's definition of technology as "the means of labour which develop within the system of social production." The problem is that Zvorikine's definition can be interpreted in two ways. On a restricted interpretation, to identify technology with production technology excludes all "medical, telecommunications, measuring and experimental technology." A broader interpretation of technology as anything that in any way contributes to the productive process makes the concept so large as to become vacuous. (On this debate, one might wish to consult two articles by Zvorikine which are not mentioned in Rapp's concluding bibliography: "Concerning a Unified Concept of Technical Progress: Zvorikine on Drucker's 'Work and Tools'," *Technology and Culture* 2, no. 3 [Summer 1961]: 249–253, followed by Peter Drucker's reply; and "The Laws of Technological Development," in Carl F. Stover, ed., *The Technological Order* [Detroit: Wayne State Univ. Press, 1963], pp. 59–74.)

Tondl himself proposes no simple definition of technology. Instead, he argues that technology involves both subjective elements (purposes and knowledge) and objective ones. With regard to the objective elements he clearly states that "one single definition is not sufficient" to encompass the complexity of technology. Objectively, technology can be described as "everything which man in his activity *puts between himself and the objective world*," as "the *resources which increase the efficiency* of human activity," as a *"synthesis of causal nets,"* or as various types of technological devices (i.e., tools, machines, automatons, etc.). But Tondl makes no attempt either to indicate which of these is the more nearly adequate objective description or to formulate a comprehensive description of his own.

The second part of Tondl's essay is concerned to distinguish technological (or engineering) sciences from the natural sciences. Analyzing both natural and technological sciences as task-solving methodologies, Tondl distinguishes the two on the basis of the tasks pursued. The task of natural science is to reveal the laws of the "behavior" of systems as well as "the limit of the possible 'behavior' or systems under given or postulated conditions." The task of technological science concerns how to synthesize a certain "behavior" from given elements or how to increase the efficiency of existing synthetic processes.

The next seven essays—and, in fact, one third of this volume—are the analytic papers from the symposium "Toward a Philosophy of Technology," *Technology and Culture* 7, no. 3 (Summer 1966), with the supplementary exchange between Wisdom and Agassi in *Technology and Culture* 8, no. 1 (January 1967). These are: Mario Bunge's "Technology as Applied Science," Joseph Agassi's "The Confusion Between Science and Technology in the Standard Philosophies of Science," J. O. Wisdom's "The Need for Corroboration: Comments on Agassi's Paper," Agassi's "Planning for Success: A Reply to J. O. Wisdom," Wisdom's

"Rules for Making Discoveries: Reply to Agassi," Henryk Skolimowski's "The Structure of Thinking in Technology," and I. C. Jarvie's "The Social Character of Technological Problems: Comments on Skolimowski's Paper." Having already been available in English for some time here and in various reprintings, and widely recognized as important by those concerned with the philosophy of technology, these papers need no further commentary. (One might hypothesize, however, that the inclusion of these essays was one factor in driving up the cost of this volume; the University of Chicago Press has in the past been quite exorbitant in its own demands for permissions fees.)

Rapp's own contribution to this volume follows as "Technology and Natural Science—A Methodological Investigation." This is translated from "Technik und Naturwissenschaften—Eine methodologische Untersuchung," in Hans Lenk and Simon Moser, eds., *Techne, Technik, Technologie: Philosophische Perspektiven* (Pullach: Verlag Dokumentation, 1973), pp. 108-132. Rapp's argument is that science and technology cannot be distinguished on the basis of a differentiation between nature and artifice—science dealing with nature or the world uninfluenced by man, technology with all those things produced by human activity—because "in the modern natural sciences the phenomena to be investigated must first be isolated . . . with the aid of complicated technical apparatus and instruments, so that in an abstract sense they too can be regarded as artifacts." On the other side, technology readily adopts both the research findings and working methods of science in ways which also undermine the traditional nature-artifact distinction. Rapp is especially good at developing illustrations of this science-technology interaction. And yet, he observes, "one cannot be content . . . to treat them as constituting . . . a single complex of science-and-technology."

In order to elucidate the true difference between science and technology, Rapp breaks technology down into four basic elements: skills, engineering science, industrial production, and technical objects. He then defines technology as "the harnessing of nature for human aims, certain material objects being produced on the basis of suitable knowledge and skills in a preconceived, consistently conducted process of production, and then put to use in an appropriate manner." In light of this definition he is able to argue that the real overlap of science and technology occurs not across the board, but only in one area, the area of engineering science or technological knowledge. Even here, however, the overlap is not such as to produce unification because of the different aims of science and technology. "Thus, whereas in natural science the primary object of interest is theoretically oriented acquisition of knowledge, in technology practical application always occupies the foreground." This difference in aim or intention generates differences in procedures. Accepting Hempel's theory of scientific explanation as based on the hypothetico-deductive method, Rapp concludes that technology, by contrast, "is concerned with the realization of a *projected* object in a *pragmatically* oriented production process." This final suggestion about the "projective-pragmatical" method of technology is the most provocative point in Rapp's piece.

The next two essays, M. I. Mantell's "Scientific Method—A Triad" and Eberhard Jobst's "Specific Features of Technology in Its Interrelation with Natural Science," continue to focus on the science-technology relationship. Mantell's paper, which first appeared in *Proceedings of the American Society of Civil Engineers* 95, no. PP1 (1969): 47–53, argues that the scientific method can be applied to the solving of any problem. Where it has been thought inapplicable this was because of a failure to distinguish three types of scientific method: (1) basic research, (2) applied research, (3) systems approach. The main body of Mantell's paper elaborates this distinction by means of a detailed description of each type of method.

Jobst's paper, which is translated from "Spezifische Merkmale der technischen Wissenschaft in ihrem Wechselverhältnis zur Naturwissenschaft," *Deutsche Zeitschrift für Philosophie* 16, no. 8 (1968): 928–935, offers a Marxist understanding of the relations between technology and science. According to Jobst, science and technology form an "essential unity" on the basis of "a dialectical interrelation." The exact character of this dialectical interrelationship or mutual influence is explored by special reference to developments in the field of thermodynamics.

Dieter Teichmann's "On the Classification of the Technological Sciences," and Tadeusz Kotarbiński's "Instrumentalization of Actions" provide two other Marxist perspectives on technology. Teichmann's paper is translated from "Zur Klassifikation der technischen Wissenschaften," in *Klassifizierung und Gegenstandsbestimmung der Geowissenschaften und der technischen Wissenschaften* (Freiberger Forschungshefte D 53 [Leipzig, 1967]), pp. 199–203. In it he considers the strengths and weaknesses of five possible ways of differentiating among the technological sciences—by historical development, by relationship to natural sciences (e.g., physical, chemical, etc., technologies), by branches of production (agricultural, transportation, etc., technologies), by productive function, and by machine operation. The essay by the Polish philosopher Kotarbiński is selected from his book *Praxiology—An Introduction to the Science of Efficient Action* (New York: Oxford Univ. Press, 1965), and discusses how the instrumentalization of actions increases their efficiency.

The three succeeding articles are discussions of design methodology by engineers. These include "A Philosophy of Engineering Design," being chapter 1 of Morris Asimov's widely used textbook *Introduction to De-*

sign (Englewood Cliffs, N.J.: Prentice-Hall, 1962); R. J. McCrory's "The Design Method—A Scientific Approach to Valid Design," from an *ASME* (American Society of Mechanical Engineers) Conference Paper, no. 63-MD-4 (1963); and A. D. Hall's "Three-Dimensional Morphology of Systems Engineering," which is reprinted from the *IEEE (Institute of Electrical and Electronic Engineers) Transactions on Systems Science and Cybernetics* SSC-5, no. 2 (April 1969): 156–160. All of these papers agree that design is of the essence of engineering, and therefore attempt to describe rational procedures for solving design problems. One weakness of this section, however, is that it does not adequately represent alternative interpretations of design as fundamentally different from science in its methods of procedure.

The final two selections are concerned with the role of experimental apparatus in engineering and in science. With regard to "The Role of Experiments in Applied Science" there is an exchange of letters between A. J. S. Pippard, author of *The Experimental Study of Structures* (London: E. Arnold, 1947), W. A. Tuplin, E. McEwen, and an anonymous reviewer—an exchange occasioned by the reviewer's comments on Pippard's book. This exchange is reprinted from the *Engineer* 196 (1953): 369–370, 465–466, and 561. With regard to "The Role of Apparatus in Cognition and Its Classification," F. V. Lazarev and M. K. Trifonova, in a translation of "Rol' priborov v poznanii i ix klassifikacija," *Naucnye doklady vyssej skoly: Filosofski nauki,* no. 6 (1970): 82–89, propose to distinguish three types of apparatus—amplifiers, analyzers, and transformers. (Along the line of these last two contributions, one relevant addition to Rapp's bibliography would be Vincent Edward Smith, "Toward a Philosophy of Physical Instruments," *Thomist* 10, no. 3 [July 1947]: 307–333.)

As is readily apparent, the scope of these analytic studies is quite large—ranging from questions of the relation between science and technology to the internal structure of technology itself and the nature of specific technological practices such as engineering design and experimentation. This volume is made doubly valuable, however, by an annotated bibliography of eighty-eight items dealing with these analytic issues. Like the book itself, the bibliography surveys both Eastern and Western literature by engineers as well as philosophers; it is especially good as a key to items in Russian, German, and Polish on such specialized East European subjects as praxiology. Since no attempt has been made to duplicate this bibliographical index in the following "Current Bibliography," anyone undertaking research in this particular area is strongly encouraged to consult the Rapp bibliography.

Carl Mitcham
St. Catharine College (Kentucky)

GEORGE GRANT: A NEGATIVE THEOLOGIAN ON TECHNOLOGY

George Parkins Grant has been called Canada's leading conservative thinker. Ramsay Cook, one of Canada's important liberal politicians, has also acknowledged that "George Grant is undoubtedly one of the most important political and social thinkers of recent Canadian intellectual history."[1] Despite his political involvements, however, neither classification adequately characterizes Grant.

Grant is a member of the department of Religious Sciences and Political Science at McMaster University, Hamilton, Ontario, Canada. A Rhodes Scholar in 1939, he received his doctorate in philosophy from Oxford, and was a member of the Philosophy Department at Dalhousie University, Halifax, Nova Scotia, from 1947 to 1960. After a year as consultant to the Institute of Philosophical Research in San Francisco, Grant joined the faculty at McMaster. Presently he resides in Dundas, Ontario, with his wife and two of his six children. His articles appear in scattered journals on topics in philosophy of education, Greek philosophy, literature, political theory, psychology, technology, and religion. But it is these last two themes which are most central to Grant's thought.

The Problem of Negative Theology

The word "theology" first occurs in Plato's *Republic*. Socrates and Adimantus build in speech a city which requires that all talk about the gods have an *imprimatur*. All theology must submit to the laws of the city, in order for the city to maintain its universal justice. But what if the city becomes corrupt? Should theology submit to a regime which advocates nontheological principles? In such an empire true discourse about the gods will become "negative theology"—viz., speeches about what God is not. Negative theology offers a darkened world glimpses of the transcendent. Not a transcendence beyond that which is present, but a transcendence signified by the reality of suffering and the lack of joy in the world. In the words of Luther, "The theologian of glory says that evil is good and good evil; the theologian of the Cross says that the thing is as it is."

In the Protestant tradition, however, it is not negative but biblical theology which has been the predominant form of Christian discourse about God. Biblical theology, by reflection on the Providence of God as manifested in His special revelation, seeks to make positive statements about God's Grace and the significance of finite existence. Negative Christian theology, while it presupposes the possibility of biblical theology, and is necessarily written from within the Christian tradition, concentrates on describing the limitations of finite existence in order to disassociate these limitations from conceptions of the Divine in men's minds. The corresponding moral force of negative theology within any particular social order is to place limitations on man's powers of doing and making. In the tension between the world and the transcendent, negative theology paradoxically attaches itself to the world in speech, in order through its critique to prepare an opening for the transcendent. (Indeed, in this respect the function of negative theology is not wholly unlike that of the Frankfurt School of critical social theory within the larger framework of Marxian social theory.) Grant's writings are difficult to understand precisely because he does negative theology. He does not speak directly about what are to him the most important things. About those things which he initially appears to speak directly, he often speaks more indirectly than expected.

Even in the obliqueness of its references to God, negative theology cannot escape the perennial conflict between contemplation and charity. What has Athens to do with Jerusalem? In a world described by Heidegger as "unthinking" and by Ellul as "loveless," Grant seeks to unite thought and faith through his discourses about technique. "Those who care about charity," Grant writes, "must care about communication, and to communicate requires systematic thought.[2] For one who does not identify in any explicit way with the Christian tradition, it is still possible to profit from Grant's work. The heart of the matter rests in the task of Grant's negative theology to reflect systematically, comprehensively on technique. In his early writings Grant, influenced by the Oxford practice of analytic philosophy, thought that analysis of moral language might provide a framework for understanding the scope and limits of human action. "The study of philosophy," he argued, "is the analysis of the traditions of our society and the judgment of those traditions against varying intuitions of the Perfection of God. It is contemplation of our own and others' activity, in the hope that in understanding it better we make it less imperfect.[3] There is, however, a liberal trust in human history as the progressive incarnation of reason which lies behind the confidence of the moral language approach. By virtue of an intellectual conversion which owes its origins to the influence of Leo Strauss and Jacques Ellul, Grant abandoned that liberal trust. He became a spectator, waiting and listening to the speeches, rituals, and strivings of a society dominated by technique.

The Books of Grant

In his early years, besides a colorful career in social and political activities, Grant concentrated on the problems of modernity and moral philosophy. In his first book, *Philosophy in the Mass Age,* he assumed "there is such a study as moral philosophy, and that by reflection we can come to make true judgments as to how we ought to act."[4] Even in this early work, Grant was struggling with "the primals" of Greekness and Christianity. He writes, "I would assert that philosophic reflection can lead us to make true judgments about right action. Contemplation can teach us the knowledge of God's law."[5]

Lament for a Nation, Grant's second book, and the one which he considers his best, suddenly threw the contemplative, moral philosopher into the public world. The years between 1960 and 1965 were ones in which a Protestant politician, John G. Diefenbaker, was elected Prime Minister of Canada. This political event in some sense represented the growing concern of some Canadians over the involvement of Canada with the United States. Grant viewed the ruthless backlash by "progressive intellectuals" against Diefenbaker as a sign of the approach of the "universal and homogeneous state," as an outgrowth of the tyranny of technology.[6] The defeat of Canadian nationalism signified the dominance of technique over man. In the closing sentences of *Lament for a Nation,* Grant wrote, "Beyond courage, it is also possible to live in the ancient faith, which asserts that changes in the world, even if they be recognized more as a loss than a gain, take place within an external order that is not affected by their taking place. Whatever the difficulty of philosophy, the religious man has been told that process is not all. *'Tandebantque manus ripae ulterioris amore.'* (Virgil, *Aeneid,* Book VI: 'They were holding their arms outstretched in love toward the further shore.')."

Technology and Empire, a combination of previously published and unpublished articles, is Grant's most comprehensive statement on a variety of issues. The first chapter, "In Defence of North America," discusses the complex roots of thought and technique, and shows how indispensable the Christian tradition is to understanding these roots. The second chapter, "Religion and the State," discusses a rather classical theme of the relation between religion and the state. For a citizen of the United States, whose tradition has always upheld the separation of church and state, Grant's second chapter will provide considerable food for thought. Is this nonconformist view of separation between church and state one aspect of the influence of technique in the political arena? The third chapter, "Canadian Fate and Imperialism," is a strong statement by a Canadian describing the "American Empire to the South." "The present happenings in Vietnam," Grant argued, "are particularly terrible for Canadians. What is being done there is being done by a society which is in some deep way our own." But this article is not the myopic view of a citizen in a satellite country of the United States. Quite the contrary. For "what lies behind the small practical question of Canadian nationalism," Grant concludes, "is the larger context of the fate of Western civilization." The fourth chapter, "Tyranny and Wisdom," is a comment upon one of the central issues concerning technique, viz. the homogenous, universal state. This article exposes Grant's concern over the relation of Socrates to Jesus. Here in the context of a commentary on the debate between Leo Strauss and Alexandre Kojève,[7] Grant argues for the necessity to think more deeply about Christianity's involvement with modern technique. The fifth chapter, "The University Curriculum," is a masterful analysis of the institution which Grant knows best, i.e., the secular university. There is no difficulty for any citizen of the United States to apply this analysis to the Kent States, the Columbias, or the Universities of California. Here again technique triumphs over the love of wisdom. Finally, "A Platitude," makes the most radical statement about technique to appear in Grant's thought. "Technique," Grant writes, "is ourselves." Grant's indebtedness to Ellul, Nietzsche, Heidegger, and Strauss is implicit in every line of this last chapter. In sum, *Technology and Empire* is truly the best book for an introduction to Grant's mature thought on technique.

Time as History is a revision of Grant's 1969 Massey Lectures, sponsored by the Canadian Broadcasting Company. "To write of the conception of time as history and to think of it as an animator of our existence is not," states Grant, "to turn away from what is immediately present to all of us. When I drive on the highways around Hamilton and Toronto, through the proliferating factories and apartments, the research establishments and supermarkets; when I sit in the bureaucracies in which the education for technocracy is planned; when I live in and with the mechanized bodies and resolute wills necessary to that system; it is then that the conception of time as history is seen in its blossoming.... The words used to explicate 'time as history' may seem abstract, but they are meant to illuminate our waking and sleeping hours in technical society."[8] Time and history unite in the technological society, Grant argues, because of the way modernity conceives the world as essentially a historical process. And, as technology comes to fill up more and more of our horizon, the language of willing and doing move closer in, to intersect with the language of process. They come together, for example, "in the sermons preached by our journalists about the achievement of landing on the moon."[9] There is a modern faith that our doing can make human existence meaningful. We have, concludes Grant, placed on ourselves the burden of creation—as is now exemplified in the aspirations of genetic engineering.

As this brief overview indicates, Grant's work bearing on technology ranges from apologetic discussions of philosophy and religion to political theory and historico-philosophical studies. In all of these, however, he is motivated by a concern for his world which only becomes clear to those who have known him personally. It is hoped that this bibliographical introduction may lead some, at least, to a deeper appreciation of Grant's thinking on technology, a thinking rooted in Greek philosophy and Christian faith.

Philip J. Hanson
Wilmington College (Ohio)

FOOTNOTES

1. Ramsay Cook, "Loyalism, Technology and Canada's Fate" in *Journal of Canadian Studies/Revue d'Etudes Canadiennes* vol. 3 (August 1970): 50–60.
2. *Philosophy in the Mass Age,* p. 102.
3. "Philosophy" in *The Royal Commission Studies* (Ottawa: The King's Printer, 1951), p. 119.
4. *Philosophy in the Mass Age,* preface of 1st ed., p. 9.
5. *Ibid.* Further confirmation of Grant's interest in religion is his Oxford dissertation on the Presbyterian theologian John Wood Oman (1860–1939).
6. *Lament for a Nation,* p. 97.
7. Cf. Leo Strauss, *On Tyranny* (Glencoe, Illinois: Free Press, 1963), which contains a response by Alexandre Kojève and Strauss's "Restatement on Xenophon's *Hiero.*"
8. *Time As History,* pp. 8–9.
9. *Ibid.,* p. 6.

(Grant Bibliography for Philosophy of Technology)

"Philosophy," in *The Royal Commission Studies* (Ottawa: The King's Printer, 1951), pp. 119–133.
"Contemplation in an Expanding Economy," in *The Anglican Outlook* (Montreal: 1955), pp. 8–11.
Philosophy in the Mass Age. Toronto: The Copp Clark Publishing Co., 1959. 2nd edition, with new introduction, 1966.
"Conceptions of Health," in Helmut Schoeck and James W. Wiggins, ed., *Psychiatry and Responsibility* (Toronto: Van Nostrand, 1962), pp. 117–134.
"Value and Technology," in *Welfare Services in a Changing Technology* (Ottawa: The Canadian Conference on Social Welfare, 1964), pp. 21–29.
Lament for a Nation: The Defeat of Canadian Nationalism. Toronto: McClelland and Steward, Ltd., 1965. 2nd edition, with new introduction, 1966.
"Critique of the New Left," *Our Generation:* double issue vol. 3, no. 4 — vol. 4, no. 1 (May 1966): 46–51. Also published as "Protest and Technology," in Charles Hanly, ed., *Revolution and Response* (Toronto: McClelland and Steward, 1966), pp. 122–128.
Technology and Empire: Perspectives on North America. Toronto: House of Anansi, 1969.
Time as History. Toronto, Canada: C. B. C. Learning Systems, 1971.
"Tradition and Revolution" in *Canadian Forum* 50, no. 591-2 (April–May 1970): 88–93. Also published as "Revolution and Tradition," in Lionel Rubinoff, ed., *Tradition and Revolution* (New York: St. Martin's Press, 1971), pp. 81–95.

"Nationalism and Rationality," in *Canadian Forum* 50, no. 600 (January 1971): 336–337.
"English Speaking Liberalism and Technique," an unpublished paper delivered before the Royal Society of Canada, 1974. Part of a book to be published in 1977.
"The University Curriculum and the Technological Threat," in W. Roy Niblett, ed., *The Sciences, the Humanities and the Technological Threat* (London: Univ. of London Press, 1975), pp. 21–35. Revised edition of a chapter from *Technology and Empire*.

CURRENT BIBLIOGRAPHY IN THE PHILOSOPHY OF TECHNOLOGY: 1973–1974

INTRODUCTION

A bibliography depends on basic decisions about its subject matter; it reflects some theory, however implicit, about the character of its material. The aim of the present bibliography of the philosophy of technology is not just to carve out a new speciality somewhere among the history of technology; philosophy of science; industrial sociology; science policy studies; bioethics; the history, sociology and philosophy of medicine; etc. There are specialities enough already in existence. Rather, building on work which already explicitly calls itself philosophy of technology—i.e., work which has focused either on articulating the cognitive and active structure of technology or on discovering its ethical and political implications—this bibliography intends to point toward a comprehensive understanding of the making and using of artifacts. As such it seeks to integrate important elements from each of the areas of research mentioned above, along with other philosophical studies of artificial intelligence, cybernetics, and art. The structure of this proposed integration is adumbrated in the outline of the bibliography as follows:

  I. Comprehensive Philosophical Works. —Books and articles which take technology as a primary theme for reflection and deal with it in terms of two or more of the categories listed below. Includes important collections and symposia, and virtually all essays with titles like "philosophy of technology" or "philosophy and technology."

 II. Ethical and Political Critiques
     A. Primary.—Philosophical analyses of the ethical and political implications and problems of technology. Includes: studies of technology and values; criticisms of technological society; moral analyses of technological change, transfer and development; technology assessment; the ethical and political issues of ecology; obligations toward the future; bioethics; and some science policy studies.
     B. Secondary. —Sociological, psychological, historical and general humanistic studies of technology and civilization. Much popular literature. Some good commentary on the work of thinkers such as Marcuse and Habermas. Some technical material on: technology and economics, planning, automation and labor, futurology, technology assessment, technology and law, technology and education, etc.
     C. Appendix: Soviet and East European Materials. —Lists mainly those Soviet and East German items which could not be adequately checked and are suspected of being ideological in character.

III. Religious Critiques
     A. Primary.—Basic analyses of the religious implications and problems of modern technology. Includes religious oriented bioethics, theology of ecology, and historico-theological studies. Largely from the point of view of Christian ethics or moral theology.
     B. Secondary. —Other religious discussions of technology which tend to be one or more of the following: popular or reportorial in

character, limited in scope, less well articulated, sociological, or discussions of relevance which, however, are only small parts of larger arguments.

IV. Metaphysical and Epistemological Studies
   A. Primary. —Discussions of one or more of the following topics: the essence of technology, the science-technology relationship and foundations of the technological sciences, the methodological structure of technological knowledge and action (particularly invention and design), phenomenological description of the man-technology relationship, the ontological status of artifacts including works of art, problems of artificial intelligence and other minds, philosophical issues of cybernetics.
   B. Secondary. —Informative technical literature, particularly on artificial intelligence, cybernetics, and engineering design.

V. Appendix
   A. Background Materials. —Historical studies of technology of potential value to philosophers, plus some general philosophical analyses of human knowledge, action, science, culture, etc. which are in some way related to questions of technology.
   B. Textbooks. —Largely anthologies dealing with the social science studies of technology.

In what was perhaps the most thoughtful review of the original *Bibliography of the Philosophy of Technology,* Otto Mayr (*ISIS* 66, no. 2 [1975]), after praising the work as a whole, makes four otherwise unpublicized criticisms: a) it included too much popular literature, b) the annotations were not critical enough, c) the primary-secondary distinction was not a good one, and d) the overall divisions of the bibliography did not reflect the real distinctions in the literature. By way of commenting more specifically on the outline given above, let me respond to these well considered remarks.

As mentioned in the general introduction, what we have been after from the beginning is a just critical assessment of the literature. It was natural, however, for bibliographers who were not yet in their thirties, dealing with a field that was not yet academically well defined, to be a bit timid in their comments and to produce annotations which were "neutrally informative rather than critical." In this second installment it can be expected that a few years experience will have increased one's boldness.

Nevertheless, the bibliography continues to include more popular literature—not to say philosophical garbage—than it might. Why? For two reasons: it is, unfortunately, only too much with us; and it still seems beneficial to have an index to the popular mind as a kind of counterpoint for deeper reflection. Slowly, over the next few years, under the pressures of an increasing volume of material and a growing awareness of the fundamental issues, we can expect stronger criteria of selection to emerge from this bibliographical practice. But while in historical research second-rate studies and the popular imagination may confidently be relegated to silence, in philosophy the situation is not quite so simple. If the owl of Minerva begins its flight at night, for one reason or another one of its intermediate stops often happens to be the city dump.

With all this in mind we have felt it necessary to retain the primary-secondary distinctions which Mayr found so objectionable. "The compilers themselves seem to have lacked confidence in the scheme, for they applied it only to two chapters out of five." Here, alas, our confidence has increased; the scheme has been applied to three chapters out of five—perhaps even to four out of five, since the distinction in the Appendix between "Philosophical and Historical Background" and "Texts" strongly resembles that used in parts II, III, and IV. The reason for continuing this scheme, despite the fact that it appears to force us "to label a great many . . . peers (and . . . betters) in an unavoidably offensive manner," is that *philosophically* it really does do the work for which it was intended. All secondary sources are secondary only in the sense that they are less immediately relevant to the encouragment of philosophical reflection on technology. Now they can require mediation either because they are not philosophy in the deepest sense or because they are not about technology in a sustained and systematic way. But even if they are not true philosophy they might still be very good sociology, history, psychology, etc. And even if they are not directly about technology, still they may contain hints or suggestions which are well worth secondary considerations. The classification admittedly involves a kind of exercise in modal logic, but for purposes of students of the philosophy of technology it serves as well as anything else we were able to devise to direct those interested to the most immediate sources of inspiration, while maintaining the freedom to venture into the realms of mediation as opportunity and inclination present themselves. The most unacceptable application of this distinction (or something like it) is the use of a third classification under Ethical-Political Critiques, an appendix for the "Soviet and East European Materials." To arbitrarily suggest that all this material is ideological in a way that many other articles are not, is more than a little unjust. The category is simply a bibliographical crutch, and one may hope it will be discarded in future bibliographical supplements.

As for Mayr's criticisms of the general structure of the bibliography as a whole, here it is best to quote him in full:

Given the present state of the philosophy of technology, to obtain an effective structure for its bibliography is to create a structure for the discipline itself. From this challenge the compilers drew back. Instead they simply divided their material, apparently *after* it had been collected, into five chapters according to traditional philosophical categories. . . . This approach I find unfortunate. First, the categories chosen are diffuse. There is a hopeless amount of overlap: a reader seeking references on any topic will always have to go through the entire book. He would have been better served with only an alphabetic arrangement and no systematic structure at all. Second, I suspect that traditional philosophical categories such as ethics, metaphysics, and epistemology (to say nothing of such unhappy pairs as "Ethics and Politics," with "Religion" as a parallel category) simply do not reflect the nature of the subject.

It seems to me that the discourse in the philosophy of technology, whenever it comes to life, centers on relatively practical and specific issues, on problems like the definition of technology, the classification of machinery, the science-technology relationship, the meaning of cybernetics, machine intelligence, the notion of progress, technology assessment, futurology. Questions like these have recently generated much interesting literature which unfortunately is widely scattered. By failing to recognize these issues our bibliography not only misses valuable material, it also forgoes an important opportunity: it would have been far better to derive its structure from the problems that the philosophy of technology is actually struggling with than from traditional and academic categories that in this context seem exceedingly stale.

With due respect to Dr. Mayr, who is curator of mechanical engineering at the Smithsonian Institution and author of an important study of *The Origins of Feedback Control* (Cambridge: MIT Press, 1971), I think these two paragraphs reveal that he speaks more as a historian than as a philosopher of technology. Historians may well not be able to employ atemporal categories or perspectives when approaching new territory. (Although from this point of view, Mayr's criticism of our apparent *a posteriori* approach to classification of the literature is somewhat peculiar.) Philosophers, under the influence of historicism, are sometimes tempted to adopt a similar attitude—but they do so only at the expense of modish superficiality or the replication of considerable segments of intellectual history. As has been argued in the "Introduction: Technology as a Philosophical Problem" of *Philosophy and Technology* (a companion work to the original *Bibliography*), such issues as the definition of technology, classification of machines, the science-technology relationship, machine intelligence, etc., exhibit in a new form traditional metaphysical and epistemological problems; while questions about the idea of progress, technology assessment, futurology, etc., are intimately bound up with traditional ethical-political arguments. To ignore this is to make claims for originality which cannot be sustained, or to fail to do philosophy of technology. It is precisely because of his recognition of the way in which technology can become integral to traditional philosophical arguments that Frederich Dessauer is the true father of the philosophy of technology. Philosophy of technology is a *new* form of philosophy, to be sure; but it is still a *form* of philosophy. Part of its true development thus lies in recognizing and articulating the same in the different, the different in the same. This is not to deny, however, the need for a good subject index to cross-reference any number of these more specialized topics.

Without meaning excessively to belabor a simple review, there is one other issue hinted at by Mayr's objection to the conjunction of "Ethics and Politics" in parallel to "Religion." Presumably he means to suggest that all three terms could be profitably collected under the rubric of "technology and values." We have, however, explicitly eschewed the term "values," preferring instead the more traditional disciplinary categories, because of the way the idea of values already skews philosophical consideration. The primacy of the concept of values in discourse about the aims of human action, and the consequent implicit subjectivizing of that discourse, is part of a distinctly modern critique of the character and possibility of ethics, politics, and religion—a critique which is in important aspects linked to the development of modern science and technology and its rejection of any natural teleology. To presume this primacy thus tends to prejudge significant points at issue in the ethical, political, and religious evaluations of technology; the term "values" is not quite so neutral as is sometimes presumed.

The ideal of this ongoing bibliography of the philosophy of technology is to reference all truly philosophical analyses of technology in major Western languages, along with as much material from other languages as it is possible to locate using our limited resources and facilities. In most cases this does not include book reviews, although select mention of reviews is often made in the annotations for some more widely discussed works. Also, as can be readily appreciated, much more marginal material (on technology and civilization, technological change, technology assessment and transfer, futurology, engineering science and practice, popular works on ecology, etc.) has been included from English than from other languages.

Mechanically, preparation of the bibliography has made use of the following reference tools:

Applied Science and Technology Index—an index to a variety of technical and engineering periodicals; subject headings for "Technological Change" and "Technology—Social Aspects"

Bibliografia Filosofica Italiana (Centro di Studi Filosofici di Gallarte)—limited to works in Italian; no annotations; subject heading on "Futurologia"

Bibliographie de la Philosophie/Bibliography of Philosophy (published with the aid of UNESCO and CNRS by J. Vrin, Paris)—international in scope, with annotations in French, English, German, etc.; books only; no special category for philosophy of technology

Bibliographie Philosophie (Information und Dokumentation im Institute für Gesellschaftswissenschaften, Berlin, DDR)—mainly East European books and articles; no annotations; subject index of the topic "Philosophie und Technik"

Bibliography of Bioethics (Center for Bioethics, Kennedy Institute, Georgetown Univ., Washington, D.C.)—comprehensive for English-language materials; but the index language used to describe each document does not contain "philosophy of technology"

Bibliography of Society, Ethics and the Life Sciences (Institute of Society, Ethics and the Life Sciences, Hastings-on-Hudson, N.Y.)—sections on "Introductory Readings in Ethics and the Life Sciences," "Values, Ethics and Technology," and "Genetics, Fertilization, and Birth; I, General Readings; B, Ethical, Philosophical, and Historical Material" are all especially relevant to philosophy of technology

British Humanities Index—subject headings for "Technological Change," "Technological Progress," "Technology," "Art and Technology," "Sociology and Technology," etc.

Bulletin Signaletique (Centre de Documentation Sciences Humaines, Paris)—international; abstracts of articles and book reviews; subject index for "technique" and "technologie"

Catholic Periodical and Literature Index—indexes an international list of popular and scholarly Catholic journals and books; subject heading for "Technology (Philosophy)" as well as "Technology and Civilization," "Technolcracy," etc.

Essay and General Literature Index—indexes English collections of essays, with particular emphasis on humanities and social sciences; small subheading under "Technology" of "Philosophy," plus large subject list under "Technology and Civilization"

International Bibliographie der Zeitschriftenliteratur aus allen Gebieten des Wissens/International Bibliography of Periodical Literature Covering All Fields of Knowledge—indexes an exceptionally large international list of periodicals; subject headings for "Technik," "Technokratie," "Technologie"; weaknesses: tends to be two or three years behind on many journals, and with so many journals being indexed the subject headings include too wide a diversity of materials

International Political Science Abstracts/Documentation Politique In-

ternationale (International Political Science Association)—articles from an international list of journals; index usually includes subject headings for "Technocracy," "Technology," etc.

Monthly Catalog of United States Government Publications—subject headings for "Technology," "Technology Assessment," "Technology Transfer," etc.

National Union Catalog: Subject Index (Library of Congress, Washington, D.C.)—international in scope; subject headings for "Technology," "Technology—Philosophy," "Technology and Civilization," "Technology and Ethics," etc.

Philosopher's Index (Philosophical Documentation Center, Bowling Green Univ., Bowling Green, Ohio)—international; articles only; good annotations, many provided by the authors themselves; subject headings for "Technology," "Artifacts," "Machines," "Cybernetics," "Artificial Intelligence," "Technocracy," "Technique," etc.

Reader's Guide to Periodical Literature—indexes mostly popular English language periodicals; subject headings for "Technology and Civilization," "Technology and State," "Technology Assessment," "Technology Transfer," etc., but not "Technology and Philosophy"

Répertoire Bibliographique de la Philosophie (Catholic Univ. of Louvain)—books, articles, and some reviews in Dutch, English, French, German, Italian, Latin, Portuguese, Spanish, Catalan, etc.; exhaustive with respect to books; reasonably complete on articles in Western languages; no annotations; subject heading for "Philosophie de la culture et de la technique"

Social Sciences and Humanities Index/Social Sciences Index and Humanities Index—indexes a wide range of English language academic journals; with its reorganization into two separate works in 1973, both works introduced a subheading "Technology—Philosophy" along with subject headings for "Technology and Civilization," "Technology Assessment," "Technology Transfer," etc.

Subject Guide to Books in Print (prepared and published by R. R. Bowker Co., New York, for United States publishers)—subject headings for "Technology" include everything from engineering textbooks to social science books.

As the descriptive comments readily suggest, the most important of these works are the *Philosopher's Index* and *Répertoire Bibliographique de la Philosophie*. In consequence, the present bibliography has attempted to list and independently annotate all materials which are classified by either of these works in such a way as to suggest their relevance for philosophy of technology—even when, in fact, this turns out to be minimal.

In conjunction with use of the above bibliographical works we have also directly reviewed the following journals for articles and citations dealing with the philosophy of technology:

British Journal of Aesthetics
Cross Currents
Deutsche Zeitschrift für Philosophie
The Engineer
Les Etudes Philosophiques
Futurist
Impact of Science on Society
International Philosophical Quarterly
Isis
Hastings Center Report
Hastings Center Studies
Man and World
New Scholasticism
Philosophy Today
Review for Religious
Review of Metaphysics
Review of Politics
Revue Philosophique de Louvain
Science
Science Studies
Scientia (Milan)
Social Research
Soundings (Vanderbilt Univ.)
Soviet Studies in Philosophy
Studies in Soviet Thought
Systematics
Technology and Culture
Telos
Theological Studies
Theology Digest
VDI-Zeitschrift (Verein Deutscher Ingenieure Zeitschrift)
Voprosy filosofii
Zeitschrift für Philosophie Forschung
Zygon

It is expected that in the future the number of journals directly monitored will increase. We are open to suggestions from individuals and journals themselves of others that should be so included.

Finally, although this "Current Bibliography: 1973–1974" is designed to update the *Bibliography of the Philosophy of Technology*, which covers materials up through 1972, we have picked up a number of missed items from the period 1970 on. Missed works prior to 1970 will be dealt with differently at another time.

* * *

Concluding Note: Seeing this bibliography through the galleys, and being forced by extensions in my information to alter and enlarge it considerably at this late stage—so considerably, in fact, that I have significantly retarded publication, and must apologize to the editor, publisher, and readers for the resultant delay—I have been made exceptionally aware of the limitations of this work. There is important material, especially in Japanese, which has been totally ignored. Spanish, Italian, Portuguese, and Scandinavian works are included on a too casual basis. The Soviet and East European survey is inadequate. And something will have to be done in future editions of this work to improve organization. My mind, at this point, is inclined toward some material subdividing of secondary sources into more specific topics such as technological forecasting, technology and economics, technology assessment, technology and law, technology transfer, etc. I would welcome comments and suggestions from users of this work. — C.M.

I. COMPREHENSIVE PHILOSOPHICAL WORKS

"Architecture, Technology, and the City," *Soundings* (Vanderbilt Univ.) 57, no. 3 (Fall 1974). A symposium which includes the following: J. J. McDermott's "Space, Time and Touch: Philosophical Dimensions of Urban Consciousness" (pp. 253–274), D. P. Billington's "Structures and Machines: The Two Sides of Technology" (pp. 275–288), T. R. LaPorte's "Beyond Machines and Structures: Bases for the Political Criticism of Technology" (pp. 289–304), and J. W. Cook's "Structures and Meaning: Values Apparent in Architectural Design" (pp. 305–317).

Beaune, Jean-Claude. *La technologie*. Paris: Presses Universitaires de France, 1972. Pp. 96. A short analysis making use of forty-six brief selections from a large number of historical, literary, technical, and philosophic authors—e.g., D. Santos, R. Queneau, S. Acquaviva, L. Mumford, J. Ladriere, P. Ducasse, A. Espinas, R. S. Merrill, R. J. Forbes, A. R. Hall, J. Crank, etc.—preceded by laconic commentary. Chapter titles: "First Approaches to Technology," "Science and Technology," "Cybernetics," "Sociological Points of View," "Economic Activities and Technological Development," "Ethnology and Technology." Very impressionistic.

Beck, Robert N. "Technology and Idealism," *Idealistic Studies* 4, no. 2 (May 1974): 181–187. "The purpose of this brief paper is to show that the primary problem of technology, like all problems related to possibilities and actions, is the conceptual adequacy of the intentions and values it implies, and not, as many critics have suggested, its social effects" (p. 181). Prepared for the XVth World Congress of Philosophy.

Billington, David P. "Structures and Machines: The Two Sides of Technology," *Soundings* (Vanderbilt Univ.) 57, no. 3 (Fall 1974): 275–288. Structures are static, machines are dynamic. These two elements thus exemplify two views of technology. Through reflection upon reaction to technology as machines in America and as structures in the Netherlands and Spain, argues that structures are primary and machines secondary technology. See also T. R. LaPorte's "Beyond Machines and Structures: Bases for the Political Criticism of Technology," *Ibid.*, pp. 289–304, which argues that technology-as-physical-entity has been emphasized at the expense of technology-as-social-organization.

Durbin, Paul T. "Technology, Philosophy of," *New Catholic Encyclopedia* Vol. 16 (Supplement 1967–1974), pp. 446–448. A short overview.

Ganovski, Sava, ed. *Science, Technology, Man*. Sofia: Publishing House of the Bulgarian Academy of Science, 1973. Pp. 316. An advance vanity publication of the XVth World Congress of Philosophy in which "Bulgarian philosophers outline their approach to a number of important and topical problems related to contemporary science, technical progress and man. Along with the positive discussion attention has been paid to criticism of various unscientific conceptions on the above-mentioned problems." Contents: Introduction by Sava Ganovski, "The Philosophical Problems of Science, Technique and Man." Part I, Philosophy and Science, includes: Nikolai Iribadjakov's "Philosophy and Anti-philosophy," A. Polikarov's "Nature, Knowledge, Dialectics," Stephan Vassilev-Vassev's "Man, Knowledge, Practice," Nikola Stephanov's "On the Question of Building a Structural Configurator of Scientific Activity," Svetoslav Slavkov's "Mathematics, Knowledge, Technological Advancement," Dobrin Spassov's "Philosophy, Language, Logic, Praxis," Boris Chendov's "On Formalized Systems in Scientific Knowledge," Bogdan Dyankov's "The Polysemantic Structure of the Natural Languages," Elena Panova's "Once Again: Is It Possible for Metaphysics to Be a Science?," Kiril D. Darkovski's "Bertrand Russell's Philosophical Development and the Crisis of Modern Positivism." Part II, Philosophy, Humanism, Creation, includes: Stephan Angelov's "The Revolutionary Humanism of Socialist Society," Vassil Momov's "Social Environment and Education," Nikolai Mizov's "Man, Religion, Social Progress," Atanas Stoykov's "On Some Aspects of the Relationships Between Art and Technics," Todor Stoychev's "Art and Mythology," Pancho Roussev's "Philosophical Anthropology?" Part III, Philosophy and Society, includes: Peter Velchev's "Philosophy and Revolution," Niko Yahiel and Yulian Minkov's "Scientific Activity and Social Progress," Marko Markov's "The Scientific and Technical Revolution and the Management of Big Social Systems," Todor Vulov's "The Interaction between Society and the Personality," Velichko Dobriyanov's "Technological Determinism and Technological Futurology—A Distorted Reflection of Technological Progress."

"The Historiography of Technology." *Technology and Culture* 15, no. 1 (Jan. 1974): 1–48. A symposium which includes: R. P. Multhauf's "Some Observations on the State of the History of Technology" (pp. 1–12), E. S. Ferguson's "Toward a Discipline of the History of Technology" (pp. 13–30), E. T. Layton, Jr.'s "Technology as Knowledge" (pp. 31–41), and D. S. Price's "On the Historiographic Revolution in the History of Technology: Commentary on the Papers by Multhauf, Ferguson, and Layton" (pp. 42–48). Considerable discussion of the proper definition of technology. Ferguson and Layton papers annotated separately under "Ethical-Political" and "Metaphysical-Epistemological" sections, respectively.

Hüber, Kurt, and Albert Menne, eds. *Natur und Geschichte* [Nature and History], X

Deutscher Kongress für Philosophie, Kiel, October 8-12, 1972. Hamburg: Meinen, 1973. Contains three papers of special note: Hans Lenk's "Zur neueren Ansätzen der Technik-Philosophie" (pp. 236–260) and one each by Siegfried Maser and Günter Ropohl listed under IV, Metaphysical and Epistemological Studies.

Huning, Alois. *Das Schaffen des Ingenieurs; Beiträge zu einer Philosophie der Technik* [The activity of engineers; contributions to a philosophy of technology]. Düsseldorf: VDI-Verlag, 1974. Pp. 203. A good overview and introduction to the philosophy of technology. Begins with a general discussion of the problems of the philosophy of technology and the ideas of Kapp, Zschimmer, Schröter, Dessauer, Tuchel, H. Sachsse, M. Scheler and P. Wust, Jaspers, Heidegger, Habermas and Marcuse, and Marxist philosophy of technology. The author then considers a number of specific problems: the definition of technology, theoretical and practical elements in technology, methods of technical work, the creativity of engineers, technology and economics, politics, and ethics. Good notes and brief, unannotated bibliography. By the chairman of a VDI study group concerned with the philosophy of technology. See also the author's *Ingenieurausbildung und Soziale Verantwortung* (Düsseldorf: 1974).

Jonas, Hans. *Philosophical Essays: From Ancient Creed to Technological Man.* Englewood Cliffs, N.J.: Prentice-Hall, 1974. Pp. xviii, 349. Includes "Technology and Responsibility: Reflections on the New Tasks of Ethics," "Seventeenth Century and After: The Meaning of the Scientific and Technological Revolution," and "Philosophical Reflections on Experimenting with Human Subjects." For more detail see review.

Lenk, Hans, and Simon Moser, eds. *Techne, Technik, Technologie; Philosophishe Perspektiven* [Techne, technique, technology; philosophical perspectives]. Pullach: Verlag Dokumentation, 1973. Pp. 247. Contents: Simon Moser's "Kritik der traditionellen Technikphilosophie" (pp.11–81), Hans Rumpf's "Gedanken zur Wissenschaftstheorie der Technikwissenschaften" (pp. 82–107), Friedrich Rapp's "Technik und Naturwissenschaften—eine methodologische Untersuchung" (pp. 108–132), Kurt Hüber's "Philosophische Fragen der Technik" (pp. 133–151), Günter Ropohl's "Prolegomena zu einem neuen Entwurf der allgemeinen Technologie" (pp. 152–172), Ulf Niederwemmer's "Programmatische Skizze für eine Sozialphilosophie der Technik" (pp. 173–197), Hans Lenk's "Zu neueren Ansatzen der Technikphilosophie" (pp. 198–231). Unannotated bibliography. Name index. Notes on contributors. Moser's essay which takes up one third of the volume and is a slightly revised version of his "Zur Metaphysik der Technik," from *Metaphysik einst jetzt* (Berlin: De Gruyter, 1958), is a critical examination of the theories of D. Brinkmann, J. Conant, C. Weisacker, F. Dessauer, H. Schmidt, M. Heidegger, Plato, and Aristotle. This essay has been translated into English as "Toward a Metaphysics of Technology," *Philosophy Today* 15, no. 2 (Summer 1971): 129–156. Rapp's essay is also available in English as "Technology and Natural Science—A Methodological Investigation," in F. Rapp, ed., *Contributions to a Philosophy of Technology* (Boston: D. Reidel, 1974), pp. 93–114. Rumpf's paper offers a detailed analysis of science and technology in terms of their subject matter, aims, methodology, and the structure of their propositions, as well as the common traits of both fields which transcend their differences. Ropohl's essay proposes a general theory of technology on the basis of an input-output scheme of matter, energy, and information.

Medawar, Peter. "What Is Human about Man in His Technology," *Smithsonian* 4, no. 2 (May 1973): 22–28. Using Karl Popper's theory of the three worlds as a framework, Medawar, a Nobel laureate in medicine and physiology, forecasts an optimistic outlook for technological man. World One, the physical world, evolves haphazardly in Darwinian fashion. World Two, the mental world, is reflected in our technological instruments (organ projections) and our knowledge about those instruments. The evolution of World

Two is distinguished from the physical world's random variations by the transfer of acquired characteristics and knowledge. World Three, the objective world of ideas themselves, is the greatest source for hope in man's evolution because it is not purely an autonomous world but partially a product of the human mind. Science and technology, when subordinated to this evolving third world, "offer the only means of escaping the misfortunes for which they are responsible" (p. 69). Condensed in *Intellectual Digest* 3, no. 12 (Aug. 1973): 68–69. Subsequently published as "Technology and Evolution," pp. 101–109, in *Technology and the Frontiers of Knowledge*. Garden City, N.Y.: Doubleday, 1975. See also the author's *The Hope of Progress* (London: Methuen, 1972), which argues that science can solve the problems it creates.

Pirsig, Robert. *Zen and the Art of Motorcycle Maintenance: An Inquiry into Values*. New York: William Morrow, 1974. Pp. 412. An unusually philosophical novel. The plot concerns a man's attempt to relate to himself before he underwent shock therapy and to his son who also seems headed for a mental breakdown. But the reason for this split between parts of himself and his son is, in the protagonist's mind, bound up with a cultural hiatus between the rational or technological and the romantic or aesthetic attitudes toward the world. The struggle to bring these together centers around a careful analysis of the man-machine relationship, particularly as manifested in the care and maintenance of a motorcycle. G. Basalla, reviewing the book in *Science* 187 (Jan. 24, 1975): 248–250, states: "Pirsig's philosophical insights into the man-machine relationship . . . might well serve as a preliminary guide for those who in charting this territory will aid man's attempt to redirect technology toward the realization of broadly based social goals."

Prior, Aldo. "La filosofía y el impulso tecnológico" [Philosophy and the technological impulse], pp. 248–450 in *II° Congreso Nacional de Filosofía, Actas*, Tomo I. Buenos Aires: Editorial Sudamericana, 1973.

Proceedings of the XVth World Congress of Philosophy; Varna, Bulgaria, September 17–22, 1973. Vol. I. *Philosophy and Science; Mortality and Culture; Technology and Man*. Sofia: Sofia Press Production Centre, 1973. Pp. 395. "The first two volumes of the Proceedings of the Congress belong to the materials which will be distributed to all duly registered participants. They include reports which have been submitted to the plenary sessions and colloquia." Contents of the third section of this volume include: J. A. Passmore's "Man and Technology" (pp. 247–250). R. Rixta's "Čelovek i texnika" (pp.251–260), V. G. Afanas'ev's "Nauka, texnika, upravlenie" (pp. 261–268), H. Beck's "Seinserfahrung und Gesellschaftsbildung in der technischen Herausforderung" (pp. 269–272), D. A. Crocker's "The Humanistic Value of a Science of Human Action" (pp. 273–280), V. Dobriyanov's "Technology in Society—An Attempt at a Definition" (pp. 281–284), I. T. Frolov's "Contemporary Science and Humanism" (pp. 285–292), K. Fuch-Kittowski, R. Tschirschwitz, and B. Wenzlaff's "Mensch und Automatisierung" (pp. 293–296), S. Ganovski's "The Philosophical Conception of Man" (pp. 297–300), A. Gedö's "Technik und Wissenschaft als 'Ideologie' " (pp. 301–304), R. T. DeGeorge's "Technology and Reason" (pp. 305–310), E. Hahn's "Technik und Subjektivität" (pp. 311–318), N. S. Illarionov's "Technological Revolution and Traditional Values in Modernizing Social Structures" (pp. 319–320), S. F. Kasprzyk's "On the Concept of Technology and its Relation to Science and Technic" (pp. 321–326), R. Kitchener's "Behavioral Technology or Science?" (327–330), G. Köhler's "Mensch und Technik im Wechselspiel der Kräfte" (pp. 331–336), A. Kosing's "Wissenschaftlich-technische Revolution und sozialistischer Humanismus" (pp. 337–342), M. M. Kuri's "Man and Technics" (pp. 343–346), M. B. Mitin's "Problema gumanizacii texniki i social'nyj prolress" (pp. 347–354), W. D. Nietmann's "Technology as a Fact of Life" (pp. 355–360), St. Panou's "Die Logik der Technik und der

Mensch" (pp. 361–364), H. L. Parson's "Technology and the Crisis of Dehumanization in the United States Today" (pp. 365–370), H. Sachsse's "Die Technik in der Sicht Herbert Marcuses und Martin Heideggers" (pp. 371–376), J. R. Sanabria's "L'homme et la technique" (pp. 377–382), W. Schirmacher's "Heideggers Radikalkritik der Technik als gesellschaftlicher Handlungsentwurf" (pp. 383–388), J. N. Theodoracopoulos' "L'homme et la technique" (pp. 389–392), Mourad Wahba's "The Twentieth-century Man-ism" (pp. 393–395). One other paper from section one—G. Van Riet's "Science et philosophie, technique et moralité" (pp. 123–128)—plus four papers from section two—L. N. Kogan's "Naučno-texničeskaja revoljucija i duxovnaja kul'tura" (pp. 179–184), F. V. Kanstantinov's "Naučno-texničeskaja revoljucija i problemy nravst-vennogo prolressa" (pp. 185–192), R. Routley's "Is There a Need for a New, an Environmental Ethic?" (pp. 205–210, and J. Wróblewski's "Technics, Instrumental Values and Ethical Relativism" (pp. 239–246)—are also relevant to the theme of technology.

Proceedings of the XVth World Congress of Philosophy; Varna, Bulgaria, September 17–22, 1973. Vol. II. *Reason and Action in the Transformation of the World; Philosophy in the Process of the Scientific and Technological Revolution; Knowledge and Values in the Scientific and Technological Era; Structure and Methods of Contemporary Scientific Knowledge*. Sofia: Sofia Press Production Centre, 1973. Pp. 392. Relevant contents include: Ivan Gobry's "La culture comme transformation du monde" (pp. 29–32), Fred Kohlsdorf's "Zur Rolle gesellschaftlicher Ideen und Theorien in unserer Zeit" (pp. 39–42), Assen Kojarov's "Forecasting and Transformation of the Contemporary World" (pp. 43–46), J. O. Cotten's "Note sur les contradictions actuelles de la conception bourgeoise de la connaissance objective" (pp. 79–81), Wolfgang Eichhorn's "Wissenschaftlich-technischer Fortschritt und philosophische Weltanschauung" (pp. 83–86), Wilhelm Ettelt's "Philosophische Weltperspektive und Technik" (pp. 87–90), Günter Kröber's "Die wissenschaftlich-technische Revolution und das Problem der Entwicklungsgesetzmässigkeiten der Wissenschaft" (pp. 91–94), Joseph Meurers' "Das philosophische und wissenschaftliche Denkbemühen im Fortschritt der Technik" (pp. 95–99), T. I. Oiserman's "Technophobie als Phänomenon des gesellschaftlichen Bewusstseins" (pp. 101–104), Giuseppe Prestipino's "La philosophie au sein de la revolution scientifique et technique" (pp. 105–108), Friedrich Rapp's "Zur Methodologie der Technik" (pp. 109–112), V. Stoljarow's "Weltanschauliche Auseinandersetzung im Wirkungsfeld der wissenschaftlich-technischen Revolution" (pp. 119–121), Toncho Trendafilov's "The Technical as a Form of Movement of the Social" (pp. 133–135), Gene G. James's "The Place of Value in a World of Reason" (pp. 219–221), Alexandre Tillmann's "Vers de nouvelles valeurs? Au delà de la technique" (pp. 261–264).

[Note: The quality of papers in these two volumes of *Proceedings* does not require that they be annotated separately; two further volumes are, however, scheduled for publication. For commentary on the Congress itself, see: Luce Fontaine-De Visscher, "Le XVe Congrès mondial de philosophie," *Revue Philosophique de Louvain* 71, 4th series no. 12 (November 1973): 765–768; J. R. Seibold, "Ciencia, tecnología y hombre, Reflexions sobre el XV Congreso Mundial de Filosofía," *Stromata* 30 (1974), 337–342; and Norris W. Clarke, "Reflections on the XVth World Congress of Philosophy and the 1st International Congress of Metaphysics," *International Philosophical Quarterly* 14, no. 1 (March 1974): 115–124. Archie Bahm's "The XVth World Congress of Philosophy: A Brief Report," *International Philosophical Quarterly* 13, no. 4 (December 1973): 553–554, gives a non-evaluative description.]

Pucciarelli, Eugenio. "Ciencia y filosofía en el mundo de la tecnica" [Science and philosophy in the world of technology], *Cuadernos de Filosofía* 12 (January–June):

225–242. Philosophy as knowledge presupposes contemplation, or a disinterested attitude toward reality, and is the activity of theoretical man. The technological age is the result of a pragmatic attitude, which measures knowledge by success, and is the work of *Homo faber*. By moulding a new world, technology brings about a change in the conception and practice of philosophy.

Rapp, Friedrich, ed. *Contributions to a Philosophy of Technology: Studies in the Structure of Thinking in the Technological Sciences*. Dordrecht, Holland, and Boston: D. Reidel, 1974. Pp. xv, 228. Reprints the epistemological papers from the symposium "Toward a Philosophy of Technology," *Technology and Culture* 7, no. 3 (Summer 1966), along with some more recent work by European thinkers and practicing engineers. Annotated bibliography. See review for more detail.

Schuurman, Ir. Egbert. *Techniek en Toekomst: Confrontatie met wijsgerige beschouwingen* [Technology and the future: a confrontation of philosophical views]. Assen: Van Gorcum, 1972. Pp. 568. A philosophical analysis of modern technology into tools (technological objects), execution (technological form-giving), and preparation (technological design), by a student of H. Van Riessen. Includes an examination of the theories of F. G. Jünger, M. Heidegger, J. Ellul, H. Meyer, N. Wiener, K. Steinbuch, and G. Klaus. Concludes by arguing for the need of a universal religious community as a basis for technology—this religious community understood from the point of view of Protestant Christian theology. English language summary pp. 399–425. Extensive unannotated bibliography, pp. 465–533. Name index.

———. "Over Anti-Techniek en Totale Technocratie" [Anti-technology and total technology], *Philosophia Reformata* 37 (1972): 156–173.

Technology and the Frontiers of Knowledge. The First Frank Nelson Doubleday Lectures, at the National Museum of History and Technology, Smithsonian Institution, 1972–1973. Garden City, N.Y.: Doubleday, 1975. Pp. 134. Contents: Foreword by Daniel J. Boorstin, Saul Bellow's "Literature in the Age of Technology" (pp. 3–22), Daniel Bell's "Technology, Nature, and Society" (pp. 25–70), Edmundo O'Gorman's "History, Technology, and the Pursuit of Happiness" (pp. 75–96), Sir Peter Medawar's "Technology and Evolution" (pp. 101–109), and Arthur C. Clarke's "Technology and the Limits of Knowledge" (pp. 113–133). Bellow, observing the decline of the authority of the imagination in an age of technology, argues that the novelist reacts either by trying to create a total world of his own or by surrender. In detailing the first alternative, he presents James Joyce's *Ulysses* as "*the* account of human life in an age of artifacts" (p. 11). O'Gorman's essay restates in particularly cogent form the view of technology of Ortega y Gasset. Clarke's piece is a typical praise of technology as producing new knowledge which scholars in their ignorance reject as impossible. The papers by Bell and Medawar are annotated separately under Ethical-Political Primary and Comprehensive, respectively.

Verein Deutscher Ingenieure, ed. *Mensch und Technik* [Man and Technology]. Düsseldorf: VDI-Verlag, 1963–1974. Between 1963 and 1974 the VDI study group on man and technology published eight volumes of proceedings covering discussions in the years 1961–1962, 1963–1964, 1965–1966, 1967–1968, 1969–1970, and 1971–1972. These volumes contain a number of papers by such thinkers as A. Huning, H. Rumpf, R. W. Goering, F. Rapp, P. Wittich, etc.

II. ETHICAL AND POLITICAL CRITIQUES

A. Primary Sources

Alibrandi, T. "Tecnica, natura e politica" [Technology, nature and politics], *Il Mulino* (Bologna) 21, no. 221 (May–June 1972): 407–423. Problems of opposition between technology and nature are able to be remedied by a politics of pluralism.

Albrecht, Ulrich. "Die Werturteilsfrage in der Technik" [The question of value judgment in technology], *Zeitschrift für Allgemein Wissenschafts Theorie* 1 (1970): 161–172.

Anton, John P. "The Paradox of Urban Aesthetics in Technological Society" (in Greek), *Annales D'Esthetique* 8 (1969): 148–161. The city as a product of "techne" is both the outcome and the matrix of art. As matrix or container of art it witnesses to an abundance of goods and efficiency, yet as work of art itself it appears defective in aesthetic form.

Augenstein, Leroy G. *Come, Let Us Play God*. New York: Harper and Row, 1969. Pp. ix, 150. A plea to the public to set up a "new decision-making apparatus" to deal with the growing problems in medical ethics. The book mainly has a question-and-answer format. Apparently intended to be a primer for initiating discussion and preparing the public for the awesome responsibility of making life-and-death decisions "that cannot be sidestepped." This book, however, may quite safely be sidestepped. Augenstein is chairman of the department of Biophysics at Michigan State University.

Axelos, Kostas. *Arguments d'une recherche* [Points of inquiry]. Paris: Éditions de Minuit, 1969. Pp. 216. Contains a number of articles on the theme of planetary man, including: "Vers une pensée planétaire" (pp. 168–170); "Art, technique et technologie planétaires" (pp. 173–176); "Qui est donc l'homme planétaire?" (pp. 181–186); "Schéma du jeu de l'homme et du jeu du monde; Vers la formation de l'homme planétaire; onze thèses de base" (pp. 200–204).

Barbour, Ian G., ed. *Western Man and Environmental Ethics; Attitudes Toward Nature and Technology*. Reading, Mass.: Addison-Wesley, 1973. Pp. 276. A well-focused anthology which includes the following sixteen articles: Lynn White's "The Historical Roots of our Ecologic Crisis," Lewis Moncrief's "The Cultural Basis of our Environmental Crisis," René Dubos's "A Theology of Earth," L. White's "Continuing the Conversation" (an original contribution for this volume), H. Paul Santmire's "Historical Dimensions of the American Crisis," Leo Marx's "Pastoral Ideals and City Troubles," Gabriel Fackre's "Ecology and Theology," Wendell Berry's "A Secular Pilgrimage," W. Murdoch and J. Connell's "All About Ecology," Ian McHarg's "The Place of Nature in the City of Man," N. J. Faramelli's "Ecological Responsibility and Economic Justice," James Carroll's "Participatory Technology," Paul Goodman's "Can Technology Be Humane?", "The Power to Destroy, the Power to Create" by Ecology Action East, J. Randers and D. Meadows, "The Carrying Capacity of Our Global Environment: A Look at the Ethical Alternatives." Although designed as a textbook, this work is listed here because of the editor's good analytic introduction and the presence of some valuable essays which are not otherwise readily available.

Bell, Daniel. *The Coming of Post-Industrial Society*. New York: Basic Books, 1973. Pp. 507. A comprehensive sociological analysis of the ascendency of a new elite—i.e., scientists, professionals, technocrats—based on the increase of knowledge and its manipulation by computers. This is related to other features of contemporary technological society: the transition from a goods-producing to a service-producing economy; divorce between ownership and managerial control of corporations; the disjunction between cultural trends and social structure (i.e., the so-called adversary or counterculture vs. the established economy); and the problem of equality vs. meritocracy, "the allocation of positions in a knowledge society." Widely reviewed. See particularly T. Tilton, with reply by Bell, *Social Research* 40, no. 4 (Winter 1973): 728–760; and Trent Schroyer, *Telos* no. 19 (Spring 1974): 162–176; Dennis Little, "Post-Industrial Society and What It May Mean," *Futurist* 7, no. 6 (Dec. 1973): 259–262; and Michael Marien, "Daniel Bell and the End of Normal Science," *ibid.*, pp. 262–268. Marien's review is followed by a useful bibliographical survey of the origins of the concept of "post-industrial" society.

———. "Technology, Nature and Society; The Vicissitudes of Three World Views and the Confusion of Realms," *American Scholar* 42, no. 3 (Summer 1973): 385–404. Defines nature as "the organic and inorganic realms of the earth that are changed by man" and "the order of things, which is discerned by man." "Technology is the instrumental ordering of human experience within a logic of efficient means, and the diremption of nature to use its powers for material gain." "Society is a moral order, defined by consciousness and purpose, and justified by its ability to satisfy men's needs, material and transcendental." With these distinctions Bell tries to distinguish in what ways modern man is different from ancient man. Subsequently published, pp. 25–70, in *Technology and the Frontiers of Knowledge* (Garden City, N.Y., Doubleday, 1975).

Beres, Louis René. "The Errors of Cosmopolis; World Order and the Vision of Human Oneness," *Philosophy Today* 18, no. 3 (Fall 1974): 234–247. A critique of the dangerous lack of realism in some utopian arguments for a unified world state, with reference to Richard A. Falk's *This Endangered Planet* (New York: Random, 1972), DeChardin, and the modern origins of some of these ideas in Kant, Goethe, Rousseau, etc.

Berger, Peter L. *Pyramids of Sacrifice: Political Ethics and Social Change.* New York: Basic Books, 1974. Pp. xiv, 242. The argument of this book is built around the image of the great pyramid at Cholula, Mexico, where in the course of history a number of different kinds of sacrifice have taken place. First, Olmec and Toltec priests directed the laborious construction of the pyramid. Next, Aztecs sacrificed thousands of victims to the god Quetzalcoatl. Later, Catholic priests forced people to build hundreds of churches on and around the pyramid. And today the descendants of these slaves and victims toil again under the supervision of national archaeologists to resurrect the great pyramid. The argument can be summed up by quoting from the twenty-five theses at the beginning of the text: 1) "The world today is divided into ideological camps. The adherents of each tell us with great assurance where we're at and what we should do about it. We should not believe any of them." 4) "Capitalist ideology, as based on the myth of growth, must be debunked." 8) "Socialist ideology, as based on the myth of revolution, must be debunked." 20) "Modernity exacts a high price on the level of meaning. Those who are unwilling to pay this price must be taken with utmost seriousness, and *not* be dismissed as 'backward' or 'irrational'." 25) "We need a new method to deal with questions of political ethics and social change (including those of development policy). This will require bringing together two attitudes that are usually separate—the attitudes of "hard-nosed" analysis and of the utopian imagination. What this book is finally all about is just this—some first steps toward a *hard-nosed utopianism.*"

Berger, Peter L., Brigitte Berger, and Hansfried Kellner. *The Homeless Mind; Modernization and Consciousness.* New York: Random, 1973. Pp. ix, 258. As the authors summarize this book in the preface: "Part I . . . attempts to isolate certain crucial elements of modern consciousness and relate these to the institutional processes with which they are linked. Part II analyzes the process of modernization, that is, the diffusion of modern consciousness in what is commonly called the Third World today. Part III takes up a number of phenomena in the advanced industrial societies that appear to be protests against modernity and that . . . we have called 'demodernization.' The conclusion briefly discusses the pragmatic and political implications of this argument."

Blackstone, William J. "Ethics and Ecology," *Southern Journal of Philosophy* 11, nos. 1–2 (Spring–Summer 1973): 55–71. Concerned with implications of ecology for ethics and a theory of rights. Analyzes concepts of equilibrium and homeostasis, and the idea of access to a livable environment as a moral, human, or legal right. Argues for caution in ethical application of ecological concepts. Includes an explicit note on the general way technology complicates ethics.

Boorstin, Daniel J. "Technology and Democracy," pp. 97–123, in Taylor Littleton, ed., *Our Secular Cathedrals: Change and Continuity in the University*. Univ. of Alabama Press, 1973. Develops some thoughts in relation to his *The Americans: The Democratic Experience* (1973). Discusses four consequences of "the use of technology to democratize our experience." One, "attenuation," describes the "thinning-out or the flattening of experience." For example, mass photography makes the real experience of special or foreign places duller. Two, the "decline of congregation" refers to how the centralization of all forms of production leads to the privitization of experience. No one meets "at the well" because of indoor running water; entertainment is more private because of TV. Three, "the rising sense of momentum" means the inability to stop certain processes due to their velocity, as with atomic research or space exploration. Four, the success of technological solutions to technical problems leads to an illusory "belief in solutions" for all human problems.

Brooks, Harvey. "Technology and Values: New Ethical Issues Raised by Technological Progress," *Zygon* 8, no. 1 (March 1973): 17–35. "Living with technology is like climbing a mountain along a knife-edge which narrows as it nears the summit. With each step we mount higher, but the precipices on either side are steeper and the valley floor farther below. As long as we can keep our footing, we approach our goal, but the risks of a misstep constantly mount. Furthermore, we cannot simply back up, or even cease to move forward. We are irrevocably committed to the peak" (pp. 34–35). And on into thin air? See also the author's "Technology and Zero Growth," *Daedalus* 102 (Fall 1973): 139–152.

Burhoe, Ralph W. "What Specifies the Values of the Man-made Man?" *Zygon* 6, no. 3 (September 1971): 224–246. Argues that technology is powerless to change the nature of things. "You cannot make water out of hydrogen and sulfur, you must use hydrogen and oxygen . . ." (p. 242). Ultimately men's errors are harmless.

———. "Evolving Cybernetic Machinery and Human Values," *Zygon* 7, no. 3 (September 1972): 188–209. Concludes that the danger to man from any kind of machine—whether virus, lion, atom bomb, or computer—is not as great as the failure of the machine in man's head to see the imperatives in cosmic nature for human values.

Burns, Chester R., and H. Tristram Engelhardt, Jr., guest eds. "The Humanities and Medicine," *Texas Reports on Biology and Medicine* 32 (Spring 1974). Contains thirty-two articles dealing with issues like amniocentesis, abortion, and human experimentation. Includes Engelhardt's "Explanatory Models in Medicine" which examines the definition of "normal" and an article on behavior control by Joe E. Tupin which examines the problem of free will.

Callahan, Daniel. "Living with the New Biology," *Center Magazine* 5, no. 4 (July–August 1972): 4–12.

———. *The Tyranny of Survival*. New York: Macmillan, 1973. Pp. 284. Using Freud's theory of culture as a framework, argues for the need to overcome the two tyrannies of individualism and survival, and to develop a social ethic for the limiting of technology. Important work.

———. "Bioethics as a Discipline," *Hastings Center Studies* 1, no. 1 (April 1973): 66–73.

———. "Science: Limits and Prohibitions," *Hastings Center Report* 3, no. 5 (November 1973): 5–7. Accepts the fact that man is a technological animal, but argues for the necessity of a science of limits based on a Freudian "reality principle." "This is essentially a task for the cultural superego, which could curb the desire for technological infinity and transcendence by scaling down the emotional and visionary demands made upon technology in the first place—by putting in their place a sense of radical finiteness . . ." (p. 6). Thoughtful article.

———. "Medicine's 'New' Ethics: A Challenge to Your Daily Decisions," *Prism* 2 (April 1974): 43–46.

Cassell, Eric J. "Dying in a Technological Society," *Hastings Center Studies* 2, no. 2 (May 1974): 31–36. Dying has become a technical rather than a moral problem, with resultant confusion between the mechanical events of a body ceasing to function and the passing of a person. Thus the need to restore a balance between the technical and the moral. See also the author's "Making and Escaping Moral Decisions," *Hastings Center Studies* 1, no. 2 (1973): 53–62, on why doctors have the right to make moral decisions.

Cruz Cruz, Juan. "Strukturelle Grundlagen des Kulturellen und des Technischen" [Fundamental structure of cultures and of technologies], *Philosophia Naturalis* 15 (1974): 1–14.

Dacal Alonso, Jose Antonio. "Marginacion y arte" [Margination and art], *Logos* (Mexico) 1 (May–August 1973): 219–237. The assignment of art to a marginal role in culture is unique to advanced civilizations, and one of the primary causes of this phenomenon derives from technical-scientific inventions.

Denninger, Erhard. "Die Herausforderung der Technik an das Recht in der technologischen Gesellschaft" [The challenge of technology and law in technological society], *Universitas* 25, no. 11 (1970): 1135–1158.

———. "Diritto e Societa tecnologica" [Law and technological society], *Rivista Internazionale di Filosofia del Diritto* 50 (October–December 1973): 651–674. An examination of problems of jurisprudence posed by rapid technological change. Includes a critical discussion of antitechnological judicial theory. Domination, security, and equality are held to be common to both the technological and political activities. Considers the adequacy of "technological planning law."

Easlea, Brian. "Who Needs the Liberation of Nature?" *Science Studies* 4, no. 1 (January 1974): 77–92. A critical review of William Leiss's *The Domination of Nature* (New York: Braziller, 1972). Leiss is critical of the counterideology of "the liberation of nature" as undermining scientific and technological progress and deflecting energy from a necessary socialist reform of political institutions. Easlea argues in reply "that a world in which *everything* is perceived in terms of use-value, as mere instrumentality, can never be a socialist one." And he calls for the socialist development of an image of man as one "who speaks on behalf of all those who have no voice" (Victor Serge), including nature whose only "chance to speak [is] through the minds of men" (Horkheimer).

Eulau, Heinz. "Skill Revolution and Consultative Commonwealth," *American Political Science Review* 67 (March 1973): 169–191.

———. "Technology and the Fear of the Politics of Civility," *Journal of Politics* 35, no. 2 (May 1973): 367–385. "Technologists and antitechnologists are rightly concerned with the social consequences of technology. Their fear of civil politics is also quite logical. . . . [The] fundamentals of civil politics are alien to technologists and antitechnologists alike as they seek to cope with the political environment. They are unwilling to concede that no social order can be a perfect order—a recognition that . . . is the mark of political maturity" (p. 385).

Faramelli, Norman J. "Computers and Modeling: Reflections on Possibilities, Limits, and Mythologies," *Soundings* (Vanderbilt Univ.) 55, no. 2 (Summer 1972): 178–199. Attempts an ethical assessment of the strengths and weaknesses of computer modeling and decision-making dealing with social problems.

Fekete, John. "McLuhanacy: Counterrevolution in Cultural Theory," *Telos*, no. 15 (Spring 1973): 75–123. A long, critical analysis of McLuhan from a New Left perspective. "In regard to McLuhan, we must insist that reified passivity is not necessary to technology, and that the genuinely new period in human history has not yet begun, but begins only with a qualitative break in the continuity of capital" (p. 123).

Feld, Bernard T. "Human Values and the Technology of Weapons," *Zygon* 8, no. 1 (March 1973): 48–58. General description of the problem of "military pollution" with some suggestions for bringing it under control through individual and professional group action.

Ferguson, Eugene S. "Toward a Discipline of the History of Technology," *Technology and Culture* 15, no. 1 (January 1974): 13–30. Opens with a description of the state of the history of technology; argues why this discipline can fill an academic need; and finally suggests a new organizing theme to help meet this need. The last part contains a strong argument for the nonneutrality of technology, with good notes to relevant historical literature.

Ferkiss, Victor. "Man's Tools and Man's Choices: The Confrontation of Technology and Political Science," *American Political Science Review* 67, no. 3 (September 1973): 973–980. A review of ten recent books.

———. *The Future of Technological Civilization*. New York: George Braziller, 1974. Pp. 369. A critique of classical liberalism coupled with the formulation of an alternative political theory called "ecological humanism." This new theory of "'technological man,' the creature who both creates and controls his technology" (p. 88) is "naturalistic in that it is rooted in the assumption that man is part of nature and his salvation lies in acting in accordance with this fact . . . holistic in that it is based on the realization that everything in man's world—the physical planet he lives on, the society he lives in, and himself—is closely interrelated in a single system . . . immanentist in recognizing that the reordering of human society and man's nature can never come from outside or 'above,' nor can it be blueprinted in advance; it can only grow out of whatever already exists" (p. 89). Attempts to draw out the consequences, both theoretical and practical, of these principles which were first stated in the author's earlier *Technological Man* (New York: Braziller, 1969). Large but unannotated bibliography. Favorably reviewed by R. Benne, *Zygon* 9, no. 4 (December 1974): pp. 352–356; and S. Carpenter, *Technology and Culture* 16, no. 4 (October 1975): 696–699.

———. "Ecological Humanism and Planetary Society," *Humanist* 34, no. 3 (May–June 1974): 24–36. Chapter 19 in the author's *The Future of Technological Civilization* (New York: Braziller, 1974).

———. "Political Philosophy and the Facts of Life," *Zygon* 9, no. 4 (December 1974): 272–287. Condensed version of chapter 7 of the author's *The Future of Technological Civilization* (New York, Braziller, 1974).

Fernández-Lomana Del Río, Ramón. "Dos principios para una axiología de la técnica" [Two principles of an axiology of technology], *Anuario Filosófico* 5 (1972): 155–169.

Field, G. Lowell. "The Impending Agenda for Political Philosophy," *Ethics* 83, no. 4 (July 1973): 322–326. Argues that technology has not solved the perennial questions of political philosophy. Our growing awareness of the failures of technology leads to the conclusion that we have not solved the "ancient dilemmas."

Fletcher, Joseph. *The Ethics of Genetic Control; Ending Reproductive Roulette*. Garden City, N.Y.: Doubleday, 1974. Pp. xxi, 218. The Anglican theologian of *Situation Ethics* (1966) extends the argument of that book into the field of genetic control. "Jesus talked about bringing a more abundant life—a life of quality, not mere life. Raw nature is often painfully indifferent to human values. It is unmoral" (p. 131). Fletcher persuasively attacks the traditional objections to genetic control. But his vision does not go beyond the abolition of disease and suffering. His discussion of health in any positive sense (other than as the absence of disease) is limited to a few superficial statements about the "quality of life." Fletcher has no conception of philosophy as anything other than calculation.

———. "Indicators of Humanhood: A Tentative Profile of Man," *Hastings Center Report* 2, no. 5 (Nov. 1972): 1–4. See also Fletcher's update of this article in *Ibid.* 4, no.6 (December 1974): 4–7.

Gabor, Dennis. *The Proper Priorities of Science and Technology.* Southampton: Univ. of Southampton, 1972. Pp. 16. The eighteenth Fawley Foundation lecture. "I have tried to sketch out an idealistic programme for science and technology: work towards a stable equilibrium state through the intermediate steps of pollution elimination and waste avoidance" (p. 15). Pessimistic about the ability of democracy "to respond to what appears as a small or remote danger" (p. 16).

George, François. "Forgetting Lenin," *Telos* no. 18 (Winter 1973–1974): 53–88. An extended critique of Lenin's worship of mechanization and efficiency. First two sections originally published in *Les Temps Modernes* (April 1973): 1735–1772; third section previously unpublished.

Geymorat, L. "Neutrality Is Impossible," *Scientia* (Milan) 107, nos. 9–10 (1972): 759–763. On the basis of an analysis of science as a part of culture which therefore affects society, argues against the supposed neutrality of sciences and technology. By a professor of the philosophy of science at the University of Milan.

Glaser, Wilhelm R. *Soziales und instrumentales Handeln; Probleme der Technologie bei Arnold Gehlen und Jürgen Habermas* [Social and instrumental actions; problems of technology according to Arnold Gehlen and Jürgen Habermas]. Stuttgart: W. Kohlhammer, 1972. Pp. 214. Brief bibliography.

Golding, Martin. "Obligations to Future Generations," *Monist* 56 (January 1972): 85–99.

Grant, George. "Tradition and Revolution," *Canadian Forum* 50, no. 591–592 (April–May 1970): 88–93. Also published as "Revolution and Tradition," in Lionel Rubinoff, ed., *Tradition and Revolution* (New York: St. Martin's Press, 1971): 81–95. "It is clear to anybody who lives in advanced technical society that initiation of the new and the different is now subordinated to the regular occurrence of the unvarying same. The overcoming of chance by technique may have made us oblivious to the true eternity; but it more and more opens to us an immanent sempiternity of the same. . . . These words about technique are simply to illustrate that in the word revolution, the prefix 're' remains although we would like to lose it" (p. 85).

Gustafson, James M. "Basic Ethical Issues in the Biomedical Fields," *Soundings* (Vanderbilt Univ.) 53, no. 2 (Summer 1970): 151–180. Discusses nine pertinent issues in the field. "I shall develop these issues and propositions in such a way that some of the reasons for contrasting judgments become clear. In some instances, the reasonable position appears to be a dialectical one between two *prima facie* opposing propositions" (p. 153). Subsequently included in the author's *Theology and Christian Ethics* (Philadelphia: United Church Press, 1974), along with his "Genetic Engineering and the Normative View of the Human," which first appeared in Preston Williams, ed., *Ethical Issues in Biology and Medicine* (Cambridge, Mass.: Schenkman, 1973).

Hamilton, Michael, ed. *The New Genetics and the Future of Man.* Grand Rapids, Mich.: Wm. B. Eerdmans, 1972. Pp. 242. Three "state of the field" papers followed by response. The first is Leon Kass (biologist), "New Beginnings in Life," with response by F. Grad (lawyer), J. Fletcher (theologian), and D. Callahan (philosopher). The second is W. French Anderson (research biochemist), "Genetic Therapy," with response by A.

Motulsky (clinical geneticist), A. Capron (lawyer), and P. Ramsey (theologian). The third is Samuel Epstein (research chemist), "Pollution and Health," with response by J. Johnson (industrial chemist), L. Bickwit, Jr. (Senate counsel), and C. Powers (theologian). In a favorable review of this book in *Theological Studies* 34, no. 4 (December 1973): 747–749, Robert Springer, S.J., criticizes Kass's implicit longing for an older theology. "Such a theology is no longer viable."

Hardin, Garrett. *Exploring New Ethics for Survival; The Voyage of the Spaceship Beagle.* New York: Viking Press, 1972. Reprint Baltimore: Penguin, 1973. Pp. xii, 273. Serious attempt by a microbiologist and geneticist to develop an ecological "lifeboat" ethics. Sees overpopulation as the primary problem. Reprints the author's original "The Tragedy of the Commons" essay from *Science* (December 13, 1968), of which this is an elaboration. But cf. also the author's *Nature and Man's Fate* (New York: Rinehart, 1959), and *Mandatory Motherhood: The True Meaning of "Right to Life"* (Boston: Beacon Press, 1974).

———. "Living on a Lifeboat," *BioScience* vol. 24, no. 10 (October 1974). For a negative criticism of Hardin's basic position see Daniel Callahan, "Doing Well by Doing Good: Garrett Hardin's 'Lifeboat Ethic'," *Hastings Center Report* 4, no. 6 (Dec. 1974): 1–4.

Harrison, Frank R., III. "What Kind of Beings Can Have Rights?" *Philosophy Forum* 12, nos. 1–2 (September 1972): 113–129. An argument, in response to Joel Feinberg's "The Rights of Animals and Future Generations" (paper delivered to the 4th Annual Conference in Philosophy, Univ. of Georgia, Feb. 18–20, 1971), to the effect that robots can have rights. See also Harrison's "The Pains of R-George, Robot," *Southern Journal of Philosophy* 9, no. 4 (Winter 1971).

Heilbroner, Robert L. "Growth and Survival," *Foreign Affairs* 51, no. 1 (October 1972): 139–153. A critique of the "anti-growth school" (see "A Blueprint for Survival," *Ecologist* [January 1972]; and D. H. Meadows, *et al., The Limits to Growth* [New York: Universe Books, 1972]). Heilbroner is more optimistic about our technical capacities to respond to the ecological crisis than this "school," but is more pessimistic than the "school" about our social capacities to adapt to the crisis.

———. *An Inquiry into the Human Prospect.* New York: W. W. Norton, 1974. Pp. 150. Considers man's greatest external threats to be the population explosion, nuclear war, and environmental pollution. The consequence of the population explosion in the underdeveloped countries and the ensuing chaos will be "the eventual rise of 'iron' governments, probably of a military-socialist class" (p. 39). In the hands of such a desperate government, nuclear weapons could quite conceivably become a means to blackmail the wealthier nations into a massive transfer of goods. As regards the environment, Heilbroner sees the emission of heat due to energy reactions as the main long-range problem. "The limit on industrial growth therefore depends in the end on the tolerance of the ecosphere for the absorption of heat" (p. 50). He is skeptical about the capacity of capitalism and even socialism to respond adequately to these problems. "Each has been marked with serious operational difficulties; each has overcome these difficulties with economic growth" (p. 75). But it is the limit of growth that ultimately confronts us. Likewise, the limit of "human nature" confronts us. How much manipulation and social transformation can man endure in overcoming these threats? The problems with establishing birth-control programs in Third World countries is a case in point. Concludes that there is no hope without the "payment of a fearful price" (p. 136). Selections first published in *The New York Review of Books* 20, nos. 21–22 (January 24, 1974): 21–34. For review see "The Human Prospect: Heilbroner's Challenge to Religion and Science," *Zygon* 10, no. 3 (September 1975), which includes articles by L. Gilkey, D. Campbell, E. Dunn, J. Caggiano, V. Ferkiss, R. Burhoe, and S. Katz.

Hermann, Rolf-Dieter. "Art, Technology, and Nietzsche," *Journal of Aesthetics and Art*

Criticism 32, no. 1 (Fall 1973): 95–102. Analyzes contemporary art and technology projects in light of Nietzsche's notions of "nihilism" and "the eternal return of the same."

Hilton, B., D. Callahan, M. Harris, P. Condliffe, B. Berkeley, eds. *Ethical Issues in Human Genetics*. New York: Plenum Press, 1973. Pp. xi, 455. A series of papers and follow-up discussions. See especially D. Callahan's "The Meaning and Significance of Genetic Disease: Philosophical Perspectives," J. Gustafson's "Genetic Counseling and the Uses of Genetic Knowledge—An Ethical Overview," P. Ramsey's "Screening: An Ethicist's View," L. Kass's "Implications of Prenatal Diagnosis for the Human Right to Life," C. Fried's "The Need for a Philosophical Anthropology."

Illich, Ivan. *Tools for Conviviality*. New York: Harper & Row, 1973. Pp. xiii, 135. A critique of advanced industrial society which argues that in its conquest of nature, its bureaucratic structure, its educational system, its social mobility, and its rate of technological change, this society has become unbalanced. "There are two ranges in the growth of tools: the range within which machines are used to extend human capability and the range in which they are used to contract, eliminate, or replace human functions. In the first, man . . . can exercise authority on his own behalf and therefore assume responsibility. In the second, the machine takes over . . ." (pp. 91–92). Uses *convivial* "as a technical term to designate a modern society of responsibly limited tools."

Kass, Leon. "Making Babies—the New Biology and the 'Old' Morality," *The Public Interest* no. 26 (Winter 1972): 18–56. Argues that *in vitro* fertilization and cloning are dehumanizing. "Let us simply look at what we have done in our conquest of nonhuman nature. We shall find there no grounds for optimism as we now consider offers to turn our technology loose on human nature" (p. 55).

Keller, J. A. "Types of Motives for Ecological Concern," *Zygon* 6, no. 3 (1971): 197–209. The three motives: 1) crass self-interest, 2) enlightened self-interest, and 3) altruism. The third motive is the only one which admits that nature has an intrinsic value apart from man. Argues that there is a contemporary need to restore this ancient attitude toward nature. Provocative essay.

Kreilkamp, Karl. "Towards a Theory of Science Policy," *Science Studies* 3, no. 1 (January 1973): 3–29. A theoretically oriented assessment of science policy studies by reviewing Harvey Brooks, *The Government of Science* (Cambridge, Mass.: MIT Press, 1968); Dan K. Price, *Government and Science* (New York: Oxford Univ. Press, 1962); Dan K. Price, *The Scientific Estate* (Cambridge, Mass.: Harvard Univ. Press, 1965); Jerome R. Ravetz, *Scientific Knowledge and Its Social Problems* (Oxford: Clarendon Press, 1971); Edward Shils, ed., *Criteria for Scientific Development: Public Policy and National Goals* (Cambridge, Mass.: MIT Press, 1968).

Kwee, Swan Liat. *De mens tussen mythe en machine* [Man between myth and machine]. Amsterdam: Wetenschappelijke Vitgererij, 1974. Pp. 240. By a professor of philosophy at the Technological Univ. of Eindhoven, The Netherlands. He is also the co-editor of *Growing Against Ourselves: The Energy-Environment Tangle* (Lexington, Mass.: D. C. Heath, 1972), in which he has authored a paper entitled "Evaluation and Outlook" (pp. 207–220).

Lapati, Americo D. "Skinner and the Nature of Man," *New Scholasticism* 47, no. 4 (Autumn 1973): 501–515. Skinner's behavior modification can be accepted as *a* method, but because of his limited view of the nature of man, not as *the* method for improving learning processes or treating mental patients.

Lasch, Christopher. "Birth, Death, and Technology: The Limits of Cultural Laissez-faire," pp. 294–307 in the author's *The World of Nations: Reflections on American History, Politics, and Culture*. (New York: Knopf, 1973). First appeared in *Hastings Center Report* 2 (June 1972): 1–4.

Laszlo, Ervin. *A Strategy for the Future; The Systems Approach to World Order.* New York: Braziller, 1974. Pp. ix, 238. Part one conceptualizes a systems view of world order; part two tries to outline a strategy for producing this desired order. R. R. Rossini, in a review in *Technology and Culture* 16, no. 4 (October 1975): 699–700, says that "The strategy strikes me as a long exercise in optimism." An appendix summarizes the basic ideas of general systems theory.

Lenk, Hans. "Technocracy and Scientism? Remarks Concerning an Ideological Discussion," *Man and World* 5, no. 3 (August 1972): 253–271. "I want to discuss how, and how far, the increasing influence of scientific and technological developments on society and government leads to a political system in which democracy is losing 'its substance,' insofar as esoteric decisions of experts cannot any longer be controlled by the public and cannot be legitimated in a representative democratic manner within the 'free interplay of political powers'" (p. 253). Distinguishes a variety of issues associated with technocracy, suggesting various analytic and practical ways philosophers can contribute to the reduction of "technocratic tendencies." Followed by a discussion of scientism or "the epistemological correlate of technocracy" and its relation to the problems of technocracy. Conclusion: "On the one hand, problems of the future will *only* be solved by means of an extended application of scientific knowledge and methods of technology, that is, *only* with the help of experts and scientists. But on the other hand, those problems and the ways to cope with them will *only* be prepared for solution and decision with the help of philosophical and cultural scientific universalists as well. That is, *only* by means of comprehensive, integrated, systematic planning; but *only* by *political* decisions between genuine alternatives will it be possible to solve them in a relatively optimal way" (pp. 269–*270*).

———. ed. *Technokratie als Ideologie; sozialphilosophische Beitrage zu einem politischen Dilemma* [Technocracy as ideology; social philosophical contributions to a political problem]. Stuttgart: W. Kohlhammer, 1973. Pp. 238. Fourteen papers by moderately progressive social philosophers arguing against any flat rejection of "technocracy." Although technocracy tends to be a pejorative term, what is really needed are studies of the extent to which basic technological structures are unavoidable in the administration of an advanced industrial society. One paper, for instance, advocates the need for "technological enlightenment," arguing that modern society is amazingly uninformed concerning technology in spite of educational attempts by engineers. Also considers the problem of cooperation between engineers and social scientists and suggests the establishment of an institute for society and technology. Contents include: Hans Lenk's "'Technokratie' als gesellschaftskritisches Klischee" (pp. 9–20), Ulf Niederwemmer's "Versuch einer historisch-systematischen Ortsbestimmung des Technokratiegedankens (der Saint-Simonismus)" (pp. 21–44), Günter Ropohl's "Zur Technokratie-Diskussion in der Bundes-Republik Deutschland" (pp. 58–76), Günter Gebauer's "Der Mythos der Technokratie und seine Realität" (pp. 77–93), Hermann Lübbe's "Bemerkungen zur aktuellen Technokratie Diskussion" (pp. 94–104), Hans Lenk's "Technokratie und Technologie; Notizen zu einer ideologischen Diskussion" (pp. 105–124), Simon Moser's "Technologie und Technokratie; zur Wissenschaftstheorie der Technikwissenschaften" (pp. 125–136), Hans Lenk's "Technokratie und Szientismus" (pp. 154–172), Simon Moser's "Hochschul- und Wissenschaftsplanung zwischen Technokratie und Ideologie" (pp. 173–192), and Peter-Jörg Jansen's "Systemtechnik und Technokratie-Vorwurf" (pp. 215–222). Lenk's second paper is available in English translation as cited immediately above.

Lindsay, R. B. "The Scientific and Technological Revolutions and Their Implications for Society," *Zygon* 7, no. 4 (December 1972): 212–243. Includes discussion of the nature of science, the evolution of technology, and a criticism of K. Popper's thesis concerning the impossibility of predicting the future. Stresses the need for continual development in science and technology.

Long, H. "The Paradox of Technocracy: Mechanism, Wholeness, and Freedom Reconsidered," *Cross Currents* 23, no. 1 (September 1973): 92–106. Mechanistic thinking emphasizes external and deterministic factors. Absolutist thinking emphasizes internal factors and absolute subjectivity. What is needed, as "an alternative to the polarization of technocrats and liberationists" (p. 106), is a dynamic thinking which emphasizes the interdependence of external and internal factors.

Luria, S. E. "Slippery When Wet; Being an Essay on Science, Technology, and Responsibility," *Proceedings of the American Philosophical Society* 116 (October 1972): 351–356. The Penrose Memorial Lecture.

McCormick, Richard A. "To Save or Let Die: The Dilemma of Modern Medicine," *Journal of the American Medical Association* 229 (July 8, 1974): 172–176. Once a deformed child has been born should we try to save it? McCormick sets up the guideline of potential for human relationships. If there is hope that the infant will "be able to experience our caring and love," he argues for maximum care. Otherwise there is reason to refrain from treatment.

McGinn, Thomas. "Ecology and Ethics," *International Philosophical Quarterly* 14, no. 2 (June 1974): 149–160. Argues that "developed countries have an obligation to restore a disturbed balance of nature," but that "economic and technological development looked at on a world scale raises the different problem of distributive justice." That is, it is "against distributive justice for the West to try to halt economic development when the exploitation of nature has benefited the West exclusively."

Mailer, Norman. *Of a Fire on the Moon.* Boston: Little, Brown, 1970. Pp. vii, 472. Serious attempt by a novelist to discover the philosophical meaning of the American landing on the moon. For review see Thomas Werge, "An Apocalyptic Voyage: God, Satan, and the American Tradition in Norman Mailer's *Of a Fire on the Moon,*" *Review of Politics* 34, no. 4 (October 1972): 108–128.

Mayz Vallenilla, Ernesto. "Hombre y tecnica" [Man and technology], *Cuadernos de Filosofia* 12 (1972): 199–203.

Meyer, Hermann Josef. "El tecnologismo; análisis de una actitud espiritual" [Technologism; analysis of a spiritual attitude], *Folia Humanistica* 11 (1973): 57–74.

Mitcham, Carl. "Calder on Democracy and Technology," *International Philosophical Quarterly* 13, no. 2 (June 1973): 277–286. A critical essay-review of Nigel Calder's *Technopolis* (New York: Simon and Schuster, 1971).

"Modern Civilization." *Philosophy Forum* 12, nos. 1–2 (September 1972); 12, nos. 3–4 (December 1972); 13, nos. 1–2 (1973); 13, nos. 3–4 (1973). These four issues of *Philosophy Forum*, ed. by Ervin Laszlo (SUNY at Geneseo) and published by Gordon & Breach, on the common theme of "Modern Civilization: Achievement or Blunder?" contain the following articles: H. J. Muller's "Modern Civilization and Human Survival" (vol. 12, pp. 1–27), Alfred McClung Lee's "Modern Civilization and Human Survival: A Social-Scientific View" (vol. 12, pp. 29–66), Rubin Gotesky's "Modern Civilization: Its Demise" (vol. 12, pp. 66–111), F. R. Harrison III's "What Kind of Beings Can Have Rights?" (vol. 12, pp. 113–129), Kuang-Ming Wu's "Hope and World Survival" (vol. 12, pp. 131–147), Hans Sachsse's "Modern Civilization and Scientific Knowledge" (vol. 12, pp. 217–243), Lancelot Hogben's "Modern Civilization and Scientific Knowledge" (vol. 12, pp. 245–271), A. C. Genova's "Modern Civilization and Scientific Knowledge" (vol. 12, pp. 273–299), H. M. Kallen's "Some Reflections on Rhetoric and Newness" (vol. 12, pp. 301–313), C. Wilson's "Civilization and Individual Fulfilment" (vol. 13, pp. 1–27), A. Etzioni's "Social Science Perspectives on the Dynamics of our Crisis" (vol. 13, pp. 29–57), W. Nietmann's "The Civilized Individual" (vol. 13, pp. 59–84), W. R. Brown's "Tribal Morality and Civilization" (vol. 13, pp. 85–94), W. J. MacKinnon's "An Evolutionary Inquiry into Future Human Systems" (vol. 13, pp. 95–109), R. W. Burhoe's "The Civilization of the Future: Ideals and Possibility" (vol. 13, pp. 149–177), D. Riepe's "The Civilization of the Future: Ideals and Possibility" (vol. 13, pp. 179–208), W. Earle's "The Future of Civilization?" (vol. 13, pp. 209–231), R. Sheehan's "Self-Actualizing Persons and the Ideal Society" (vol. 13, pp. 233–247).

Morison, Robert S. "Death: Process or Event?" *Science* 173, no. 3998 (August 20, 1971): 694–698. See also Leon Kass's "Death as an Event: A Commentary on Robert Morison," *Ibid.*, pp. 698–702.

Pagello, E., and G. Pasqualotto. "Artificial Intelligence: A New Ideology or a New Technology?" *Scientia* (Milan) 108, nos. 9–12 (1973): 711–721. "We shall make some sociological remarks about A.I. by relating it to the philosophy of mind by Descartes. We shall clarify that A.I. not only can propose new scientific paradigms to society and more efficient machinery to factories but can represent an effort of general control over society by the ruling classes." Also in Italian, pp. 697–710.

París, Carlos. *Mundo técnico y existencia auténtica* [The technological world and authentic existence]. Madrid: Revista de Occidente, 1973. Pp. 199.

Passmore, John. *Man's Responsibility for Nature: Ecological Problems and Western Traditions*. New York: Scribner's, 1974. Pp. x, 213. Correlates a historical examination of the roots of Western attitudes toward nature with methodological analysis of current attempts to find a precise solution and an ethic for ecological crises. "And all the while I have had in mind my overarching intention: to consider whether the solution of ecological problems demands a moral or metaphysical revolution" (p. x). "To sum up, I find it impossible to sum up, to arrive at any neat, tidy, quotable conclusions. As I have written, I have become more and more conscious of the complexity of every problem I have touched upon.... What in general I have emphasized is that, if the world's ecological problems are to be solved at all, it can only be by that old-fashioned procedure, thoughtful action.... The modern West, I have argued, leaves more options open than most other societies.... Admittedly, its central Stoic-Christian traditions are not favorable to the solution of its ecological problems.... But they, I have sought to show, are not the only Western traditions and their influence is steadily declining" (pp. 194–195). Favorable review by F. S. Strath, *Technology and Culture* 16, no. 4 (October 1975): 685–687.

Potter, Van Rensselaer. *Bioethics: Bridge to the Future*. Englewood Cliffs, N.J.: Prentice-Hall, 1971. Pp. 205. The concept of bioethics employed in this book is not that . . . now strongly associated with the term, bioethics as ethics applied to problems of biomedical engineering. Instead, for Potter, bioethics is a synthesis of ethical values and biological facts. "We are in great need of a Land Ethic, a Wildlife Ethic, a Population Ethic, a Consumption Ethic, an Urban Ethic, an International Ethic, a Geriatric Ethic, and so on. All of these problems call for actions that are based on values *and* biological facts. All of them involve Bioethics, and survival of the total ecosystem is the test of the value systems" (pp. vii–viii). The main themes of this volume, which includes nine previously published papers, are "(a) the relation between order and disorder, (b) the concept of dangerous knowledge, (c) human progress and human survival, (d) the obligation to the future, (e) the control of technology, and (f) the need for interdisciplinary effort" (p. viii). By a professor of oncology and well-known cancer researcher. Favorably reviewed by H. Hoagland in *Zygon* 6, no. 3 (September 1971): 251–254.

———. "The Ethics of Nature and Nurture," *Zygon* 8, no. 1 (March 1973): 36–47. Accepts Callahan's argument that what we need is a culture (i.e., a "nurturing context") which has its roots in nature and grows according to what it has learned from nature. Distinguishes between a constructive cultural pluralism and a destructive cultural laissez faire.

Prini, Pietro. "Idee per una filosofia ecologica" [Ideas for a philosophical ecology], *Proteus* 3, no. 7 (1972): 3-10.

Pursell, Carroll. "Belling the Cat: A Critique of Technology Assessment," *Lex et Scientia: The International Journal of Law and Science* 10, no. 4 (October-December 1974): 130-145. This first Mellon Lecture is followed by two other papers by Pursell on the same theme: "'A Savage Struck by Lightning': The Idea of a Research Moratorium, 1927-37" (pp. 146-161); and "'Who to Ask Besides the Barber'—Suggestions for Alternate Assessments" (pp. 162-177). There is also a brief index for all three papers, p. 178. The first lecture gives a good history of the development of the idea of technology assessment (TA) in the United States, along with a critique of its limitations. The critique cites the four assumptions of TA as formulated by the National Academy of Sciences: 1) technological change will continue, 2) on balance the benefits far outweigh the injuries of technological advance, 3) the problem is not technology itself but how to "repair the deficiencies in the process and institutions by which our society puts the tools of science and technology to work," and 4) TA must "foster a climate that elicits the cooperation of business." Then argues that only the third of these assumptions is correct, but that technology is as much a "process and institution" as anything else. The second lecture is the case history of a debate a generation ago in which the apprehensions of humanists were countered by the defensiveness of technicians, resulting in a victory (by default) for the status quo. The third lecture suggests ways to expand the meaning and practice of TA.

Ramsey, Paul. "Death's Pedagogy," *Commonweal* (Sept. 20, 1974): 497-502. Contrasts two recent books on death—Marya Mannes's *Last Rights* (1974) and Stewart Alsop's *Stay of Execution* (1973).

Rapp, Friedrich. "Über die soziale Verantwortung des Ingenieurs" [On the social responsibility of engineers], *Humanismus und Technik* 16 (1972): 131-137.

Riessen, I. H. Van. *De maatschappij der toekomst* [The society of the future]. 5th revised edition, Franeker: T. Wever, 1973, Pp. iv, 374. First published in 1952, this volume was translated as *The Society of the Future* (Philadelphia: Presbyterian Reformed Pub. Co., 1957). Chapter 5 of the translation,"Science and Technique in Modern Society," contains the best available discussion in English of Van Riessen's theories of the nature of science & technology and their interrelationships.

Roatti, M. "Etica ed estetica dell'ecologia: aspettando il dies irae" [Ethics and aesthetics of ecology: waiting for the days of wrath], *Il Mulino* (Bologna) 21, no. 222 (1972): 710-736.

Sachsse, H. *Technik und Verantwortung; Probleme der Ethik im technischen Zeitalter* [Technology and responsibility; problems of ethics in the technical age]. Freiburg: Rombach, 1972. Pp. 156.

―――. "Ethische Probleme des technischen Fortschritts" [Ethical problems of technical development], *Chemie-Ingeneure-Technik* 44, no. 1 (1972): A23-A27.

―――. "Modern Civilization and Scientific Knowledge," *Philosophy Forum* 12, nos. 3-4 (December 1972): 217-243. Since nature has been conquered man must turn his concern toward himself in an intense and communal effort to realize the meaning of life.

―――. *Die Technik in der Sicht des Marxismus* [Technology from the perspective of Marxism]. Berlin: Siemens-Aktiengesellschaft, 1974. Pp. 52.

Sagasti, Francisco R. "Underdevelopment, Science and Technology: The Point of View of the Underdeveloped Countries," *Science Studies* 3, no. 1 (January 1973): 47-59. "The central thesis is that science and technology were closely related to the emergence of underdevelopment as it has so far arisen in the twentieth century, and that to a certain extent they are contributing to the maintenance and persistence of underdevelopment." A theoretical and practical analysis with especially good references to Latin American literature.

Schoenmakers, C. L. J. *Techniek en ethiek; Een bezinning op de verantwordelijkheid in de natuurwetenschappelijke techniek* [Technology and ethics; a consciousness of responsibility under natural scientific technology]. Antwerpen/Utrecht: Uitgeverij De Nederlandsche Boekhandel, 1971. Pp. 248. Brief bibliography. Summary in English, pp. 241–46. Name index. Chapter 1 argues that technology is not morally neutral; on balance it is good. Chapter 2 considers the relationship between technology and society, while chapter 3 is about harmful by-products. Chapter 4 is on the function of experiments in science and technology. Chapter 5 deals with the responsibility of scientists and engineers. Chapter 6 is on war and armaments. Chapter 7 is concerned with revolution and utopia, chapter 8 with technology and democracy.

———. *Samenleving en techniek* [Society and technology]. Antwerpen: De Nederlandsche Boekhandel, 1973. Pp. 163. A historico-philosophical study.

Schroyer, Trent. "Marx and Habermas," *Continuum* 8, no. 1 (Spring–Summer 1970): 52–64. "Habermas's systematic elaboration of the techne-praxis distinction results in a new framework for critical theory. . . . In this theory resides both the unique contribution of Habermas to critical theory and, perhaps, his sublation of Marx and Marcuse" (p. 53).

———. "The Critical Theory of Late Capitalism," in George Fischer, ed., *The Revival of American Socialism* (New York: Oxford Univ. Press, 1971): 297–321. Argues that the belief in the "value neutrality of the new policy sciences" conceals the reality of current power structures. Schroyer claims the Frankfurt School is the only school of critical scientists in the West. "It is our thesis that the scientific image of science is the fundamental false consciousness of our epoch. If the technocratic ideology is to loose its hold on our consciousness, a critical theory of science must lay bare the theoretical reifications of the scientistic image of science" (p. 301).

———. *The Critique of Domination; The Origins and Development of Critical Theory*. New York: Braziller, 1973. Pp. 282. Schroyer joins the critique (by Husserl, Heidegger, Habermas) of the "modern identification of knowledge and/or reason with the products of strict science." This critique "points to a growing cultural regression where social values and norms are replaced by technical rules which mystify the social world (e.g., I.Q. scores which reify the process of education). Interaction of persons is determined more by technical rules and less by the spontaneity of human subjectivity." What appears as a cultural crisis is "more seriously a crisis of human subjectivity itself" (pp. 28–29). Part One begins with examples of ancient attempts to deal with the problem of alienation, followed by a discussion of Marx's theory of the alienation of work. Part Two "attempts to build a nonscientific philosophy of science," originating in Kant's method of transcendental reflection. Part Three seeks to reconcile the crisis theories of Marx and contemporary cultural Marxists. Ends with "a first approximation of the application of a unified critical theory to the dynamics of American capitalism" (p. 173). Chapter one originally appeared in a slightly different form in *Abraxis* 1, no. 2 (Winter 1970).

Schuberth, Albert. *Zur Soziologie und Technik aus lebensphilosophischer Sicht* [On sociology and technology from the viewpoint of life philosophy]. Bonn: Bouvier, 1971. Pp. 17.

Shinn, Roger. "Perilous Progress in Genetics," *Social Research* 41, no. 1 (Spring 1974): 83–103. Gives a good short description of the new advances in genetics; 1) fetal diagnosis and selective abortion, 2) selective breeding, and 3) genetic engineering. Concludes with a mediocre discussion of the ethical issues involved in genetics.

Shriver, Donald W., Jr. "Technological Change and Multi-Valued Choice—A Personal Inquiry," *Soundings* (Vanderbilt Univ.) 53, no. 1 (Spring 1970): 4–19. Reflections on how to take ethical responsibility for technological change. Argues for "a sense of extended system-relations," "a sense of strategic confusion," and "a sense of extended obligation."

———. "Invisible Doorway: Hope as a Technological Virtue," *Zygon* 8, no. 1 (March 1973): 2–16. Technological creativity is not value-neutral, but ordinarily it does serve not one but a multiplicity of values. A technology-policy maker cannot avoid making value-choices. As to what ought to be the first-priority value in technological planning, argues for the ideal of love of humankind as an end-value.

Sperry, R. W. "Science and the Problem of Values," *Zygon* 9, no. 1 (March 1974): 7–21. Uses recent developments in the concept of mind and mind-brain relations to challenge the separation of science and values.

Stone, Christopher D. *Should Trees Have Standing? Toward Legal Rights for Natural Objects*. Los Altos, Calif.: W. Kaufman, 1974. Pp. xvii, 102. Foreword by Garrett Hardin. The author's influential article of the same title from the *Southern California Law Review* 45, no. 1 (Spring 1972): 450–501, which was part of a symposium on "Law and Technology," along with the opinion of the U.S. Supreme Court in the case of *Sierra Club v. Morton* in which Justice William O. Douglas adopted Stone's basic argument that natural objects should in certain circumstances be treated, like corporations, as legal persons.

"Technology and the Human Future." *Zygon* 8, no. 1 (March 1973). A symposium which includes D. Shriver's "Invisible Doorway: Hope as a Technological Virtue," H. Brooks's "Technology and Values: New Ethical Issues Raised by Technological Progress," V. R. Potter's "The Ethics of Nature and Nurture," and B. Feld's "Human Values and the Technology of Weapons." See annotations on individual papers under appropriate classifications.

Toulmin, Stephen E. "Achievements . . . and Responsibilities," pp. 271–294 in A. L. C. Bullock, ed., *The Twentieth Century: A Promethean Age*. New York: McGraw-Hill, 1971.

Tribe, Laurence. *Channeling Technology Through Law*. Chicago: Bracton Press, 1973. Pp. 644. An important work which goes beyond casebook questions to issues of jurisprudence and philosophy. Contains seven major parts: "The Concept of Technology Assessment," "The Role of Law: An Overview," "The Supersonic Transport as a Test Case for Legal Control," "Biomedical Technology and Asexual Reproduction," "Electronic Monitoring and Neurological Manipulation," "Computer Systems for Enhancing the Potentialities of the Individual," "Institutional Arrangements for Improved Technology Assessment in the Federal Government."

———. "Technology Assessment and the Fourth Discontinuity: The Limits of Instrumental Rationality," *Southern California Law Review* 44 (June 1973): 617–660. The "fourth discontinuity" is the gap between people and machines. Tribe discusses the present limitations of technology assessment in bridging this gap. Argues that technology assessment should include factors like intuition, unease, etc.

Truitt, Willis H., and T. W. Graham Solomons, eds. *Science, Technology and Freedom*. Boston: Houghton Mifflin, 1974. Pp. xii, 272. Divides the articles into four categories: 1) Science, Technology, and Culture, 2) The Origins and History of Science, 3) Science and Technology: The Moral and Political Issues, and 4) Science, Technology, and the Environment. Includes contributions from J. Ellul, C. P. Snow, D. Struik, T. Kuhn, L. Kass, H. Marcuse, A. Huxley, and B. Commoner.

Uribe, Hector Gonzalez. "El hombre, dueno de la tecnica o mediatizado a ella" [Man, master of technology or dominated by it], *Revista de Filosofia* (Mexico) 6 (May–August 1973): 249–257. Science and technology are not sufficient by themselves to remedy the maladies of humanity. It is necessary to find their meaning and value in an axiological and humanistic perspective. In conjunction with the technological revolution of automation comes the danger of alienation. This situation cannot be solved by technical remedies. Argues that philosophers must fight for authentic human liberation, reaching to the interior of man. It is the work of philosophy to carry man to a rediscovery of himself by means of education and new community of justice and love.

Vahrenkamp, Richard, ed. *Technologie und Kapital* [Technology and capital]. Frankfurt a/M: Suhrkamp, 1973. Pp. 234. Contents: Alfred Sohn-Rethel's "Technische Intelligenz zwischen Kapitalismus und Sozialismus," Hans-Dieter Bahr's "Die Klassenstruktur der Maschinerie Anmerkungen zur Wertform," Frank Deppe's "'Alte' und 'neue' Arbeiterklass," André Gorz's "Technische Intelligenz und kapitalistische Arbeitsteilung," Wolfgang Neef and Rainer Morsch's "Veränderungen im Arbeitsprozess, ihre Auswirkungen auf das Bewusstsein von Naturwissenschaftlern und Ingenieuren sowie Folgerungen für die Ausbildung der technischen Intelligenz," Hellmuth Lange's "Gewerkschaftliche Aktion und politisches Bewusstsein der wissenschaftlich-technischen Intelligenz in Frankreich," Manuel Bridier's "'Neue Arbeiterklasse' oder 'neue Bourgeoisie'?," and Richard Vahrenkamp's "Entwicklungsmöglichkeiten der Technologie als Produktionsverhältnis."

Veatch, Robert M. "'Doing What Comes Naturally'," *Hastings Center Report* 1 (September 1971): 1-2.

———. "Medical Ethics: Professional or Universal?" *Harvard Theological Review* 65 (1972): 531-559. Attacks the concept of "professional ethics."

Walters, LeRoy. "Technology Assessment and Genetics," *Theological Studies* 33, no. 4 (December 1972): 666-683. Sees "technology assessment" as a "useful analytical tool" for coping with the problems in the field of human genetics. Good bibliographic notes.

———. "Ethical Issues in Experimentation on the Human Fetus" in "Selected Issues in Medical Ethics," *Journal of Religious Ethics* 2 (Spring 1974): 3-98.

Weil, Simone. *Oppression and Liberty*. London: Routledge and Kegan Paul, 1958. Reprint Amherst, Mass.: Univ. of Massachusetts Press, 1973. Pp. xvi, 195. In-depth critique of Marxism and the religion of progress in general. Of special relevance is the section "Sketch of Contemporary Social Life." Contains essays from before and after her conversion to Christianity. For two previously unpublished letters of Weil which deal explicitly with technology see P. Guillerme's "Deux lettres inédites de Simone Weil," *Dialogue* (Canada) 12, no. 3 (September 1973): 454-464.

Weinberg, Alvin M. "Technology and Ecology—Is There a Need for Confrontation?" *Bioscience* 23, no. 1 (1973): 41-46.

Williams, Preston N., ed. *Ethical Issues in Biology and Medicine*. Cambridge, Mass.: Schenkman, 1973. Pp. vii, 296. Proceedings of a symposium on the Identity and Dignity of Man, held at Boston University in December, 1969. Contents include: Bernard D. Davis's "Threat and Promise in Genetic Engineering," L. Harold Dewolf's "Organ Transplants as Related to Fully Human Living and Dying," James M. Gustafson's "Genetic Engineering and the Normative View of the Human," Francis D. Moore's "Social Investment and Patient Welfare in Organ Transplantation," Hudson Hoagland's "Biological Considerations of Aggression, Violence and Crowding," and Roger L. Shinn's "Population and the Dignity of Man." Also includes reports on panels, workshops, and recommendations.

Williams, Roger. *Politics and Technology*. London: Macmillan, 1971. Pp. 80. An examination of industrial society, futuristic interpretations of post-industrial society, utopian views of technology, dystopian views, and the "political imperatives of technology."

Winner, Langdon. "On Criticizing Technology," *Public Policy* 20, no. 1 (Winter 1972): 35-59. A brief survey of the criticisms of McLuhan, Mumford, Mesthene, Marcuse, etc., and current technology assessment programs. "The need for technology criticism beyond technology assessment is manifest in many of the crucial dilemmas which the future places in our path" (p. 58).

"World Famine and Lifeboat Ethics: Moral Dilemmas in the Formation of Public Policy."

Soundings (Vanderbilt Univ.) 59, no. 1 (Spring 1976). A special issue which includes the following articles: G. R. Lucas, Jr.'s "Political and Economic Dimensions of Hunger" (pp. 1–28), S. W. Hinds's "Relations of Medical Triage to World Famine: A History" (pp. 29–51), J. Fletcher's "Feeding the Hungry: An Ethical Appraisal" (pp. 52–69), H. Tristram Engelhardt, Jr.'s "Individuals and Communities, Present and Future" (pp. 70–83), W. Harrelson's "Famine in the Perspective of Biblical Judgements and Promises" (pp. 84–99), J. Sellers's "Famine and Interdependence (pp. 100–119), G. Hardin's "Carrying Capacity as an Ethical Concept" (pp. 120–137).

B. Secondary Sources

Adams, Margaret. "Science, Technology, and Some Dilemmas of Advocacy," *Science* 180, no. 4088 (May 25, 1973): 840–842. Social implications of biomedical research.

Alberoni, Francesco. "Technical Progress and the Dialectics of Existence," *Human Context* 4, no. 2 (1972): 285–305. Considers future alternatives for liberation. The same article is available in Italian, pp. 264–284.

Alfven, H. "Science, technocratie et le pouvoir politico-économique" [Science, technocracy and the power of political economics], *Impact: Science et Société* (Paris) 22, nos. 1–2 (January–June 1972): 91–99.

Alviset, Lucien. *Suicide ou renouveau d'une civilisation?* [Suicide or renewal of civilization?]. Brussels: Vander, 1974. Pp. ix, 176.

Anderson, S. E. "Science, Technology, and Black Liberation," *Black Scholar* 5 (March 1974): 2–8.

Ashby, Eric. *Adapting Universities to a Technological Society*. San Francisco: Jossey-Bass Publishers, 1974. Pp. xvi, 158. Argues that the state and private foundations have the right to influence vocational higher education, but not nonvocational higher education. "I suggest that if nonvocational higher education is to serve its real purpose (which is to civilize people) it ought not to attract people who only want to be certified, not civilized" (p. 140). By the Chancellor of Queen's University, Belfast, England. See also the author's *Reflections on Technology in Education* (Haifa: Technion-Israel, Institute of Technology, 1967), a lecture delivered on January 3, 1967. The author's earlier *Technology and the Acadamies* has also been translated into Spanish as *La tecnologia y los académicos, ensay sobre las universidades y la revolución científica* (Caracas: Monte Avila, 1970).

Asimov, Isaac. "The Art of the Tomorrow Seekers," pp. 30–43 in *1969 Britannica Yearbook of Science and the Future*. Chicago: Encyclopaedia Britannica Inc., 1968. A historical review of science-fiction literature.

Aultman, Mark. "Technology and the End of Law," *Natural Law Forum* 17 (1972): 46–79.

Bahman, Abdur. *Trimarti: Science, Technology and Society; A Collection of Essays*. New Delhi: People's Publishing House, 1972. Pp. xii, 315. Foreword by Joseph Needham.

Baram, Michael S. "Technology Assessment and Social Control," *Science* 180, no. 4085 (May 4, 1973): 465–473. "It is now time to . . . develop a coherent framework for the social control of technology, and ensure that forthcoming processes of technology assessment and utilization will be systematic and humane" (p. 472). Baram proposes such a framework.

Bateson, Mary Catherine. *Our Own Metaphor: A Personal Account of a Conference on the Effects of Conscious Purpose on Human Adaptation*. New York: Knopf, 1972. Pp. xiv, 324. The author is the daughter of anthropologists Gregory Bateson and Margaret Mead. This book is a report on a conference organized by G. Bateson to consider how Western thought "might contain systematic distortions of view which, when implemented by modern technology, become destructive of the balances between individual man, human society, and the ecosystem of the planet" (pp. 12–13). For a critical review see W. D. Lewis, *Technology and Culture* 15, no. 1 (January 1974): 146–150.

Baudrillard, Jean. *Le Système des objets* [The system of objects]. Paris: Denoël-Gonthier, 1972. Pp. 253.

Baxter, William F. *People or Penguins: The Case for Optimal Pollution.* New York: Columbia Univ. Press, 1974. Pp. 110. By a law professor of Stanford Law School. "Continuance of our environmental quagmire, he suggests, is caused by weakminded politicians who interfere in the free-market economy by imposing legal sanctions against specific polluters for their rational behavior. Instead of this futile approach, we should modify the market forces that motivate pollution. He recommends (1) promoting class-action suits that would force polluters to compensate injured individuals and (2) instituting an effluent tax that would compensate society for injury to the environmental commons."—G. H. Sewell, review in *Technology and Culture* 16, no. 4 (October 1975): 687–689.

Beckmann, Klaus Martin. *Menschlichkeit und Technologie; ein Beitrag zur ökumenischen Humanum-Studie* [Humanity and technology; a contribution to the ecumenical human studies]. Wuppertal: Jugenddienst-Verlag, 1971. Pp. 32. Technology and ethics.

Bendix, Reinhard. *Work and Authority in Industry; Ideologies of Management in the Course of Industrialization.* Second revised edition. Berkeley: Univ. of California Press, 1974. Pp. L, 464. First published in 1956, this comprehensive historico-sociological study is obviously a labor of many years' dedicated research. Analyzes entrepreneurial ideologies in eighteenth-century England, eighteenth- and nineteenth-century Russia, and those which grew out of the American experience.

Benko, François. *Las fuentes de la civilización tecnologica* [The sources of technological civilization]. Caracas: Universidad Central de Venezuela, Ediciónes de la Biblioteca, 1974. Pp. 407.

Berger, Harry, Jr. "Outline of a General Theory of Cultural Change," *Clio* 2, no. 1 (October 1972): 49–63. A theory based on three key elements: transcendence, transposition, and technology. Relevant to theories of technology transfer.

Berlin, Isaiah. "The Divorce Between the Sciences and the Humanities," *Salmagundi* no. 27 (Summer–Fall 1974): 9–39. A study in the history of ideas, with special attention to the Enlightenment and Vico. This issue of *Salmagundi* also includes "Art and Technology: A Dialogue Between Harold Rosenberg and Benjamin Nelson" (pp. 40–56), and thus apparently comprises what could be published after difficulties arose of a previously announced special issue on "Technology and Civilization" which was to include works by Max Weber and Martin Heidegger.

Bermudo, José Manuel. *El McLuhanismo: ideología de la tecnocracia* [McLuhanism: ideology of technocracy]. Barcelona: Picazo, 1972. Pp.219.

Bernard, H. R., and Pertti Pelto, eds. *Technology and Social Change.* New York: Macmillan, 1972. Pp. vii, 354. Illustrated. Bibliography. "The study of technological innovation and its effects on social and cultural systems remains one of the most neglected areas in anthropological research." This collection of field studies ranges from an analysis of the effects of snowmobiles on Lapp reindeer herdsmen to the problems of resettlement in Northern Rhodesia due to the Kariba Dam project. Favorably reviewed by S. Lieberstein, *Technology and Culture* 15, no. 1 (January 1974): 144–146.

Bierman, A. K. *The Philosophy of Urban Existence: A Prolegomenon.* Athens: Ohio Univ. Press, 1973. Pp. x, 194. An eccentric essay with a number of marginally relevant chapters: e.g., ch. 2, "The Vocation of Man and How to Go to Bed with the Machine in Eden"; ch. 9, "You, as Transcendental Artificer." Pop ideology.

Blair, J. M. "Centrifugal Technology: The New Industrial Revolution," *Nation* 214 (June

12, 1972): 749–754. Abridged as "New Industrial Revolution," *Current* 145 (November 1972): 58–64.

Blumenberg, Hans. *Säkularisierung und Selbstbehauptung* [Secularization and self-assertion]. Frankfurt: Suhrkamp, 1974. Pp. 293. This constitutes parts 1 and 2 of a work which the author calls a contribution to the legitimatization of the new age. Part 3 was published earlier as *Der Prozess der theoretischen Neugierde* [The process of theoretical curiosity] (Frankfurt: Suhrkamp, 1973). Pp. 309.

Borgosz, Jozef. "Herbert Marcuse's 'Homo Novus' as an Expression of the Crisis of the Consumption-Oriented Personality Model," *Dialectics and Humanism* 1 (Winter 1974): 163–171.

Bourg, Carroll J. "Work and/or Job in Advanced Industrial Societies," *Soundings* (Vanderbilt Univ.) 57, no. 1 (Spring 1974): 113–125. "By job I mean that human activity for which there is financial remuneration. By work I mean that human activity which is a mode of self-expression and a matter of personal claim" (p. 123). A good brief introduction to some sociological issues.

Bronk, Detlev W. "The Human Values of Science and Technology," *1971 Britannica Yearbook of Science and the Future*. Chicago: Encyclopaedia Britannica, Inc., 1968, pp. 424. Popular defense of science and technology.

Brooks, Harvey. "The State of the Art: Technology Assessment as a Process," *International Social Science Journal* 25, no. 3 (1973): 247–256. The same article follows in French.

Broudy, Harrry S. "Humanism in Education," *Journal of Aesthetic Education* 7, no. 2 (April 1973): 67–77. Distinguishes between classical (discipline oriented) humanism and modern (freedom oriented) humanism. Argues that classical humanistic education provides the self-cultivation needed to exploit the resources of technological culture for new forms of individuality and freedom.

Brown, Lester R. "Issues of Human Welfare," *Humanist* 33, no. 6 (November–December 1973): 17–21. Argues for a new social ethic to direct technology for global good.

Bryen, Stephen David. *The Application of Cybernetic Analysis to the Study of International Politics*. The Hague: Nijhoff, 1971. Pp. 135. Proposes a model of international relations based on cybernetics, a branch of general systems theory.

Bugliarello, George. "Re-Thinking Technology: Steady-State Earth or Biosoma," *World Development* 1 (August 1973): 45–51.

Burrage, Michael. "Democracy and the Mystery of the Crafts: Observations on Work Relationships in America and Britain," *Daedalus* 101 (Fall 1972): 141–162.

Cazenobe, J. "Technologie et linguistique" [Technology and linguistics], *Revue de Synthèse* 94 (1973): 211–229.

Cernea, Mihail. "Traditional culture and technological progress," *Revue Roumaine des Sciences Sociales—Série de Philosophie et Logique* 17 (1973): 105–115.

Cerroni, Umberto. *Tecnica e Libertà*. Bari: De Donato, 1970. Pp. 116.

Cetron, Marvin J., and Bodo Bartocha, eds. *Technology Assessment in a Dynamic Environment*. New York: Gordon and Breach, 1973. Pp. xiv, 1036.

Charbonneau, Bernard. *Le Système et le chaos; critique du développement exponentiel* [The system and chaos; critique of exponential development]. Paris: Anthropos, 1973. Pp. 413.

Chen, Kan, with the assistance of George J. Karl, eds. *Technology and Social Institutions*. New York: Institute of Electrical and Electronics Engineers, 1974. Pp. 224. Bibliography, pp. 213–215. Proceedings of the Engineering Foundation Conference on Technology and Social Institutions, Pacific Grove, Calif., 1973.

Civilization and Science: In Conflict or Collaboration? New York: Elsevier, 1972. A Ciba Foundation symposium which includes H. Bloch's introduction and summary, P. B.

and J. S. Medawar's "Some Reflections on the Theme of Science and Civilization," S. Toulmin's "The Historical Background to the Anti-Science Movement," E. Shils's "Anti-Science: Observations on the Recent 'Crisis' of Science," H. Thiemann's "Science: A Consequence of Science Policy or an Expression of Culture?," G. Pelletier's "Science, Technology, and the Political Response," A. Weinberg's "Science and Trans-Science," F. A. Long's "Science and the Military," M. Roche's "Science in Spanish and Spanish-American Civilization," H. G. Johnson's "Some Economic Aspects of Science," G. Mathur's "Scientific Research and Long-Term Growth," M. Toda's "The Need for a Science of Civilization." Reviewed by I. S. Spiegel-Rösing, *Technology and Culture* 14, no. 4 (October 1973): 678–681.

Clouser, K. Danner. "What Is Medical Ethics?" *Annals of Internal Medicine* 80 (May 1974): 657–660.

Coates, Vary T. "Technology in the Balance," *Futurist* 7, no. 2 (April 1973): 66–68. Despite the U.S. Congress's creation of the Office of Technology Assessment, Coates claims that the government is doing relatively little in this crucial area. See also J. Mendell's "Technological Forecasting Comes of Age," *Ibid.*, pp. 69–70. A good, brief introduction to assessment and forecasting. Includes small bibliography.

Cottrell, William F. *Technology, Man, and Progress*. Columbus, Ohio: Charles E. Merrill, 1972. Pp. 194. Raises more questions that it answers. See also the author's earlier *Energy and Society* (New York: McGraw-Hill, 1955), reprinted Westport, Conn.: Greenwood Press, 1970.

Crosser, Paul K. *War Is Obsolete; The Dialectics of Military Technology and Its Consequences*. Amsterdam: Grüner, 1972. Pp. v, 245. The first half of the book argues that thermonuclear bombs and missiles mark a qualitative change in the evolution of weaponry technology, thus making all nuclear war strategy futile. The second half analyzes the consequences of general disarmament and demilitarization policy.

Crowe, Michael J. "A New Age in Science and Technology?" *Review of Politics* 34, no. 4 (Oct. 1972): 141–153. "Science and especially technology are frequently taken to mean the products of science and technology. . . . But science and technology are also systems of exciting and enriching ideas that will not of themselves pollute or corrupt or dehumanize." Calls for cultural appreciation of these ideas and social science study *about* science and technology.

Daddario, Emilio Q. "Priorities for the Future: Guidelines for Technology Assessment," pp. 292–311 in *1971 Britannica Yearbook of Science and the Future*. Chicago: Encyclopaedia Britannica, Inc., 1970.

Dahlberg, K. A. "The Technological Ethic and the Spirit of International Relations," *International Studies Quarterly* 17, no. 1 (March 1973): 55–88.

Darby, William J. "Acceptable Risk and Practical Safety: Philosophy in the Decision Making Process," *Journal of the American Medical Association* 224 (May 21, 1973): 1165–1168.

Dasgupta, Subhayu. *Hindu Ethos and the Challenge of Change*. Calcutta: Minerva, 1972. Pp. 277. Bibliography. Discusses technology's influence on Hindu civilization.

Davisson, William I. "Technology and Social Change," *Review of Politics* 34, no. 4 (October 1972): 172–184. A brief analysis of the effect of technology on society in the case of automobiles, economic growth and pollution, and the computer. "I have come to the conclusion that technological change becomes autonomous as a result of the requirements of the delivery system." Cf. also Thomas J. Musial and Julian R. Pleasants, "Mendelian Evolution and Mandalian Involution: Speculations About the Foundations of Cultural Change," *Ibid.*, pp. 154–171. no. 4 (Oct. 1972): 154–171.

De Finetti, B. "Prejudice and Liberty," *Scientia* (Milan) 107, nos. 9–10 (1972): 777–787. A mathematician and economist of the University of Rome argues that the scientist has the same commitment to the common good as other citizens, although with more obligation to educate while not imposing his own ends.

Dessel, Norman F., Richard B. Nehrich, Jr., and Glenn L. Voran. *Science and Human Destiny.* New York: McGraw-Hill, 1973. Pp. x, 318. Illustrated.

Dreitzel, H. P. "Social Science and the Problem of Rationality: Notes on the Sociology of Technocrats," *Politics and Society* 2, no. 2 (Winter 1972): 165–182. Argues that there is no final answer to the question of the reasonableness of rationality. Topic approached from three angles: 1) methodology of social sciences, 2) application of social science knowledge, 3) relationship between social class and individual behavior.

Dubos, René. "The Promises and Perils of Science," pp. 210–221 in *1969 Britannica Yearbook of Science and the Future.* Chicago: Encyclopaedia Britannica Inc., 1968.

———. "Humanizing the Earth," *Science* 179, no. 4075: (February 23, 1973): 769–772.

Eaton, John. *Technology and the State.* Nottingham: Bertrand Russell Peace Foundation, 1973. Pp. 8. Spokesman pamphlet no. 36.

Edge, David O., and J. N. Wolfe, eds. *Meaning and Control: Essays in Social Aspects of Science and Technology.* London: Tavistock Publications, 1973. Pp. x, 274. Papers from a seminar on Social Aspects of Science and Technology, University of Edinburgh, 1970.

Ellul, Jacques. "Search for an Image," *Humanist* 33, no. 6 (November–December 1973): 22–25.

Etzioni, Amitai. *Genetic Fix.* New York: Macmillan, 1973. Pp. 276. Reflections by this important sociologist in "journal form." Mainly about an international conference on the social and ethical implications of biology and medicine. Lively introduction to the field. Appendix includes some documents in this field.

———. "Humane Technology," *Science* 179, no. 4077 (March 9, 1973): 959. "The task before us is to marshal more of technology to the service of human purposes, not to put technology into a self-destruct, reverse-thyself gear." For two other editorials with a positive attitude toward technology see B. Davis's "Genetic Engineering: How Great Is the Danger?" *Science* 186, no. 4161 (October 25, 1974): 309; and R. Davis's "Technology as a Deterrent to Dehumanization," *Science* 185, no. 4153 (August 30, 1974): 737.

Feld, B. T., T. Greenwood, G. W. Rathjens, and S. Weinberg, eds. *Impact of New Technologies on the Arms Race.* Cambridge, Mass.: MIT Press, 1971. Pp. 371. Contains the proceedings from the 10th International Pugwash Symposium, 1970. For an informative review see K. Tsipis's "Pugwash on the Arms Race," *Science and Public Affairs; Bulletin of the Atomic Scientists* 28, no. 1 (January 1972): 47–48.

Ferre, A. "Technology and Freedom: The Communication Revolution," *Vital Speeches* 38 (January 15, 1972): 217–220.

Florman, Samuel C. "Anti-Technology: The New Myth," *Civil Engineering* 42 (January 1972): 68–70.

Flowers, Brian. *Technology and Man.* Liverpool: Liverpool Univ. Press, 1972. Pp. 29. Argues that we need a broader and less specialized education in order to come to terms with technology.

Foster, George McClelland. *Traditional Societies and Technological Change.* 2nd ed. New York: Harper and Row, 1973. Pp. 286. Bibliography. First published in 1962 under the title *Traditional Cultures and the Impact of Technological Change.*

Frank, Helmar. "Bildungstechnologie und Lehrplanung" [Technological development and the rule of planning], *Grundlagestudien aus Kybernetik und Geisteswissenschaft* 14 (1973): 73–84.

Frankel, Charles. "The Specter of Eugenics," *Commentary* (March 1974): 25–33. Between the extremes of certain eugenic proposals by scientists and the uncompromising negative attitude of some religious authorities, Frankel supports a more balanced response to the issues in biomedicine.

Freeman, David M. *Technology and Society: Issues in Assessment, Conflict, and Choice*. Chicago: Rand McNally, 1974. Pp. xx, 181.

Frosini, Vittorio. "I Calcolatori Elettronici e il Nuovo Mondo Civile" [Electronic computers and the new world civilization], *Rivista Internazionale di Filosofia del Diritto* 50 (October–December 1973): 704–711. The computer is a new "person" in the legal sense, with an "artificial reason" different from the "natural reason" of human beings. This calls for a sociology of cybernation to understand the present technological revolution.

Fuller, Watson, ed. *The Biological Revolution; Social Good or Social Evil?* Garden City, N.Y.: Doubleday, 1971. Pp. xi, 345. A discussion of the social implications of modern biological discoveries by prominent scientists. Includes articles by M. H. F. Wilkins, R. G. Edwards, J. D. Watson, J. Bronowski, etc.

Furlow, Thomas W., Jr. "Tyranny of Technology," *Humanist* 34, no. 4 (July–August 1974): 6–9. Article by a physician on the ethical problems of euthanasia. Argues for a process definition of death and the limited use of euthanasia. Part of a symposium on euthanasia of a more popular nature which included articles by M. Kohl, J. Fletcher, D. Maguire, B. Sherwin, and O. Russel.

Garaudy, Roger. "A Role for Aesthetic Education in an Age of Great Scientific and Technical Mutation," *Teilhard Review*, vol. 7, no. 3 (October 1972).

García, Guadalupe, and Carlos Sabino. *Dictaduras de la tecnocracia* [Dictatorships of technocracy]. Buenos Aires: Proyección, 1974. Pp. 199.

Garcia-Pelayo, M. "Burocracia y tecnocracia" [Bureaucracy and technocracy], *Politeia* 2 (1973): 9–89. Technocracy is an unavoidable ideology. Included in the author's *Burocracia y tecnocracia y otros escritos* (Madrid: Alianza, 1974).

Gellen, Martin. "Der Oeko-Industrielle Komplex in der USA" [The economic-industrial complex in the USA], *Kursbuch* 33 (1973): 125–133.

Ghali, Mirrit Boutros. *Tradition for the Future*. Oxford: Alden Press, 1972. Pp. 288. Reviewed by H. J. Muller, *Technology and Culture* 15, no. 1 (January 1974): 82–86.

Gironella, Juan Roig. "Traduccion al nivel humano del tecnicismo de la ciencia y filosofia" [Translation to the human level of the technical vocabulary of science and philosophy], *Espiritu* 20 (January–June 1971): 16–28. Despite classification as such by the *Philosopher's Index* this is not really related to the philosophy of technology. Concerned more with how any abstract language raises difficulties for popular interpretation.

Goldman, J. E. "Toward a National Technology Policy," *Science* 177, no. 4054 (September 22, 1972): 1078–1080. Laments the fact that our technical priorities have changed little despite the enormous changes technology has effected on the world. To correct this imbalance a national technology policy must encourage a periodic reassessment of goals.

Gottschling, Ernst. "Zur Kritik technokratischer Ideologien" [On the critique of technocratic ideologies], *Staat und Recht* 22, no. 3 (1973): 393–408.

Gronemeyer, Reimer. "Friedlicher Wandel und Übergangstrategien" [Peaceful change and the strategy of transition], *Internationale Dialog Zeitschrift* 6 (1973): 236–242.

Günther, Hans. "Sozialpolitik und Post-Industrielle Gesellschaft" [Social politics and post-industrial society], *Soziale Welt* 24 (1973): 1–24.

Haggerty, Patrick. *The Productive Society*. New York: Carnegie Press, 1974. Pp. 173. Illustrated. Chapter 1 is concerned with the question "Is growth obsolete?" Chapter 2 asks "Can we have growth and freedom?" Chapter 3 discusses the importance of education and R&D. Chapter 4 presents suggestions for a National Development Act.

"Haggerty's analysis and proposals are essentially conservative. They are the views of a successful business executive . . ."—A. Weintraub, review in *Technology and Culture,* 16, no. 4 (October 1975): 684.

Hamilton, David. *Technology, Man and the Environment.* New York: Scribner's, 1973. Pp. 357. A survey, from a British perspective, of "the overall pattern of technology: what technology is, what its effects are on everyday life, how it is changing the world, and the problems it brings" (p. 7). Weak philosophically and socially; e.g., defines technology as "the means by which Man extends his power over his surroundings" (p. 17). But contains a comprehensive nontechnical although rather conventional picture of technological developments in electronics, materials, computers, energy, space and oceans exploration, construction, medicine, etc. Also contains chapters on the relation between technology and war (pp. 301–315) and technology and politics (pp. 316–331).

Harris, N. "We and Our Machines: 200 Years of Love and Hate," *New Republic* 171 (November 23, 1974): 24–29ff.

Hartley, John. "Work-how-you-like Modules for Suiting Operators to Output," *Engineer* 239, no. 6176 (July 25, 1974): 38–39. Pilot scheme in Renault factory aims to free workers from the tyranny of automation.

Hetman, François. *Society and the Assessment of Technology; Premises, Conceptions, Methodology, Experiments, Areas of Application.* Paris: Organization for Economic Co-operation and Development, 1973. Pp. 420. Bibliography, pp. 391–413. As the title indicates, a rather eclectic grab-bag. But one of the most useful available.

Hirsch, Joachim. *Wissenschaftlich-technischer Fortschritt und politisches System* [Scientific-technological progress and political system]. Frankfurt: Suhrkamp, 1970. Pp. 292.

Hollingum, Jack. "Self-Determination Is Moving Along the Assembly Line," *Engineer* 237, no. 6139 (November 8, 1973): 64–67. Allowing groups of workers to decide how to split up the job, not tearing out the assembly line, is the message from Saab. Concerning worker self-determination, see also J. Pullin's "Worker Participation, Ethics, and the Need for Industry to Act," *Ibid.,* p. 54.

Hortleder, Gerd. *Das Gesellschaftsbild des Ingenieurs* [The social image of engineers]. Frankfurt: Suhrkamp, 1970. Pp. 226. Bibliography.

Ihde, Don. "Back to Rock, A Musical Odyssey," *Music and Man* 1, no. 1 (1973): 1–10.

Il'Enkov, E. V. *L'uomo e i Miti Della Tecnica* [Man and the myths of technology]. Edited by Ignazio Ambrogio. Roma: Editori Riuniti, 1971. Pp. 273.

"Implications of Science-Technology for the Legal Process, The," *Denver Law Journal* 47, no. 4 (1970): 549–680. This conference includes: Wilbert E. Moore's "Science and Technology v. Law, or A Plague on Both Your Houses" (pp. 553–558) with comment by Harold P. Green, Melvin Kranzberg, and Shirley Johnson; Michael S. Baram's "The Social Control of Science and Technology" (pp. 567–578) with comment by John G. Welles and Louis H. Mayo; Arthur Selwyn Miller's "The Law as a Center for Policy Analysis" (pp. 587–608) with comment by Raymond Bowers, John Gilmore, and Michael Baram, and a reply by Miller; Franklin P. Huddle's "Political Adaption to a Technology-Surfeited Soceity" (pp. 629–643) with comment by John A. Weese, Mason Willrich, and Edward Schwartz; James W. Curlin's "Saving Us from Ourselves: The Interaction of Law and Science-Technology" (pp. 651–663) with comment by Ernest M. Jones, Joseph F. Coates, Philip L. Bereano, and Wilbert E. Moore; plus an "Introduction" and "Summary" by Robert B. Yegge. Assumptions of the conference: "1) that there is a need for external monitoring of science and technology and 2) that the law should play some role in that enterprise."

James, Bernard J. *The Death of Progress.* New York: Knopf, 1973. Pp. xix, 166.

Jantsch, Erich. *Technological Planning and Social Futures.* New York: Wiley, 1972. Pp.

xiv, 256. Illustrations. The purpose of this book is to bring into focus the institutional and corporate roles in planning the world's social and technological futures.

Katz, Jay, ed. *Experimentation with Human Beings.* New York: Russell Sage Foundation, 1972. Pp. xlix, 1159. Massive volume using the legal "case method" approach in dealing with all the complex ethical issues involved in human experimentation. Includes a myriad of actual case studies, reports from the *Congressional Record,* and excerpts from relevant articles and books.

Katz, Solomon H. "The Dehumanization and Rehumanization of Science and Society," *Zygon* 9, no. 2 (June 1974): 126–138. "I shall discuss three themes: rapid technological change and its associated effects upon man in the twentieth century, the probability of national and even international revitalization movements, and, finally, the need for a new science of man" (p. 126).

Kelly, Kevin. *Youth, Humanism, and Technology.* New York: Basic Books, 1972. Pp. vi, 182.

Kemeny, John G. *Man and the Computer.* New York: Scribner's, 1972. Pp. viii, 151. Forecasts an interaction between man and computers that will undermine man's distrust of computer systems. First chapter contains a good history of computers.

Kleinberg, Benjamin S. *American Society in the Postindustrial Age; Technocracy, Power, and the End of Ideology.* Columbus, Ohio: Merrill, 1973. Pp. vi, 279.

Koch, Claus, and D. Senghaas, eds. *Texte zur Technokratiediskussion* [Texts on the discussion of technocracy]. Frankfort: Europäische Verlagsanst, 1970. Pp. 354. Bibliography.

Krieger, Martin H. "What's Wrong with Plastic Trees?" *Science* 179, no. 4072 (February 2, 1973): 446–455. Discusses reasons for preserving rare natural environments.

Landheer, B., J. H. M. M. Loenen, and Fred. L. Polak, eds. *Worldsociety.* The Hague: Nijhoff, 1971. Pp. 211. Two papers of interest: F. L. Polak's "Balance of Technology," and B. Landheer's "Industrial Society as the Basis of Worldsociety."

Leach, Gerald. *The Biocrats.* New York: McGraw-Hill, 1970. Pp. 317. Revised edition Baltimore: Penguin Books, 1972. Pp. 365. Not philosophical. Vaguely ethical. Contains basic information, charts, and statistics dealing with such problems as population control, test-tube reproduction, fetal medicine, transplants, etc.

Lefringhausen, Klaus. *Der Mensch im Sog der Technik* [Man in the suction of technology]. Wuppertal: Jugenddienst-Verlag, 1971. Pp. 32. On technology and ethics.

Leonard, N. "Economic Technology and the Liberal-Humanist Dream," *Social Science* (Winfield, Kansas) 48, no. 3 (Summer 1973): 142–151. Uses decision theory to try to rescue the liberal-humanist ideal from its conflicts with economic technology.

Leprince-Ringuet, Louis. *Science et bonheur des hommes* [Science and the good of man]. Paris: Flammarion, 1973. Pp. 264.

Lévy-Leblond, J.-M. "Is There a Crisis in Science or in Society," *Scientia* (Milan) 107, nos. 9–10 (1972): 806–809. "A science *for* the people can only be a science *by* the people." By a physicist at the University of Paris.

Linde, Hans. *Sachdominanz in Sozialstrukturen* [Dominance of things in social structure]. Tübingen: Mohr 1972. Pp. 86. Summaries in English and French. "This discussion is devoted to answering the twofold question, (a) what place profane objects (artifacts of the category of implement) have in the structuring of societies and also what place they can (b) correspondingly be given in the theoretical context of sociology."

Logan, Scott. "Man and Technology: The Responsibilities of Philosophy and the Philosopher in the Brave New World," *Dialogue* (Phi Sigma Tau) 15 (May 1973): 70–74. A distinctly popular argument to the effect that philosophers should be concerned about technology.

Lorenz, Konrad. *Civilized Man's Eight Deadly Sins.* New York: Harcourt Brace Jovanovich, 1974. Pp. 107. Favorably reviewed by Jay Martin Anderson in *American*

Scientist 62 (November–December 1974): 740–741. Discusses "eight separate but causally connected processes that are threatening to destroy . . . mankind as a species." These processes are named as: overpopulation, pollution, industrial development, entropy of feelings, genetic decay, the break with tradition, indoctrinability, and nuclear weapons. Despite the fact that these are all closely related to the ethical and political issues of technology, and the book is by a respected ethologist, the work is only marginally relevant. Journalistic.

Loth, David G., and Morris L. Ernst. *The Taming of Technology*. New York: Simon and Schuster, 1972. Pp. 256. Popular introduction to the issues of technology and law.

Magalhães Gomes, Francisco de Assis. "Humanismo e technologia" [Humanism and technology], *Kriterion* 20, no. 67 (1973–1974): 223–241.

Martinescu, Theodor. *Technocratie-democratie in capitalismul contemporan* [Technocratic democracy in contemporary capitalism]. Bucharest: Editura politica, 1973. Pp. 64.

Martins, Carlos Estevan. *Tecnocracia e captialismo: a politica dos tecnicos no Brasil* [Technocracy and capitalism: the politics of technology in Brazil]. São Paulo: Editora Brasiliense, 1974. Pp. 214.

Massari, Teresa. "La 'Teoria Sociale Critica' di Juergen Habermas" [The "critical social theory" of Jürgen Habermas], *Rivista de Filosofia* 63 (July 1973): 455–488.

Mattick, Paul. *Critique of Marcuse*. New York: Herder and Herder, 1972. Pp. 107. Argues against Marcuse's pessimistic assessment of capitalism's ability to retain power through the beneficent use of advanced technology.

Medford, Derek. *Environmental Harassment or Technology Assessment?* New York: Elsevier, 1973. Pp. xv, 358.

Mendelsohn, Everett, Judith P. Swazey, and Irene Taviss, eds. *Human Aspects of Biomedical Innovation*. Cambridge: Harvard Univ. Press, 1971. Pp. 234. Papers which grew out of the Harvard Program on Technology and Society. Two major subjects: social policy for biomedical science; and the interaction of science, technology, and medicine.

Mesarovic, Mihajlo, and Eduard Pestel. *Mankind at the Turning Point*. New York: E. P. Dutton and Reader's Digest Press, 1974. Pp. 210. This is the second report to the Club of Rome. Like the first report, *The Limits to Growth* (1972), this volume contains the results of a number of sophisticated technical projections of the future. "Our scientifically conducted analysis of the long-term world development based on all available data points out quite clearly that . . . a passive course leads to disaster. It is most urgent that we do not avert our eyes from the dangers ahead, but face the challenge squarely and assess alternative paths of development in a positive and hopeful spirit" (p. vii). Argues for "organic growth" as one possible solution to the problems it raises. Given the determinedly up-beat tone, and the fact that organic growth is a slightly less radical idea than the no-growth implications of *The Limits to Growth*, this volume has been more favorably received than its predecessor by *Time* magazine and friends.

Millichamp, David. "Maintaining Ethical Standards Cuts Out Antisocial Effects," *Engineer* 237, no. 6123 (July 19, 1973): 34–39. Use of increasingly sophisticated techniques may enable designers to boost output without allowing time for social considerations.

Montalenti, G. "Prometheus," *Scientia* (Milan) 107, nos. 9–10 (1972): 795–799. Given the fact that science and technology have both good and bad consequences, what should be the attitude of the scientist toward his work? Freedom of pure research needs to be preserved. But scientists and technologists should become aware of their responsibilities toward society, and politicians should accept the guidance of scientists about the application of science to human welfare. By a geneticist at the University of Rome.

Montgomery, John D. *Technology and Civic Life: Making and Implementing Development Decisions*. Cambridge, Mass.: MIT Press, 1974. Pp. 239. Examines the problems in

Third World countries produced by the interactions between technology and government. Favorably reviewed by D. G. Montgomery, *Technology and Culture* 16, no. 4 (October 1975): 689–691.

Moore, Wilbert Ellis, ed. *Technology and Social Change*. New York: Quadrangle, 1972. Pp. vii, 236. A collection of articles from *The New York Times*. Of note: W. F. Ogburn's "Can Science Bring Us Happiness?," Bertrand Russell's "The Science to Save Us from Science," Kenneth Keniston's "Does Human Nature Change in a Technological Revolution?," Eric Hoffer's "Automation Is Here to Liberate Us," etc.

Morais, José Xavier Pessoa de. *Communicação, tecnologia e destino humano (uma filosofía para o impasse tecnológico)*. Rio de Janiero: Civilização Brasileira, 1972. Pp. 256.

Motulsky, Arno G. "Brave New World?" *Science* 185, no. 4152 (August 23, 1974): 653–663. Describes the recent discoveries in biology and the ethical problems involved in each. States that each problem must be considered separately, instead of blanket condemnation or acceptance of the "new biology."

Muller, Herbert J. *Uses of the Future*. Bloomington: Indiana Univ. Press, 1974. Pp. iv, 264. Rather diffuse ruminations about the character of the future and the cultural forces of the present, with a good deal of reference to technology. Not hard-core philosophy.

Murchland, Bernard. *The New Iconoclasm; Reflections for a Time of Transition*. New York: Doubleday, 1972. Pp. xiii, 151. The new "iconoclasm," in the author's sense, "refers to the destruction of past meanings in the name of a reality that has not yet come into being" (p. 3). A literary analysis. The chapter on the machine as image of alienation (pp. 66–76) is the most relevant to philosophy of technology.

Murphy, Thomas P. "Technology and Political Change: The Public Interest Impact of COMSAT," *Review of Politics* 33, no. 3 (July 1971): 405–424. A case study of "how the dividend arising out of the efficiencies of new technology developed with government funding should be distributed."

Nabseth, L., and G. F. Ray, eds. *The Diffusion of New Industrial Processes; An International Study*. Cambridge: Cambridge Univ. Press, 1974. Pp. xvii, 324. Bibliography, pp. 316–319.

Nebbia, Giorgio. "Premesse culturali dell'attuale crisi ecologica" [Cultural premises of the current ecological crisis], *Proteus* 2, no. 4 (1971): 25–63.

Neville, Robert T. "The Limits of Freedom and the Technologies of Behavior Control," *Human Context* 4, no. 3 (1972): 433–446. On how psychotropic drugs and psychosurgery can enhance or diminish freedom, with a distinction between the subjective experience of freedom and the objective enjoyment of things. Part of a symposium "On Human Freedom, Behavior Control, Values and Technologies" which includes an article by E. Laszlo and a review of B. F. Skinner's *Beyond Freedom and Dignity* by A. Comfort.

Nicklin, D. J. *In Defence of Technology*. St. Lucia: Univ. of Queensland Press, 1971. Pp. 13. Illustrated. From the University of Queensland inaugural lectures. A chemical engineer argues that technology is not the cause of world problems, but their cure. Technology, by increasing productivity, does away with poverty.

Nieburg, Harold L. "Reversing Technological Innovation," *Dissent* 16, no. 1 (January–February 1969): 59–63. Describes what the author calls a "principle of reversal." "Every new technological advance either contains or provokes its own negation; e.g., "every new offensive weapon begets a defensive system." In response the "tech-fixers" superimpose "superior, next-generation ... control systems ... and so aggravate the problem." Although "expensive and exasperating" this reversal principle nevertheless "promotes a healthy ... human reaction that finds ways of subverting the megamachine to more humane, if primitive realities."

Olsen, Fred A., ed. *Technology: A Reign of Benevolence and Destruction*. New York: MSS Information Corp., 1974. Pp. 250.

Paillet, Marc. *Marx contre Marx: la société technobureaucratique* [Marx against Marx: the technobureaucratic society]. Paris: Denoël–Gonthier, 1972. Pp. 259. On the problems bureaucracy creates for socialism.

Pallascio-Morin, Ernest. *La machine dans le destin de l'homme* [The machine in human destiny]. Montreal: Librairie Beauchemin, 1974. Pp. 111.

Pasqualotto, Giangiorgio. *Avanguardia e tecnologia; Walter Benjamin, Max Bense e i problemi dell'estetica tecnologica* [The avant-garde and technology; Walter Benjamin, Max Bense and the problems of aesthetic technology]. Rome: Officina, 1971. Pp. 224.

Patanè, Leonardo R. *Ragazzi, invenzioni e tecnica* [Children, inventions and technology]. Catania: N. Giannotta, 1973. Pp. 279. Illustrated. Bibliography.

Paty, M. "Science et Humanisme" [Science and Humanism], *Scientia* 108, nos. 3–4, 5–6 (1973): 205–220. A tribute to Paul Langevin and his theories of science and technology.

Paul, Leslie. *Eros Rediscovered*. New York: Association Press, 1970. Pp. 191. Chapter 5, "Sexual Alienation and Technology" (pp. 70–92), surveys theories of the relationship between civilization and sexual gratification.

Pelto, Pertti. *The Snowmobile Revolution: Technology and Social Change in the Arctic*. Menlo Park, Calif.: Cummings Press, 1973. Pp. 225. Illustrated. Short bibliography. A field study of the effects of the snowmobile on the Skolt Lapps in northeastern Finland. "The argument that I wish to present in this book rests on the assumption that, whatever their origins, technical changes that shift production processes (in this case reindeer herding) from local autonomous sources of energy to a dependence on outside sources (for example, gasoline) will almost certainly have wide-ranging consequences on the social and cultural patterns of the affected people . . ." ("Preface"). Pelto concludes that the effects of the snowmobile were harmful and recommends some alternatives to the current situation of these Lapps.

Pereira, Luiz Carlos Bresser. *Technoburocracia e contestação* [Technobureaucracy and protest]. Petropolis: Editora Vozes, 1972. Pp. 306.

Perez Guerra, Alfonso. *Los tecnicos y el poder de las pasiones* [Technicians and the power of passions]. Barcelona: SHEPSA, 1972. Pp. 272. Brief bibliography.

Peterson, Richard A. *The Industrial Order and Social Policy*. Englewood Cliffs, N.J.: Prentice-Hall, 1973. Pp. x, 159. Bibliography, pp. 120–154.

Pfeiffer, Werner. *Allgemeine Theorie der technischen Entwicklung als Grundlage einer Planung und Prognose des technischen Fortschritts* [General theories of technical development as the foundation for planning and predicting technical progress]. Göttingen: Vandenhoeck und Ruprecht, 1971. Pp. 152. Bibliography.

Philo, Greg, and Paul Walton. "Max Weber on Self-interest and Domination," *Social Theory and Practice* 2, no. 3 (Spring 1973): 335–346. Argues "that Weber's equation of self-interest and domination in a technologically advanced society, by ignoring the mass of that society, leads to the kind of conflict between formed and substantive rationality that may increasingly render calculations and thus bureaucratic domination inefficient and irrational" (p. 344).

Piel, Gerard. *The Acceleration of History*. New York: Knopf, 1972. Pp. 369. A collection of essays most of which appeared first in places like *Scientific American, Science and Public Affairs,* and *Nature*. The chapter on "The Heritage of Science in a Civilization of Machines" (pp. 51–63) and the section on "Technology and Democratic Institutions" (pp. 99–139) deserve notice.

Pizzuli, Francis C. "Asexual Reproduction and Genetic Engineering: A Constitutional Assessment of the Technology of Cloning," *Southern California Law Review* 47 (February 1974): 476–584. A ban on cloning could be founded on the state's right to preserve the nature of the nuclear family or equally fundamental cultural ideas.

Plattel, Martin G. *Utopian and Critical Thinking*. Pittsburgh: Duquesne Univ. Press, 1972.

Pp. 156. Good review of utopian attitudes in Part I, ch. 4, "The History of Utopian Thinking" (pp. 27-41).

Plebe, Armando. "Il terricidio: mito e futurologia" [Terracide: myth and futurology], *Proteus* 2, no. 4 (1971): 157-163.

Polak, Frederik Lodewijk. *The Image of the Future; Enlightening the Past, Orientating the Present, Forecasting the Future.* 2 vols. Translated by Elise Boulding. Leyden: A. W. Sythoff; and New York: Oceana Publications, 1961. From *De toekomst is uerleden tijd; Culture-futuristische uerkenningen* (Utrecht: W de Haan, 1955). A comprehensive historical-philosophical study of futurology. Vol. 1, *The Promised Land, Source of Living Culture* (Pp. 456), contains a couple of short sections dealing explicitly with "Pure and Applied Natural Science" (p. 298) and "Technology" (pp. 327-328), plus a good discussion of utopian and anti-utopian speculation. Vol. 2, *Iconoclasm of the Images of the Future; Demolition of Culture* (Pp. 368), continues the analysis of anti-utopian thought in conjunction with a discussion of "The Broken Image of Western Culture" in terms of religious faith, philosophy, science, depth-psychology, and modern art. Abridged English translation (Amsterdam and New York: Elsevier, 1973). Pp. viii, 319. By a philosopher who first studied law and economics and was a successful business executive. A major European futurologist.Cf. also the author's *Prognostics; A Science in the Making Surveys and Creates the Future* (Amsterdam and New York: Elsevier, 1971). Pp. xxviii, 425. This is an English translation of the abridged edition of *Prognostics* (1969), which was first published as a two-volume work (Denventer: AE. E. Kluwer, 1968).

——. *De nievwe wereld der automatie; Een industriëlle en sociaal-culturele revolutie* [The new world of automation; industrial and social-culture revolution]. 2nd ed. Hilversum: W. de Haan; Antwerpe: Standard Wetenschappelijke Vitg., 1966. Illustrated with 20 pages of photographs. A popular work which updates the author's earlier *Automatie: Industriëlle en culturele revolutie* (Zeist: W. de Haan, 1958). Pp. 165.

Poppi, Antonio. "L'Emancipazione etico-politica dalla razionalita positivistica e tecnocratica secondo J. Habermas" [Ethical-political emancipation from positivist and technocratic rationality according to J. Habermas], *Rivista di filosofia neo-scolastica* 64, no. 4 (July-September 1972): 471-484. Summarizes Habermas's ethical thought from his more important theoretical essays. Argues that in the moral tension developed in his theory of "critical reasons" Habermas approximates classical elements in the foundations of ethics.

Prenting, Theodore O., and Nicholas T. Thomopoulos. *Humanism and Technology in Assembly Line Systems.* Rochelle Park, N.J.: Hayden Book Co., 1974. Pp. 404. This book provides a comprehensive view of the information, strategies, and techniques now available regarding assembly line work. Striking a balance between technical advances and optimum worker productivity and satisfaction, the authors provide an invaluable guide to developing the most productive assembly systems. A book that really understands the worker!

Rama Rao, P. S. S. "The Structure of a Non-Violent Society: An Analysis of Gandhian Thought," *Journal of Thought* 9 (January 1974): 39-46. Points out that Gandhi's non-violent society presupposes a rural, nontechnological and economically self-sufficient social order.

Raskin, Marcus. "A Subjective/Objective Rendering of Technology." *Humanist* 33, no. 5 (September-October 1973): 25-27. Out of some pointed subjective descriptions of technological life, the author makes a strong argument for the objective limitation of technology. "Technology and its forms become questions of value and purpose to be judged accordingly."

Riccio, Stefano. "Valori culturali e diritti naturali nell'età technologica" [Cultural values

and natural rights in the age of technology]. *Incontri Culturali* 6 (1973): 349–356.
Richardson, Robert A. "Dissent From the Imagery of Spaceship Earth," *North American Review* 258 (Summer 1973): 3–7.
———. "Some Social Consequences of Technological Success," *Hastings Center Report* 4, (November 1974): 4–6.
Ridgeway, James. *The Politics of Ecology*. New York: E. P. Dutton, 1970. Pp. 222. Liberal account of the government's involvement (or rather noninvolvement) in the ecological crisis in the 1960s, by the editor and founder of the radical weekly *Hard Times*.
Rinehart, Kenneth L., Jr., William O. McClure, and Theodore L. Brown, eds. *Wednesday Night at the Lab: Antibiotics, Bioengineering, Contraceptives, Drugs, and Ethics*. New York: Harper & Row, 1973. Pp. xii, 226. Based on a series of lectures at the University of Illinois at Urbana.
Ritterbush, Philip C., with the assistance of Martin Green, eds. *Technology as Institutionally Related to Human Values*. Washington: Acropolis Books, 1974. Pp. 198. Bibliography by Ritterbush and Belinda Barrington on "Technology as an Institution: Social and Cultural Aspects" pp. 157–190. Index.
Roberts, R. W. "Mankind and the Technological Imperative," *Vital Speeches* 41 (November 15, 1974): 68–70.
Romanell, Patrick. "A Philosophic Preface to Morals in Medicine," *Bulletin of the New York Academy of Medicine* 50 (January 1974): 3–27. Argues that the purpose of medical ethics has never been adequately examined. States that all previous medical codes skirt the issue which is central to ethics: "the daily frustration over choices between right and right."
Ropohl, Günter. "Die Systemtechnik und das gesellschaftliche Bewusstsein des Ingenieurs" [Systematic technology and the social consciousness of engineers], *Technica* (1970): 303–305.
———. "Thesen zur technologischen Aufklärung" [Theses on technological enlightenment], *Dortmunder Hefte* 2, no. 1 (1971): 19–22.
Rostenne, Paul. "Le piège de la société technologique" [The trap of technological society], *Giornale di Metafisica* 29 (1974): 9–44.
Rousseau, G. S. "The Peril of Princes: What Ever Happened to Those Two Other Cultures?" *Denver Quarterly* 7, no. 2 (Summer 1972): 20–45. A discussion of C. P. Snow's theory of the two cultures—science and the humanities.
Rubinoff, Lionel. "The Crisis of Modernity: The Implicit Barbarism of Technology," pp. 3–24, in L. Rubinoff, ed., *Tradition and Revolution*. New York: St. Martin's Press, 1971. Pp. 173. The introduction to this book, which includes articles by I. Illich, G. Grant, and R. J. Lifton.
Ruhnow, Martin, "Technik und Bildung" [Technology and development], *VDI-Zeitschrift* 114, no. 17 (December 1972): 1249–1256.
Sachsse, Hans, ed., with Hubert Fein, et al. *Technik und Gesellschaft* [Technology and society]. Pullach: Verlag Dokumentation, 1974. Pp. 413. This is volume one of a proposed three-volume work.
Salomon, Jean-Jacques. *Science and Politics*. Cambridge, Mass.: MIT Press, 1973. Pp. xxii, 273. "In sum, *Science and Politics* is a flawed work, and the treatment of the central topic is not fresh or compelling. The work may be of interest, however, for the bits and pieces of first-rate history and philosophy that are sprinkled throughout."—B. Bozemon, review in *Technology and Culture* 16, no. 3 (July 1975): 506–508. Favorably reviewed by J. R. Ravetz, *Science Studies* 4, no. 3 (July 1974): 295–297.
Savigear, P. "Some Political Consequences of Technocracy," *Journal of European Studies* (London) 1 (1971): 149–160.

Schischkoff, Georgi. "Kybernetik und Geisteswissenschaften" [Cybernetics and the humanities], *Schopenhauer-Jahrbuch* 53 (1972): 339–360.

Schlecht, Otto. "Technik—Wirtschaft—Gesellschaft" [Technology—economy—society], *VDI-Zeitschrift* 115, no. 16 (November 1973): 1243–1248. Technology, economics and society are defined by an ongoing interaction in which the engineer too is involved. New products and technologies influence a country's economic position, by means of trade to an even greater extent if economics is understood to be an open and dynamic system. This system also contains the challenge that every economic development should be measured in the light of the quality of life, because every technological development should make life easier for mankind. Schlecht is a secretary of the West German ministry of economics.

Schmookler, Jacob. *Patents, Invention, and Economic Change: Data and Selected Essays.* Edited by Zvi Grilliches and Leonid Hurwicz. Cambridge, Mass.: Harvard Univ. Press, 1972. Pp. xvii, 292. The first of the selected essays points out wide variations in the role of technological advance and rejects the myth that technological progress occurs at an ever-increasing rate. The second argues that "as a firm gets bigger it has to spend more to get an invention, the probability that it will use the inventions it makes declines, and so does the probability that its inventions will be significant" (p. 37). The third essay is on research science policy. The last two argue that economists should pay more attention to technological change as an economic activity. The last half of the volume, the "data," contains over 400 time series on patent activity in industry, providing raw material for future empirical research in this area.

Schroeder, Oliver. *The Dynamics of Technology: From Medicine and Law to Health and Justice.* Cleveland: Law-Medicine Center, 1972. Pp. 72.

Schumacher, E. F. "Economics Should Begin with People, Not with Goods," *Futurist* 8, no. 6 (December 1974): 274–275. Includes a bibliography to "intermediate technology" institutions and periodicals. Followed by a Schumacher interview. This entire issue of the *Futurist* is devoted to intermediate technologies.

Schwitzgebel, Ralph K. "Aesthetic Directions for Technology," *Soundings* (Vanderbilt Univ.) 53, no. 3 (Fall 1970): 293–302. Argues that "the humane disciplines, particularly art and literature" need to provide "positive alternatives for the future."

"Science, Culture and Society." *Scientia* (Milan) 107, nos. 9–10 (1972): 751–809. A symposium stimulated by an editorial of the same title in *Scientia* vol. 107, nos. 5–6 (1972). Contains the following articles, first in Italian then in English: L. Geymonat's "Neutrality Is Impossible," B. De Finetti's "Prejudice and Liberty," G. Montalenti's "Prometheus," J.-M. Lévy-Leblond's "Is There a Crisis in Science or in Society?"

"Science, Culture and Society," *Scientia* (Milan) 107, nos. 11–12 (1972): 939–972. The second in a series of symposia, containing the following articles, first in Italian then in English: M. Aloisi's "Science and Power," L. Bulferetti's "The Political Conditioning of Science," A. Pala's "Class and Science."

"Science, Culture and Society," *Scientia* (Milan) 108, nos. 1–2 (1973): 7–40. The third in a series of symposia, containing the following articles, first in Italian then in English: G. Giorello's "Objectivity and Non-Neutrality," B. Suchodolski's "Science and the Value of Life," L. Villa's "An Accusation Against Science?" Giorello teaches mathematics and philosophy at the University of Milan; Suchodolski is director of the Institute of Pedagogical Sciences at the University of Warsaw; Villa is emeritus professor of medicine at the University of Milan.

"Science, Culture and Society," *Scientia* (Milan) 108, nos. 7–8 (1973): 471–534. The fourth in a series of symposia, containing the following articles, first in Italian then in English: R. Canestrari's "Ideology and Research," G. Ciccotti and G. Jona Lasinio's "The Project of Research," G. R. Feiwel's "On the Relevance of Economics" (English only),

A. Visalberghi's "Science, Social Classes and Education." Canestrari lectures in psychology at the University of Bologna Medical School; Ciccotti and Lasinio are both theoretical physicists; Feiwel is an economist; and Visalberghi is with the philosophy department of the University of Rome.

"Science, Culture and Society," *Scientia* (Milan) 108, nos. 9–12 (1973): 663–696. The fifth in a series of symposia, containing the following articles, first in Italian then in English: L. M. Calabi's "Social Knowledge and Materialist Criticism," Margaret Mead's "The Social Responsibility of Anthropologists" (English only), L. Villa's "Responsibility of the Science Teacher." Calabi lectures in politics and economics; Mead is the well-known anthropologist.

Sibley, Mulford Q. *Technology and Utopian Thought*. Minneapolis: Burgess, 1971. Pp. 55. Lecture from the American Political Science Association, Chicago (September 1971).

Siebert, Charles. "A Conversation with F. J. Von Rintelen," *Listening* 8, nos. 1–3 (1973): 125–128. A silly interview with a visiting German philosophy professor at DePaul Univ. in Chicago. Some mention of technology on pp. 127–128. But every question and answer is so brief as to be philosophically worthless; the tone is pompous to boot.

Sinclair, Bruce, Norman R. Ball, and James O. Peterson, eds. *Let Us Be Honest and Modest: Technology and Society in Canadian History*. Toronto: Oxford Univ. Press, 1974. Pp. xvi, 309.

Sklair, Leslie. *Organized Knowledge: A Sociological View of Science and Technology*. St. Albans: Hart-Davis MacGibbon, 1973. Pp. 284.

Skolnikoff, Eugene B. *The International Imperatives of Technology; Technological Development and the International Political System*. Berkeley: Institute of International Studies, Univ. of California, 1972. Pp. ix, 194. Bibliography, pp. 187–194.

Slater, Philip E. *Earthwalk*. Garden City, N.Y.: Anchor Press, 1974. Pp. 230. A radical and sometimes rhetorically inclined critique of the anti-ecological aspects of modern technology. Does, however, raise some interesting points about the idea of machines as extensions of human organs.

Smith, Joe Mauk. *Innovation and Social Responsibility*. Delft: Waltman, 1972. Pp. 33. English and Dutch text.

Sousa, José Pedro Galvão de. *O estado technocrático* [The technocratic state]. São Paulo, Brazil: Saraiva, 1973. Pp. 143. Some bibliographical references.

Spier, Robert F. G. *Material Culture and Technology*. Minneapolis: Burgess, 1973. Pp. iii, 36.

Starr, Chauncey, and Richard Rudman. "Parameters of Technological Growth," *Science* 182, no. 4110 (October 26, 1973): 358–364. Commentary on D. H. Meadows, *et al.*, *The Limits to Growth* (New York: Universe Books, 1972).

Steenbergen, Bart van, and Eduard van Hengel. *Technocratie: ideologie of werkelijkheid* [Technocracy: ideology of actuality]. Groningen: Wolters-Noordhoff, 1971. Pp. 119. A discussion of the ideas of Scheisky, Freyer, Marcuse, Habermas, etc.

Steffens, Henry J., and H. N. Muller, III, eds. *Science, Technology, and Culture*. New York: AMS Press, 1974. Pp. viii, 204. The results of a dialogue between the Western Electric Company and the University of Vermont.

Steinbuch, Karl. *Mensch-Technik-Zukunft; Basiswissen für die Probleme von morgen* [Man-technology-future; basic knowledge for the problems of tomorrow]. Stuttgart: Deutsche Verlags-Anstalt, 1971. Pp. 352. Illustrated. Bibliography. Deals with technology assessment.

———. *Kurskorrektur* [Course correction]. Stuttgart and Degerlock: Seewald Verlag, 1973. Pp. 167. On technology and civilization.

Stöber, Gerhard J., and Dieter Schumacher. *Technology Assessment and Quality of Life*. New York: Elsevier, 1973. Pp. 302. This Proceedings of the Fourth General Conference of the "Salzburg Assembly: Impact of the New Technology" (SAINT) includes: G. J. Stöber's "Quality of Life—Its Scope and Elements," Clark C. Abt's "The Social Role of Technology," Francois Hetman's "Social Objectives and New Desirable Technologies," D. Schumacher's "Technology Assessment—The State of the Art," etc.

Strasser, Gabor, and Eugene M. Simons, eds. *Science and Technology Policies: Yesterday, Today, and Tomorrow.* Cambridge, Mass.: Ballinger Pub. Co., 1974. Pp. 286.

Strong, Maurice F., ed. *Who Speaks for Earth?* New York: W. W. Norton, 1973. Pp. 173. Lectures sponsored by the International Institute for Environmental Affairs, June 1972. Includes: Barbara Ward's "Only One Earth" (pp. 19–31), René Dubos's "Unity Through Diversity" (pp. 33–42), Thor Heyerdahl's "How Vulnerable Is the Ocean?" (pp. 45–63), Gunner Myrdal's "Economics of an Improved Environment" (pp. 67–105), Carmen Miró's "Population" (pp. 109–125), Sir Solly Zuckerman's "Science, Technology, and Environmental Management" (pp. 129–150), and Aurelio Peccei's "Human Settlements"(pp. 153–167. Zuckerman's essay is a negative critique of *Limits to Growth* (1972).

Sumner, Jeremy. "Philosophy-production by the Masses: Not Mass Production," *Engineer* 238, no. 6154 (February 21, 1974): 43–45. Western industrialization is usually irrelevant to the needs of developing nations. British industry should provide work in the Third World by financing small projects which use less complex, energy-saving machines.

Sun, P. Hans. "From Over-man to Omega Man," *Political Query* (Wilmington) 1, no. 2 (Fall 1974): 120–135. A brief critique of the anthropological theories of Nietzsche and Fudpucker as containing elements of technological totalitarianism, from the point of view of classical Tai-Chi theory.

Susinos Ruiz, Francisco. *La técnica, complicación del hombre* [Technology, complication of man]. Santander: Institución Cultural de Cantabria, 1974. Pp. 241.

"Symposium: Law and Technology." *Southern California Law Review* 45, no. 1 (Spring 1972). This symposium which begins with a preface by Justice William O. Douglas and an introduction by John G. Burke, includes the following: A. J. Rosenthal's "The Federal Power to Protect the Environment: Available Devices to Compel or Induce Desired Conduct," Christopher D. Stone's "Should Trees Have Standing? — Toward Legal Rights for Natural Objects," A. D. Tarlock, R. Tippy, and F. E. Francis's "Environmental Regulation of Power Plant Siting: Existing and Proposed Institutions," J. C. Oppenheimer and W. H. Lambright's "Technology Assessment and Weather Modification," J. Lederberg's "The Freedoms and the Control of Science: Notes from the Ivory Tower," and "Conditioning and Other Technologies Used to 'Treat?' 'Rehabilitate?' 'Demolish' Prisoners and Mental Patients." This is also the first in a series of symposia devoted to this general theme, others of which are listed below.

"Symposium: Law and Technology." *Southern California Law Review* 46 (June 1973). Includes: Laurence H. Tribe's "Technology Assessment and the Fourth Discontinuity: The Limits of Instrumental Rationality," and S. Breyer and P. W. MacAvoy's "The Federal Power Commission and the Coordination Problem in the Electrical Power Industry."

"Symposium: Law and Technology." *Southern California Law Review* 48 (November 1974): 209–570. Includes: R. W. Findley and S. J. Plager's "State Regulation of Nontransportation Noise: Law and Technology," M. H. Shapiro's "Who Merits Merit? Problems in Distributive Justice and Utility Posed by the New Biology," M. R. Gelpe and A. D. Tarlock's "The Uses of Scientific Information in Environmental Decision-Making," L. E. Allen's "Formalizing Hohfeldian Analysis to Clarify the Multiple Senses of 'Legal Right': A Powerful Lens for the Electronic Age," and "Guilt by Physiology: The Constitutionality of Tests to Determine Predisposition to Violent Behavior."

Taviss, Irene. *Our Tool-Making Society.* Englewood Cliffs, N.J.: Prentice-Hall, 1972. Pp. 145. Revised versions of the state-of-the-art introductions to the four Harvard University Program on Technology and Society *Research Reviews* of "Technology and the Polity" (#4), "Technology and Values" (#3), "Technology and the Individual" (#6), "Implication of Biomedical Technology" (#1), "Technology at Work" (#2), "Technology and the City" (#5). For this volume, the first three retain their states as separate chapters, while the last three are combined in one chapter on "Technology and Social Problems." Brief bibliography and index.

Technologie und Kapital. Frankfurt:Suhrkamp,1973. Pp. 234. Eight essays of which seven were originally a series of lectures at Heidelberg University in 1971.

Teixeira, Anisio. *Cultura e tecnologia* [Culture and technology]. Rio de Janeiro: Fundação Getúlio Vargas, Instituto de Documentação, 1971. Pp. 70.

Terrón, Eloy. *Ciencia, técnica y humanismo* [Science, technology and humanism]. Madrid: Gráf. Espejo, 1973. Pp. 261.

Thomas, Lewis. "Commentary: The Future Impact of Science and Technology on Medicine," *BioScience* 24, no. 2 (February 1974): 99–105.

Thompson, William Irwin. *Passages about Earth: An Exploration of the New Planetary Culture.* New York: Harper & Row, 1973. Pp. 207. "The disease, for Thompson . . . lies at the scientific and technological heart of industrial civilization. Its cure will be the emergence of a new Pythagorean science replacing the old Archimedean one. Pythagorean science, though available to technology, will find equal expression in art and religion. Archimedes was *Homo faber,* Pythagoras *Homo ludens.* In the new Pythagorean synthesis, *Homo ludens* comes into his own. Thompson sees a 'four-stage process: (1) crazies; (2) artists; (3) savants; and (4) pedants' (p. 132). Artists like C. S. Lewis and, more recently, Doris Lessing have already carried on into stage 2. Stage 3 is dawning in the work of savants like Paolo Soleri, C. F. von Weizsaeker, and others. The bulk of *Passages about Earth* is a kind of erudite travelogue, a visit to stage-3 hidden valleys . . ."—John A. Miles, Jr., in a review in *Zygon* 9, no. 3 (September 1974): 256–263.

Thrall, Charles A., and Jerold M. Starr. *Technology, Power, and Social Change.* Lexington, Mass.: D. C. Heath, 1972. Pp. 224. Reprinted Carbondale, Ill.: Southern Illinois University Press, 1974. Pp. 169. Papers from a symposium at the University of Pennsylvania. Includes essays by L. Mumford, R. Theobald, R. Boguslaw, M. Kranzberg, C. R. Dechert, A. Montagu, M. Bookchin, etc. Favorably reviewed by H. J. Muller, *Technology and Culture* 15, no. 1 (January 1974): 82–86.

Thring, Meredith Woolridge. *Machines—Masters or Slaves of Man?* Stevenage, England: Peter Peregrinus, 1974. Pp. 115. Illustrated. Examines "the misuse of technology" (chapter 2) and explores the idea of a "technology-based Utopia." "I put it quite simply — the only hope for survival of humanity with a life of decent quality into the 21st century is for the large majority of Scientists, Engineers and Technologists to accept full responsibility for all the effects of their professional activities and for explaining all these effects to the general public to understand the choices involved and insist on the choice being made which takes most account of all world inhabitants in future generations" (p. 108). By a well-known British engineer. Cf. also the author's *Man, Machines and Tomorrow* (London and Boston: Routledge and Kegan Paul, 1973).

Tonsor, Stephen. "Science, Technology and the Cultural Revolution," *Intercollegiate Review* 8, no. 3 (Winter 1973): 83–89. Criticizes the antitechnological attitudes of both the New Left and the Old Right as expressions of "satiated groups" rather than true revolutionaries (p. 87). "The price of survival is not less reason but more science . . . not less . . . but more technology" (p. 89).

Trautmann, Wolfgang. *Utopia und Technik; zum Erscheinungs und Bedeutungseandel des Phänomens in der modernen Industriegesellschaft* [Utopia and technology; on the manifestations and changes of meaning of utopian phenomena in modern industrial soceity].

Berlin: Duncker and Humblot, 1974. Pp. 150. Originally the author's thesis.
Tsanoff, Radoslav Andrea. *Civilization and Progress.* Lexington: Univ. Press of Kentucky, 1971. Pp. 376. Part one is "A Historical Review of the Idea of Social Progress." Part two, "Social Confidence and the Despair of Progress: Alternative Judgments of Civilization," contains a chapter on "Economic Values, Technology, and Human Progress." By a professor of philosophy at Rice University.
Urban, George R., with the collaboration of Michael Glenny, eds. *Can We Survive Our Future? A Symposium.* New York: St. Martin's Press, 1972. Pp. vi, 399. From interviews originally broadcast over Radio Free Europe. Contents: Arnold J. Toynbee's "Technical Advance and the Morality of Power," Philip Rieff's "The Loss of the Past and the Mystique of Change," Nigel Despicht's "Old Values and the Demands of New Technology," Werner Heisenberg's "Rationality in Science and Society," Jacques Ellul's "Conformism and the Rationale of Technology," Erich Jantsch's "For a Science of Man," Louis Armand's "The Anachronisms of Sovereignty in a Technological World," Gunnar Rander's "NATO and the Environment," Bernard Cazes's "Opportunities and Pitfalls of Future-Oriented Research," Ossip K. Flechtheim's "Marxism and the Third Road," and Brian Aldiss's "Learning to Live With a Doom-Laden Future."
Uscatescu, George. "Avventura e rischio dell'uomo nella civiltà technologica" [Adventure and danger of man in technological society], *Proteus* 2, no. 4 (1971): 147–155.
Vacca, Roberto. *The Coming Dark Age.* Trans. from Italian, *Il Medioevo prossimo venturo* (1971), by J. S. Whale. Garden City, N.Y.: Doubleday, 1973. Pp. 221. An optimistic, popular, apocalyptic interpretation of contemporary history. Very thin social philosophy.
Vafa, A., and M. Drobyshov. "Technological Progress and Spiritual Culture in the Countries of the Third World," *Asian Survey* 14 (March 1974): 207–219.
Verein Deutscher Ingenieure, ed. *Studium der Technik—Ingenieure von morgen* [Studies of technology — engineering of the future]. Düsseldorf: VDI-Verlag, 1971.
———. ed. *Wirtschaftliche und gesellschaftliche Auswirkungen des technischen Fortschrifts* [Economic and social effects of technical development]. Düsseldorf: VDI-Verlag, 1971.
Visscher, Maurice B., ed. *Humanistic Perspectives in Medical Ethics.* Buffalo: Prometheus Press, 1972. Pp. xiii, 297.
Ward, Barbara, and René Dubos. *Only One Earth; The Care and Maintenance of a Small Planet.* New York: Norton, 1972. Pp. xxv, 225.
Wettstädt, Günter. *Technik und Bildung; Zum Einfluss bürgerlicher Technikphilosophie auf die imperialististische Bildungsideologie* [Technology and development; on the influence of middle-class philosophy of technology on the imperialistic ideology of development]. Frankfurt a/M: Verlag Marxistische Blätter, 1974. Pp. 102. Also published in East Germany under the title *Ideologie im Zwielicht* (Berlin: Akademie-Verlag, 1974).
Wheeler, Harvey. "Technology: Foundation of Cultural Change," *Center Magazine* 5, no. 4 (July–August 1972): 48–57.
Whisnant, David E. "The Craftsman: Some Reflections on Work in America," *Centennial Review* 17, no. 3 (Summer 1973): 215–236. A defense of craftsmen in an efficiency-oriented age.
Wiff-Hansen, Johs. "Marxian Methodology and the Research of Futures," *Danish Yearbook of Philosophy* 10 (1973): 21–33. A sympathetic appraisal of Marxist ability to make a unique contribution to futurology.
Williams, Robert H. *To Live and To Die: When, Why, and How.* New York: Springer-Verlag, 1974. Pp. xviii, 346. A collection of original articles on a wide variety of topics, mostly in the area of bioethics. The editor, a professor of medicine, contributes "Prologue," "Metabolism, Mentation, and Behavior," "Body, Mind, and Soul," "Propaga-

tion, Modification, and Termination of Life: Contraception, Abortion, Suicide, Euthanasia," "Management of the Sick with Kindness, Compassion, Wisdom, and Efficiency," "Careers and Living," and "Epilogue." Other contents: E. Fischer's "On the Origin of Life," G. Omenn's "Genetic Engineering: Present and Future," K. Davis's "The Climax of World Population Growth," A. Dyck's "An Alternative to the Ethic of Euthanasia," Joseph Fletcher's "Ethics and Euthanasia," J. R. Elkinton's "Ethical and Moral Problems in the Use of Artificial and Transplanted Organs," E. Kübler-Ross's "Life and Death: Lessons from the Dying," J. L. Walker's "The Here and the Hereafter: Reflections on Tragedy and Comedy in Human Existence," M. Rothenberg's "Too Many and Too Few Limitations for Children," R. F. Rushmer's "Advantages and Disadvantages of Technological Achievements," S. Wolf's "Causes and Effects of Excessive Fears, Anxieties, and Frustrations," D. Farnsworth's "Causes and Management of Current Anxieties and Frustrations in Universities," D. X. Freedman's "The Social and Psychiatric Aspects of Psychotropic Drug Use," W. W. Menninger's "Causes and Management of Criminals: Psychiatric Aspects," J. Darrah's "The Criminal Justice System: Crimes, Criminal Processes, and Sentencing," L. Rieke's "Some Major Guides for Laws," E. M. Pattison's "Psychosocial and Religious Aspects of Medical Ethics," D. Mace's "Marriage: Whence and Whither?" J. Hampson's "Changing Views on Homosexuality, Transvestism, and Transsexualism," and M. Tumin's "Equality and Inequality: Facts and Values."

Winter, Gibson. "Human Rights in a Technological Society," *Philosophy in Context* 1(1972): 5–8. Distinctions made among civil, political, socio-economic, and cultural rights. The argument is that under the social conditions of high technology, cultural rights become decisive for the practice of other rights. Furthermore, the exercise of cultural rights assimilates citizens to techno-cultural domination unless tied to a countercultural base. Two other works by this author which it might be useful to consult: *Being Free; Reflections on America's Cultural Revolution* (New York: Macmillan, 1970), which is sensitive to the issue of technological alienation; and "Human Science and Ethics in a Creative Society," *Cultural Hermeneutics* 1, no. 2 (July 1973), 145–174, where the sociological tradition is interpreted as a modern version of the this-worldly theory of man. The religious importance of sociology is located in the struggle of Western Christianity to come to terms with historical existence after its inception in an eschatological faith.

Wright, Christopher. "Toward Future Guidance of Science in Human Affairs," *1970 Britannica Yearbook of Science and the Future*. Chicago: Encyclopaedia Britannica, Inc., 1969, pp. 428–436. By the director of the Columbia University Institute for the Study of Science in Human Affairs.

Wu, Kuang-Ming. "Hope and World Survival," *Philosophy Forum* 12, nos. 1–2 (Sept. 1972): 131–147. On how hope frees from reliance on past, present, and future.

Yablonsky, Lewis. *Robopaths*. Indianapolis: Bobbs-Merrill, 1972. Pp. 204. "Robots are machine-made simulations of people. I would coin the term *robopath* to describe people whose pathology entails robot-like behavior and existence" (pp. 6–7). A popular criticism of advanced industrial civilization.

Zbinden, Hans. *Der bedrothe Mensch; Zur sozialen und seelischen Situation unserer Zeit* [Threatened man; on the social and mental situation of our time]. 2nd ed. Munich: A. Franke, 1970. Pp. 309.

C. Appendix: Soviet and East European Materials

Aksenenok, G. A. "Nature and Man in the Time of Scientific and Technological Progress" (in Russian), *Voprosy filosofii* 27, no. 10 (1973). Socialism can solve the problems of environmental protection and the rational use of natural resources.

Anokhin, P. K. "Philosophical Importance of the Problem of Natural and Artificial Intellect" (in Russian), *Voprosy filosofii* 27, no. 6 (1973). Modeling of intellectual activity in computer systems produces knowledge about the mechanisms of thinking. However, neuro-cybernetics has not adequately cooperated with neurophysiology. The author proposes his concept of a "functional system" to bridge this gap.

Bachurin. A. V. "Scientific and Technological Progress and the Economic Mechanism" (in Russian), *Voprosy filosofii* 28, no. 4 (1974). Concerned with problems in the planning and forecasting of production and the improvement of methods and organizational structure of management in a way which helps orient future socio-philosophical research.

Bukharin, N. I., et al. *Science at the Crossroads*. London: Cass, 1971. Pp. 236. The historically important papers given by the Soviet delegation at the Second International Congress of the History of Science and Technology held in London in 1931, most of which have been long unavailable. Reviewed by Sarah White in *New Scientist* 52, no. 774 (1971): 180–181.

Ćetković, Vladan. *Tehnokratska ideologija* [Technocratic ideology]. Belgrade, Yugoslavia: Institut za političke studije Fakulteta Političkih nauka, 1973. Pp. 278. See also the author's *Birokratija i tehnokratija* [Bureaucracy and technocracy] (Belgrade: 1973). Pp. 45.

Csizmas, Michael. "Cybernetics, Marxism, Jurisprudence," *Studies in Soviet Thought* 11 (1971): 90–108.

Daglish, Robert, ed. *The Scientific and Technological Revolution: Social Effects and Prospects*. Moscow: Progress, 1972. Pp. 278.

Dimitrieva, M. S., and O. F. Ovčarov. "Vzaimodejstvie émpiričeskogoi teoretičeskogo v texničeskix naukax v processe progressivnogo razvitija nauki i techniki" [The interaction of empirical and theoretical (elements) in the technological sciences in the process of the progressive development of science and technology] in S. A. Kugel, et al., eds., *Problemy dejatel' nosti učenogo i naučnyx kollektivov* vol. 5 (Leningrad: 1973), pp. 102–105.

Dorosinski, W. "Wykaz ważniejszych publikacji poświeconych projektowaniu" [Select bibliography of works on design], *Prakseologia* 41 (1972): 245–256. Bibliography with 266 titles, many in Polish.

Dubinin, N. P. "The Philosophy of Dialectical Materialism and Problems of Genetics" (in Russian), *Voprosy filosofii* 27, no. 4 (1973). "The progress of genetics has brought about a revolutionary situation in biology, which calls for radical changes in biological thinking and for posing in a new way the question of relationship between the different branches of biology. The possibilities of regulating heredity, opened by modern genetics, and theoretical conditions for evolving genetic engineering have raised social and ethical problems before scientists. A correct solution of these problems is possible only if Marxist-Leninist fundamentals are consciously applied to genetics. In turn, the enrichment of the conceptual mechanism of genetics helps make more concrete the laws and categories of dialectical materialism in relation to the biological form of the motion of matter and to biological knowledge."—from the English language summary.

Edeling, Herbert, and Hans Kulow. "Wissenschaftlich-technische Revolution und Entwicklung der Arbeiter-persönlichkeit im Sozialismus" [Scientific-technological revolution and the development of worker personality in socialism], *Deutsche Zeitschrift für Philosophie* 22, nos. 10–11 (1974): 1274–1283. In the organic joining of the gains of the scientific-technological revolution with the advantages of the socialist economic system, it is more important from the beginning to structure the work process so that high efficiency and the development of the socialist personality are in increasing harmony.

Fedorov, E. K., and I. B. Novik. "Man and His Natural Environment," *Soviet Studies in*

Philosophy 12, no. 2 (Fall 1973): 3–25. A Soviet analysis of the ecological crisis. "By digging deeper into objective truth in biosphere processes, man will be able rationally to combine transformation of nature with the required harmonizing of his relationship with it" (p. 24). Translated from *Voprosy filosofii* 26, no. 12 (1972), a paper prepared for the XVth World Congress of Philosophy on "Man, Science, and Technology."

Fedoseyev, P. N. "The Social Importance of the Scientific and Technological Revolution" (in Russian), *Voprosi filosofii* 28, no. 7 (1974). A lengthy and authoritative statement of the Marxist understanding of the scientific and technological revolution, by the vice-president of the USSR Academy of Sciences, originally delivered as a plenary report for the 8th International Sociological Congress in Toronto, Canada. Differentiates use of the term "revolution" in the proper sense ("social revolution") and as applied to changes in science, technology and production. Since the 1950s changes in science-technology have pointed toward social revolutionary consequences. [Marxist "scientific and technological revolution"—what in the West is sometimes called the Second Industrial Revolution.] Discussion of why Marxism has had priority in developing the analysis of this concept, along with a critique of Western sociological analyses. Specifically argues against the idea that scientific-technological progress of itself either causes social problems or can be a solution to social problems.

Filipec, Jindřich. "Kritik bürgerlicher Interpretationen der Wissenschaftlich-technischen Revolution" [Critique of the bourgeois interpretation of the scientific-technological revolution], *Deutsche Zeitschrift für Philosophie* 21, no 8 (1973): 965–980. Increasing problems of the controllability of natural and social processes under capitalism calls for the use of more conscientious scientists in the formulation of development policies. However, the bourgeois social philosophy is doomed to practical failure, since it refuses to abandon capitalist methods of production. This refusal is common to bourgeois social philosophy in all forms—from the pseudo-radicalism of Marcuse to the conservatism of Popper.

Flerov, D. N., and V. S. Barashenkov. "Science at the Epoch of Scientific and Technological Revolution" (in Russian), *Voprosy filosofii* 28, no. 9 (1974). Discusses perspectives of development in science in conjunction with problems of world resources and the future directions of scientific research. Concentrates on fundamental questions and the possibility of comprehensive theories.

Fukász, György. "Die wissenschaftlich-technische Revolution und die Veränderungen der Arbeit" [The scientific-technological revolution and the changes in work], *Deutsche Zeitschrift für Philosophie* 21, no. 7 (1973): 820–840. The way the scientific-technological revolution is carried out is determined by existing production conditions. Thus socialism requires that in pursuing the scientific-technological revolution attention is paid to the concrete structure of the character of work

Ganovsky, Sava. "Philosophical and Sociological Problems of Science, Technology and Man" (in Russian), *Voprosy filosofii* 27, no. 7 (1973): 20–26. "Marxism-Leninism teaches that the transformation of the principal force of production—man—into a free toiler . . . is what constitutes real . . . humanism." A good exposition of the Marxist view of the nature of science and technology. English translation as "Interrelations of Science, Technology, and Man in Social Philosophical Perspective," *Soviet Studies in Philosophy* 13, no. 1 (Summer 1974): 24–36.

Gasparski, W. "Prolegomena do metodologii projektowania" [Introduction to design methodology], *Prakseologia* 41 (1972): 5–22.

Gauzner, Nikolaĭ Dmitrievich. *Social Effects of the Scientific and Technological Revolution under Capitalism.* Moscow: Novoski Press Agency Publishing House, 1973. Pp. 196.

Gorbov, F. D., and V. I. Lebedev. "Man in Technical Systems" (in Russian), *Voprosy filosofii* 27, no. 6 (1973). Technical systems free man from strenuous physical labor but make higher demands of his psyche. This should be taken into account in designing technical systems, because man should not have to adapt to technical systems, instead technical systems should be adapted to man. However, great difficulties arise because we do not yet have exact knowledge about the psychic possibilities of man who becomes part of new technical systems. His true limitations become apparent only in his practical functioning. Besides, some psycho-physiological mechanisms are psychic deviations as a result of operator stress.

Gudozhnik, G. S. *Nauchno-teknicheskii progress: sushchnost' osnovnye tendentsii* [Scientific-technical progress: essence of basic tendencies]. Moscow: Nauka, 1970. Pp. 272. "An excellent general survey of current Soviet developments in the scientific-technical fields associated with advanced technology." Primarily historical.—S. Lieberstein, review in *Technology and Culture* 14, no. 2, Part I (April 1973), 323–325.

Gvishiani, J. M. "Scientific and Technological Revolution and Social Progress" (in Russian), *Voprosy filosofii* 28, no. 4 (1974). The term "scientific and technological revolution" is often used in official communist documents; is an important extension of Marxist thought, and thus deserves careful formulation. "The scientific and technological revolution is a radically qualitative transformation of productive forces, the conversion of science into an immediate productive force and, accordingly, the revolutionary changing of the material and technical foundations of social production, its content and form, the character of labor, and the social division of labor. It has become possible only thanks to a high degree of socialization of production—a process which creates objective preconditions for transition from the capitalist to the socialist mode of production."—from the English language summary.

Hager, Kurt. *Wissenschaft und Technologie im Sozialismus* [Science and technology under socialism]. Berlin: Dietz Verlag, 1974. Pp. 77.

Hegedüs, András, and Mária Márkus. "Modernization and the Alternatives of Social Progress," *Telos* (Fall 1973): 145–157. Translation of an article from the "Budapest School" of socialism which caused the authors to lose their jobs and be expelled from the Communist Party. Preceded by introductory material.

Hronský, F. "On Problems of Forming the Socialist Way of Life Under Conditions of the Scientific and Technological Revolution" (in Russian), *Teorie a Metoda* 5, no. 3 (1973): 77–96.

Iovchuk, M. T. "The Future of Scientific Philosophy in View of Social Development and Scientific and Technological Progress in the Last Third of the 20th Century" (in Russian), *Voprosy filosofii* 27, no. 6 (1973). In light of the current scientific-technological revolution and the contemporary orientation toward the practical creation of a new future world, Marxist-Leninist philosophy is changing too; its functions are being widened and made more precise, as is its place in the interaction with particular sciences, and its ideological and methodological role in society is increasing.

Khozin, G. S. "Science and Engineering, Ideology and Politics" (in Russian), *Voprosy filosofii* 27, no. 1 (1973). An analysis of the U.S. space program which argues that scientific and technological progress continues to be influenced by the military-industrial complex. In conjunction with the inability of a capitalist state to make full use of scientific and technological progress for the good of society as a whole, this fact has given rise to antitechnocratic sentiments among some sections of the people. (If only the antitechnologists understood!) English translation: "Science and Technology, Ideology, and Politics in the U.S.A.," *Soviet Studies in Philosophy* 12 (Winter 1973–1974): 50–67.

Kobhakov, Valentin Petrovich. *Naucho-tekhnicheskii progress i nravstvennost'*. Moscow: 1973. Pp. 63. On social ethics of technical civilization.

Konstantinov, F. V. "Scientific and Technological Revolution and Problems of Moral Progress" (in Russian), *Voprosy filosofii* 27, no. 8 (1973). Discusses three views of influence of science-technology on moral progress: 1) the pessimistic view, which goes back to Rousseau; b) the optimistic view, which rests on the concepts of "industrial" and "post-industrial" society; c) the Marxist view, which sees contradictions between the scientific-technological revolution and the moral situation in capitalist countries. Argues that the problems posed by scientific-technological progress can only be solved by means of a socialist transformation of society.

Kovalev, A. M., and V. I. Kovalenko. "On the Question of the Interrelations Between Scientific-Technological and Social Revolution," *Soviet Studies in Philosophy* 10, no 4 (Spring 1972): 383–394. Translated from *Vestnik Moskovskogo universiteta, seriia filosofii* no. 2 (1971).

Krasin, Y. A. "Apologetical Nature of the Theory of Post-Industrial Society" (in Russian), *Voprosy filosofii* 28, no. 2 (1974). An extended critique of Daniel Bell's *The Coming of Post-Industrial Society* (1973), with special emphasis on his methodological "axial principle."

Kravchenko, I. I., and V. S. Markov. "Scientific-Technological Progress and the Development of the Individual Under Socialism," *Soviet Studies in Philosophy* 11, no. 1 (Summer 1972): 48–69.

Kröber, Günter. "Scientific and Technological Revolution, Science and Society" (in Russian), *Voprosy filosofii* 28, no. 3 (1974). Because bourgeois philosophy, unlike Marxist-Leninist theory, does not see science as essentially a social phenomenon, but examines it as an autonomous entity with social concomitants, it is unable to separate the accidental from the essential in considering the social consequences of science. Bourgeois theory is reduced to quantitative descriptions of scientific development. The basic laws governing the development of science are laws of the interaction of science with other spheres of life, above all production; the laws of scientific development are not immanent laws.

Kubík, J. "Complex Planning of the Development of Science, Technology and Economy—Some Problems" (in Russian), *Teorie a Metoda* 5, no. 3 (1973): 97–110. Scientific and technological development can be speeded up only by improving the management of the whole science-technology-production system.

Kukel, J. "Scientific and Technological Revolution and the System of Management of National Economy" (in Russian), *Teorie a Metoda* 5, no. 3 (1973): 111–130.

Kutta, F. "Scientific and Technological Revolution and Social Planning" (in Russian), *Teorie a Metoda* 5, no. 3 (1973): 53–76.

Lingner, Edith. *Zum Wesen und zur politischen Stossrichtung technocratischer Staatsauffassungen: eine Auseinandersetzung mit einigen imperialistischen Ideologen der DRD* [On the essence and the political push toward a technocratic conception of the state: an explanation of several imperialistic ideologies of West Germany]. Potsdam-Babelsberg: Akademie für Staats- und Rechtswissenschaft der DDR, Informationszentrum Staat und Recht, 1974. Pp. 206.

Mamykin, Igor' Petrovich. *Analogiia v tekhnicheskom tvorchestve*. Minsk: Nauka i Tekhika, 1972. Pp. 168. On the philosophy of inventions.

Markov, Nikolai Vasil'evich. *Nauchno-tekhnicheskaia revoliutsiia: analiz, perspektivy, posledstviia* [Scientific-technological revolution: analysis, outlook, effects]. 2nd edition. Moscow: Politnzdat, 1973. Pp. 239.

Marković, Mihailo. *From Affluence to Praxis: Philosophy and Social Criticism*. Ann Arbor: Univ. of Michigan Press, 1974. p. xiv, 265. Two relevant chapters by this Yugoslavian social theorist: chapter two, "Possibilities for Radical Humanization in Modern Industrial Civilization;" and chapter three, "Technostructure and Technological Innovation in Contemporary Society."

"Mensch, Wissenschaft und Technik im Sozialismus" [Man, science and technology under socialism]. *Deutsche Zeitschrift für Philosophie* 21, special issue (August 1973). Pp. 224. Contents: Günther Bohring and Reinhard Mocek's "Wissenschaft — Persönlichkeit — Fortschritt" (pp. 5 ff.), Harald Schliwa's "Gesellschaftliche Bewusstheit und Humanismus in der wissenschaftlich-technischen Revolution" (pp. 36 ff.), Heinrich Opitz's 'Wissenschaftliche Erkenntnis in der sozialistischen Gesellschaft" (pp. 48 ff.), Vitali Stoljarow's "Zu weltanschaulichen Grundfragen der wissenschaftlich-technischen Revolution" (pp. 62 ff.), Herbert Hörz's "Naturerkenntnis und Ethik" (pp. 84 ff.). Klaus Fuchs-Kittowski, Reiner Tschirschwitz and Bodo Wenzlaff's "Mensch und Automatisierung" (pp. 104 ff.), Günter Kröber's "Wissenschaft, Gesellschaft und wissenschaftlich-technische Revolution" (pp. 122 ff.), Hubert Laitko's "Wissenschaft und Praxis im Sozialismus und die wissenschaftstheoretische Abbildung ihres Zusammenhangs" (pp. 141 ff.), Eberhard Fromm's "Zur Kritik revisionistischer und opportunistischer Theorien über Wissenschaft und Technik" (pp. 171 ff.), and Hermann Ley's Entfremdungseffekte und gesellschaftliches Bewusstsein in der bürgerlichen Gesellschaft von heute" (pp. 185 ff.).

Miková, L., V. Nohavica, and L. Říha. "Effectiveness of Scientific and Technological Progress and Fixed Investments" (in Russian), *Teorie a Metoda* 5, no. 3 (1973): 131–146.

Mileikovsky, A. G. "The Scientific and Technological Revolution and the Problem of the 'Qualify of Life' in Developed Capitalist Countries" (in Russian), *Voprosy filosofii* 28, no. 7 (1974). The scientific-technological revolution has precipitated deep conflict in developed capitalist countries. The failure to integrate the working class into a "society of consumption" is revealed by working class demands concerning "the quality of life." Need for further Marxist-Leninist study of this new concept.

Mitin, M. B. "The Problem of Humanization of Technology and Social Progress" (in Russian), *Voprosy filosofii* 27, no. 4 (1973). A review of Western apprehensions about technological progress and the critique of Marxism as afflicted with a "technological Eros." Admits that the problem of humanization is not subject to any facile solution, but defends socialism, which does not make a fetish of technological progress, as attempting to place progress in the service of each man and society as a whole. Balanced treatment.

Mogilev, A. V. *Chelovek—nauka—tekhnika; opyt marksistskogo analisa nauchno-tekhnicheskoi revoliutsii* [Man—science—technology; toward a Marxist analysis of the scientific-technical revolution]. Moscow: Polizdat, 1973. Pp. 366. A comprehensive and well-balanced work by a collective body of authors. Moves from a historico-philosophical discussion of science, technology, production and their interrelations (chapters I–III by the Institute of the History of Natural Sciences and Technology of the USSR Academy of Sciences), and the socio-economic influences of the scientific and technological revolution (chapters IV–VII by the Institute of Philosophy and Sociology of the Czechoslovak Academy of Sciences), to a philosophical consideration of the impact of this revolution on art, culture, religion, science, and the future (chapters VIII–X by the Institute of Philosophy of the USSR Academy of Sciences). Many references to Western literature on this subject. Translated into English as *Man, Science and Technology: A Marxist Analysis of the Scientific and Technological Revolution* (Moscow-Prague: 1973). Pp. 387. S. Lieberstein, in a review in *Technology and Culture* 16, no. 4 (October 1975), pp. 691–693, calls this "an excellent introduction to current Marxist thinking on the social consequences of technological change." For another good introduction to Marxist thought on the scientific and technological revolution see Pavel Kovaly's review of three books: by R. Richta (1969), N. V. Markov (1971), and I. A. Kozikov (1972), in *Studies in Soviet Thought* 14 (1974): 139–148.

Moskvichov, L. N. *The End of Ideology Theory: Illusions and Reality; Critical Notes on a Fashionable Bourgeois Conception.* Moscow: Progress Publishers, 1974. Pp. 191. Two sections of special relevance: one entitled "Class Ideology and the 'Technological' Approach," the other "Bourgeois 'Social Engineering'."

Müller, K. "On the Issue of the Interaction Between the Scientific and Technological Revolution and the Process of Socioeconomic Integration" (in Russian), *Teorie a Metoda* 5, no. 3 (1973): 39–52.

Olszewski, Eugeniusz. "Les sciences et les techniques dans la période de la révolution scientifico-technique" [Science and technology in the period of the scientific and technological revolution], *Organon* vol. 8 (1971): 41–53. All sciences have become more developed and interrelated, with frequent conceptual revolutions, and increased social importance. Technology, too, has become more unified and because of pollution is threatening human life. The leading trend in modern technology is automation which liberates man from a dehumanizing dependence on machines. This article is a French translation of part of the author's "Dziś i jutro rewolucji naukowo-technicznej" in E. Olszewski, Z. Rybicki, and K. Secomski, *Czynniki naszego rozwoju* [Factors of our progress] (Warsaw: Wiedza Powszechna, 1971).

Oyzerman, T. I. "Historical Materialism and the Ideology of 'Technological' Pessimism" (in Russian) *Voprosy filosofii* 27, no. 8 (1973). Historical materialism, as a summary of human experience, presents social production not merely as the manufacture of useful objects, but as the production of the needs and abilities of man as well as numerous forms of communication. Technological pessimism rejects this principle as absolutization of productive forces. But productive forces consist of people with means of material and spiritual production. Technological pessimism is reactionary because it identifies the historical destiny of mankind with the capitalist mode of production.

Petrov, B. N. "Space Research and Scientific and Technological Progress" (in Russian), *Voprosy filosofii* 28, no. 10 (1974). A pseudo-philosophical paean to space research as a manifestation of the scientific-technological revolution and a logical step in world progress.

Richta, R. "Scientific and Technological Revolution and Social Systems" (in Russian), *Teorie a Metoda* 5, no. 3 (1973): 7–38.

———. "Impact of the Scientific and Technological Revolution," *World Marxist Review* 17 (September 1974): 133–142.

Rozhin, V. P. "Lenin and Problems of the Marxist Theory of Development," *Soviet Studies in Philosophy* 9, no. 1 (Summer 1970): 45–59. Translated from *Voprosy filosofii* 23, no. 11 (1969).

Saifulin, Murad, ed. *The Future of Society: A Critique of Modern Bourgeois Philosophical and Socio-political Conceptions.* Moscow: Progress, 1973. Pp. 375. Also published in French as *L'Avenir de la société humaine* (1973).

Schoukhardine, S. V. "La révolution scientifico-technique contemporaine: état des recherches et problemes." [The contemporary scientific and technological revolution: the state of research and problems], *Organon* vol. 8 (1971). 55–65.

Sevast'ianov, V. I., and A. D. Ursul. "New Interrelations of Society and Nature in the Space Age," *Soviet Studies in Philosophy* 10, no. 2 (Fall 1971): 158–175. The first author is a cosmonaut; the second a philosopher. Translated from *Voprosy filosofii* 25, no. 3 (1971).

Shpirt, Alexksandr. Ulianovich. *The Scientific-Technological Revolution and the Third World.* Moscow: Novosti Press, 1972. Pp. 143.

Shvarts, S. S. "Problem of Human Ecology" (in Russian), *Voprosy filosofii* 28, no. 9 (1974). Analyzes the aspects of the biological science of ecology which can become part of a human ecology. Argues that the basic point of contact is the concept of an ecosystem, and that this concept rather than any "return to nature" offers the best hope for protecting the biosphere.

Slavkov, Svetoslav. "The Role and Place of Mathematics in Scientific and Technological Revolution" (in Russian), *Voprosy filosofii* 27, no. 7 (1973). Discusses how, as a result of a higher degree of abstraction and formalization in mathematics, and progress in the other sciences, mathematical method, language and style are becoming more and more distinctive of present-day scientific knowledge.

Smolyan, G. L. "Man and Electronic Computer" (in Russian), *Voprosy filosofii* 27, no. 3 (1973). The three parts of this article discuss: 1) the general character of automation as it affects management processes; 2) the interaction between man and computer and the role of man as an agent of "strategic" thinking; 3) the possibilities of creating "an artificial intellect."

Stoskova, N. N. *F. Engels o roli tekhniki v razvitti obshchestva* [Friedrich Engels on the role of technology in the development of society]. Moscow: Nauka, 1970. Pp. 80. Bibliography, pp. 78–79.

———. "Problemy tekhniki v trudakh F. Engelsa" [Problems of technology in the works of Engels], *Voprosy istorii estestvoznaniia i tekhniki* no. 3 (1970): 39–44. Outlines Engels's interpretation of the problems and history of technology, especially the interaction between technology and science.

Strumilin, S. G., and E. E. Pisarenko. "Science and Production" (in Russian), *Voprosy filosofii* 27, no. 8 (1973). Science is an active productive force. Its effect on production, however, depends not only on its own inner depth but most important on the economic basis for introducing scientific achievements into production. Neither science nor production can exist outside of the biosphere. Thus the process of humanizing nature requires all efforts of the noosphere to be extended to preserve the biosphere.

Subas, M. L. "Some Epistemological Aspects of Technological Design" (in Russian), *Voprosy filosofii* 26, no. 9 (1972): 28–36. On how engineering drawings and models serve as a link between idea and product.

Tatarkiewicz, Wladyslaw. "The Definition of Art" (in Russian), *Voprosy filosofii* 27, no. 5 (1973). Traditionally art was understood as a creative activity governed by rules and requiring skill. In the 17th century art was defined as the creation of beautiful things. In the 20th century both views have been questioned. Art is a polysemantic notion. An alternative understanding of art would argue that "a product of man's conscious activity is an object of art only when it reproduces reality, creates forms, expresses emotions and evokes admiration, moves people or startles them."

Tessman, Kurt H. "Zur Kritik des technologischen Determinismus" [Toward a critique of technological determinism], *Deutsche Zeitschrift für Philosophie* 22, no. 9 (1974): 1089–1103. The concepts of Lefèbvres are singled out from the multitude of bourgeois variants of technological determinism and examined from the point of view of dialectical-material determinism. Briefly, technological determinism makes the conditions of production determinative of historical processes, whereas dialectical materialism sees determinism resting in the relationship of conformity or nonconformity between productive forces (which are beyond any existing productive conditions) and the productive conditions.

Timakov, V. D., and N. P. Bochkov. "Social Problems of the Genetics of Man" (in Russian), *Voprosy filosofii* 27, no. 6 (1973). Discusses the possibility that the increased longevity of people suffering from heredity diseases, and thus the increased likelihood of their passing on their genetic defects to children, plus the existence of various mutagenic factors such as radiation and new chemical compounds, will adversely affect the human gene pool. Argues that these processes will not lead to genetic degeneration because some accumulation of pathological mutations does sharply alter the relationship between normal and mutant alleles. Also, increased contacts between populations reduces the probability of mutations in homozygote states. Finally, progress in modern genetics is adding new techniques to the methods of public health care.

Trendafilov, Toncho. *Nauchno-tekhnicheska i sotsiealna revoliutsiia* [Science-technology and the socialist revolution]. Sofia: Partizdat, 1973. Pp. 247. Russian and English summaries. Bibliography pp. 229-234.

Valenta, F. and L. Riha. "The Character of Present-Day Changes in Scientific and Technological Development and Impact Upon Managing the Reproduction Process" (in Russian), *Teorie a Metoda* 5 (1973): 155-179.

Yanitsky, O. N. "The Single Posturban Way of Life—the Model and Reality" (in Russian), *Voprosy filosofii* 28, no. 10 (1974). A critical analysis of the idea, loosely associated with that of postindustrial society, of the American suburb as resolving rural-urban antagonism. Argues that the scientific and technological revolution does not resolve the contradictions of capitalist urbanization, but simply reproduces them in the new form of the megalopolis and the confrontation of its inner cities and suburbs.

Vasil'chuk, Iu. A. "The Dialectics of the Forces of Production," *Soviet Studies in Philosophy* 11, no. 1 (Summer 1972): 70-100.

Volkov, Gerikh Nikolaevich. *Man and the Challenge of Technology*. Moscow: Novoski Press Agency Publishing House, 1972. Pp. 215.

Yudin, B. G. "New Elements in the Technology of Capitalist Regulation" (in Russian), *Voprosy filosofii* 27, no. 1 (1973). A review of social theory in the United States during the 1960s, in its reaction to the realization that far from solving existing social problems, economic growth and technological progress are creating new ones. The traditional capitalist idea that the state must not intervene in the economy is undermined by the necessity for social engineering and designing.

III. RELIGIOUS CRITIQUES

A. Primary Sources

Beck, Horst Waldemar. *Im Banne des Automaten* [Under the curse of automation]. Stuttgart: Steinkopf, 1971. Pp. 63.

———. *Weltformel contra Schopfungsglaube; Theologie und empirische Wissenschaft vor einer neuen Wirklichkeitdeutung* [World formula against creation faith; theology and empirical science before a new interpretation of reality]. Zurich: Theologischer Verlag, 1972. Pp. viii, 281. Originally the author's thesis.

Brungs, Robert A. "Reconciliation: Man-the-Maker and Man-the-Made," *Theology Digest* 22, no. 4 (Winter 1974): 324-332. Brungs, director of the Institute for Theological Encounter with Science and Technology, reflects the influence of de Chardin in this discussion of the dangers of man's extending his mastery of nature into a mastery of human nature (genetic control, behavior modification, etc.). Claims that human nature is open to development, but that this development must always be seen in relation to the Lordship of Christ. Leaves one with the sense that the heart of this crucial issue has been buried under vague generalizations.

Callahan, Daniel, "What Obligations Do We Have to Future Generations?" *American Ecclesiastical Review* 164 (April 1971): 265-280.

Cobb, John B. *Is It Too Late? A Theology of Ecology*. New York: Bruce, 1972.

Curran, Charles E. *Politics, Medicine, and Christian Ethics; A Dialogue with Paul Ramsey.* Philadelphia: Fortress Press, 1973. Pp. viii, 228. Curran (a Catholic) explores the views of Ramsey (a Protestant) on power, force, just war, revolution, nuclear deterrence, abortion, artificial insemination, organ transplants, right to die, and genetic engineering.

Derrick, Christopher. *The Delicate Creation; Towards a Theology of the Environment.* Old Greenwich, Conn.: Devin-Adair, 1972. Pp. x, 129. A student of C. S. Lewis and G. K. Chesterton argues that the technological conquest of nature is a manifestation of the ancient Manichaean heresy. What is needed is a sense of "cosmic piety" and gratitude to balance the modern obsessions of discontent and rebellion. Reviewed by Schuyler Brown, S.J., *Theological Studies* 34, no. 2 (June 1973): 346–347.

Dumas, A. "Ethique et technique," *Revue d'histoire et de philosophie religieuses* 54, no. 4 (1974): 495–506.

Green, Harold P. "Human Values in a Technological Society," *Dimensions in American Judaism* 5 (Winter 1971): 19–23.

Häring, Bernard, C.S.S.R. *Medical Ethics.* Edited by G. L. Jean. Notre Dame: Fides Publishers, 1973. Pp xiii, 250. A handbook on medical ethics by a distinguished moral theologian, written with the collaboration of Sr. Gabrielle L. Jean, a specialist in biology and psychology, who edited the English text and prepared the bibliography. Anna Valentini, M.D., of the University of Bologna, checked the manuscript from the medical standpoint.

Howe, Günter (1908–1968). *Gott und die Technik; Die Verantwortung der Christenheit für die wissenschaftlich-technische Welt* [God and technology; the responsibility of Christianity for the scientific and technological world]. Hamburg: Furche-Verlag; and Zurich: Theologischer Verlag. 1971. Pp. 234. See also author's *Die Christenheit im Atomzeitalter.* [Christianity in the atomic age], edited by H. Timm, preface by C. F. von Weizsacker, which contains a bibliography of Howe's works, pp. 320–322.

Kaufman, Gordon D. "A Problem for Theology: The Concept of Nature," *Harvard Theological Review* 65 (1972): 264 ff.

Keefe, Donald. "Biblical Symbolism and the Morality of *in vitro* Fertilization," *Theology Digest* 22, no. 4 (Winter 1974): 308–323. Argues that *in vitro* fertilization "will introduce into society a destructive degradation of the symbols by which we live in history," specifically the masculine and the feminine. First appeared in the Proceedings of a conference on "*In vitro* Fertilization" sponsored by the Institute for the Theological Encounter with Science and Technology, October 1974.

Longwood, Merle. "The Common Good: An Ethical Framework for Evaluating Environmental Issues," *Theological Studies* 34, no. 3 (September 1973): 468–480. Argues that political liberalism (in the tradition of Locke and Mill), with its prime concern of balancing "interest groups" to achieve social harmony, cannot adequately deal with environmental issues that affect the whole society. Proposes a "modified version of the traditional concept of the common good" (p. 480).

Lynch, William F., S.J. *Christ and Prometheus: A New Image of the Secular.* Notre Dame: Univ. of Notre Dame Press, 1970. Pp. 163. Contemporary problems of the religious imagination come from its having to face the age-old secular project of man's search for autonomy without guilt in a sharply and quantitatively new form. Suggests that the religious solution may be to see that "autonomy is not a defiance but a grace."

McCormick, Richard A., S. J. "Genetic Medicine, Notes on the Moral Literature," *Theological Studies* (Special Issue: Genetic Science and Man) 33, no. 3 (September 1972): 531–552. Probably the best introduction available in the area of ethics and genetics. Classifies the literature under three basic approaches: 1) the consequentialist calculus of Joseph Fletcher, 2) the deontological attitude of Paul Ramsey and Leon Kass, and 3) the mediating approach of James Gustafson and Charles Curran. Ends with personal reflections of the author which favor the approach of Ramsey and Kass.

———. "The New Medicine and Morality," *Theology Digest* 21, no. 4 (Winter 1974): 308–321. Using the example of dying, McCormick surveys the major moral positions about the use of new biomedical technology from the situationist view of J. Fletcher to P. Ramsey's classical stand. Concludes with some cautious suggestions of his own.

Marsch, Wolf Dieter. *Die Folgen der Freiheit; christliche Ethik in der technischen Welt* [The future of freedom; Christian ethics in the technological world]. Edited by Michael Schibilsky and Hartmut Przybylski. Gütersloh: Gütersloher Verlagshaus Mohn, 1974. Pp. 128. Contents: "Christliche Ethik in der technischen Welt," "Kybernetik und christliches Ethos," "Christliche Anthropologie und biologische Zukunft des Menschen," "Die Stadt als Ort der Utopie," "Christliche Zukunftshoffnung und rationale Zukunftsplanung," "Verantwortung für die Folgen der Freiheit, theologische Überlegungen zum Thema Umweltschutz."

Martinez de Galinsoga de Bega, Carlos. *Hacia una teología de la téchnica* [Toward a theology of technology]. Madrid: S.M., 1972. Pp. 115. A thesis.

Marty, Martin, and Dean G. Peerman, eds. *New Theology No. 10*. New York: Macmillan, 1973. Pp. 225. On the theme "*Bios* and Theology." Contents: Part I, Creation and Nature, includes: Robert T. Osborn's "A Christian View of Creation for a Scientific Age" (from *Religion in Life* [Spring 1972]), Wolfhart Pannenberg's, "The Doctrine of the Spirit and the Task of a Theology of Nature" (from *Theology* [January 1972]), Van Rensselaer Potter's "The Ethics of Nature and Nurture" (from *Zygon* [March 1973]), Richard A. McCormick's "Genetic Medicine: Notes on the Moral Literature" (from *Theological Studies* [September 1972]). Part II, Life-Sequence, includes: Paul Ramsey's "Shall We 'Reproduce'?" (from *Journal of the American Medical Association* [June 5 and June 12, 1972]), William Vrasdonk's "Toward a Theology of Eugenics" (from *The Ecumenist* [November–December 1971]), Rachel Conrad Wahlberg's "The Woman and the Fetus: 'One Flesh'?" (from *Christian Century* [September 8, 1971]), Sister M. Romanus Penrose's "Virginity and the Cosmic Christ" (from *Review for Religious* [March 1972]), Daphne Nash's "Women's Liberation and Christian Marriage" (from *New Blackfriars* [May 1972]), José M. R. Delgado's "Psychocivilized Direction of Behavior" (from *Humanist* [March–April 1972]), Daniel Maguire's "The Freedom to Die" (from *Commonweal* [August 11, 1972]), and James M. Sullivan's "The Lord of Death and Dying (from *Listening* [Dubuque, Iowa] [Autumn 1971]). The editors also provide a helpful introduction to this literature.

Mead, Margaret. *Twentieth Century Faith; Hope and Survival*. New York: Harper and Row, 1972. Pp. xviii, 172. Collection of earlier lectures and publications dealing with topics such as Christianity and technology, spiritual issues in the problem of birth control, the right to die, and immortality.

Muller-Schwefe, Hans-Rudolf. *Technik und Glaube; Eine permanente Herausforderung* [Technology and faith; a permanent challenge]. Göttingen: Vandenhoeck und Ruprecht; Mainz: Matthias-Grünewald Verlag. 1971. Pp. 305. A comprehensive theological work. Discusses the ideas of Guardini, Dessauer, Pieper, Bernaros, Teilhard, Rahner, Metz, etc.

Nelson, James Bruce. *Human Medicine: Ethical Perspectives on New Medical Issues*. Minneapolis: Augsburg Publishing House, 1973. Pp. 207. Considers questions of abortion, artificial insemination, human experimentation, genetic engineering, death, and organ transplants—outlining the alternative Christian responses to each. By a professor of Christian ethics.

Oudin, J. M. "Hylémorphisme et civilisation technique," *La pensée catholique* no. 153 (1974): 73–87.

Rahner, Karl. "Experiment: Man," *Theology Digest* sesquicentennial issue (February 1968): 57–69. "Because of advances in technology and in the social and life sciences, man's internal and external environment is coming more and more within human control. What are the limits, the meaning, and the direction of the self-creation now in man's own hands? Although human self-creation represents a flowering of personal freedom, it seems to raise new questions about the nature or essential core of man individual and man collective. This paper discusses possible built-in limits of human self-shaping, but emphasizes the positive responsibility of theology and of Christians to make future self-creation a project worthy of man's absolute future—God himself." This essay, part of a special extra number of *Theology Digest*, is one of a series of seven lectures delivered by Rahner in the fall of 1967 which happened to be related to the St. Louis University sesquicentennial theme of "Knowledge and the Future of Man." Revised version subsequently included in the author's *Theological Investigations* vol. 9 (New York: Herder and Herder, 1972): 205–224, as "The Experiment with Man; Theological Observations on Man's Self-manipulation." One other closely related article, "The Problem of Genetic Manipulation," follows, pp. 225–252. Rahner is one of the most widely respected Catholic theologians. Other studies related to the problem of technology to be found in his collected theological papers are: "Science as a 'Confession'?" *Theological Investigations* vol. 3 (Baltimore: Helicon, 1967); "Theological Remarks on the Problem of Leisure" and "The Theology of Power," *Theological Investigations* vol. 4 (Baltimore: Helicon, 1966).

Ramsey, Paul. "The Ethics of a Cottage Industry in an Age of Community and Research Medicine," *New England Journal of Medicine* 284 (April 1, 1971): 700–706.

Shils, Edward. "Faith, Utility, and the Legitimacy of Science," *Daedalus* 103 (Summer 1974): 1–15.

Toynbee, Arnold J. "The Genesis of Pollution," *Horizon* 15, no. 3 (Summer 1973). Argues that the Christian, as opposed to the pantheist, view of the world is the source of the ecological crisis.

Vaux, Kenneth. *Biomedical Ethics*. New York: Harper and Row, 1974. Pp. xviii, 134. Part one surveys the traditions of medical ethics—Hippocratic, religious, political. Part two offers "A Model for Decision-making." Part three discusses, with case illustrations, contemporary issues in biomedical technology. Postscript: "The reader may perceive a theological bias in this book. I am searching for a new natural theology. I believe that man can discover as well as create value. . . . I believe God is good. . . . He has not deceptively or cruelly ordered nature but fashioned her vulnerable to man's perception and pliable to his will" (p. 112).

———. ed. *To Create a Different Future*. New York: Friendship Press, 1972. Pp. 144. Papers from the Houston Conference on Technology and a Human Future (1972) which "sought to place the issues of a technological future in the context of spiritual meaning and ethical value." Includes J. Randers's "Global Limitations and Human Responsibility," I. Illich's "Technology and Conviviality," R. Murray's "Ethical and Moral Aspects of Genetic Knowledge and Counseling," R. T. Francoeur's "Technology and the Future of Human Sexuality," K. Vaux's "Religious Hope and Technological Planning," and introductory and closing comments by Margaret Mead and J. Edward Carothers.

Young. R. V., Jr. "Christianity and Ecology," *National Review* 26, no. 51 (December 20, 1974): 1454–1458 ff. Christian apology in answer to the accusation that Christianity is responsible for our ecological crisis. This accusation is most eminently expressed by Lynn White, Jr.'s "The Historical Roots of our Ecologic Crisis," *Science* 155, no. 3767 (March 10, 1967), 1203–1207, and Arnold J. Toynbee's "The Genesis of Pollution," *Horizon* 15, no. 3 (Summer 1973).

B. Secondary Sources

Alvarez, Juan Jose. "Cambio social y religion" [Social change and religion]. *Estudios Filosoficos* 52 (September–December 1970): 613–628. Sociologically, religion has been a stabilizing factor in the social system, but as Weber has shown in his analysis of the Protestant ethic, it has also sometimes been an element stimulating change. In a technological and secular society religion will be forced to undergo fundamental adaptations.

Barnette, Henlee Hulix. *The Church and the Ecological Crisis*. Grand Rapids: Eerdmans, 1972. Pp. 114. Aims to "summarize the salient factors in the eco-crisis in the light of the biblical understanding of man and nature." Tries to put the wealth of knowledge concerning ecology into perspective for the believer.

Bemporad, Jack, et al., eds. *Focus on Judaism, Science, and Technology*. New York: Union of American Hebrew Congregations, 1970. Pp. 228. Bibliography pp. 227–228. An adult religious education text.

Bennett, Thomas R. *Learning to Be Human in a Push-Button World*. New York: Friendship Press, 1971. Pp. 96. "Guidance for adult groups and all who seek help in finding human meaning in machine age." Sunday school literature.

Borchert, Donald M. "On Being Human in the Cybernetic Revolution," *Christian Century* 89, no. 35 (October 4, 1972): 980–984. Argues that the Bible and Marx offer pertinent guides for humanization in a cybernetic age.

Braaten, Carl E. *"Caring for the Future: Where Ethics and Ecology Meet,"* Zygon 9, no. 4 (December 1974): 311–322. The author is a professor at the Lutheran School of Theology in Chicago. This is reprinted with minor revisions from his *Eschatology and Ethics* (Minneapolis: Augsburg, 1974).

———. *Christ and Counter-Christ: Apocalyptic Themes in Theology and Culture*. Philadelphia: Fortress, 1972.

Carothers, J. Edward. *Can Machines Replace Man?* New York: Friendship Press, 1966. Pp. 64. The eighth pamphlet in the series "Questions for Christians." "If machines do replace men, it will be due to man's failure to define a sense of life's meaning in a world liberated from toil" (p. 60).

———, Margaret Mead, Daniel McCracken, Roger Shinn, eds. *To Love or to Perish; The Technological Crisis and the Churches*. New York: Friendship Press, 1972. Pp. 152. A report by the U.S.A. Task Force on the Future of Mankind in a World of Science-Based Technology. Sponsored by both the National Council of Churches and the Union Theological Seminary in New York.

Cauthen, Kenneth. *Christian Biopolitics: A Credo and Strategy for the Future*. Nashville, Tenn.: Abingdon, 1971.

Chagas, C. "The Development of Science and the Future of Mankind," *L'Osservatore Romano* (English edition) (November 28, 1974): 8–9.

Croose Parry, R.-M. "Towards a Global Consciousness," *Teilhard Review* 8 (February 1973): 17–23.

Donnelly, T. G. "In Defence of Technology," *Christian Century* 90, no. 3 (January 17, 1973): 65–69. Somewhat rambling and emotional response to the "paranoid" haters of technology, by a computer technologist. Special blame for the spread of this "paranoia" is aimed at Jacques Ellul's *The Technological Society* and "The Center for the Study of Democratic Institutions." Discussion, *Ibid.* (March 21, 1973): 345–347; (June 27, 1973): 706–707; (September 12, 1973): 895–896.

Dupré, Louis. "A New Approach to the Abortion Problem," *Theological Studies* 34, no. 4 (December 1973): 481–488.

Faramelli, Norman J. *Technethics: Christian Mission in an Age of Technology*. New York:

Friendship Press, 1971. Pp. 160. "Machines are extensions of man; it is madness to think of man being dominated by his machines" (p. 153). "We must not ask where science and technology are taking us, but rather how we can manage science and technology so that they can help us get where we want to go" (p. 154). By a chemical engineer and graduate of divinity school who has also been director of the Boston Industrial Mission.

Fletcher, John C. "Dialogue Between Medicine and Theology," pp. 150–163 in Claude A. Frazier, ed., *Should Doctors Play God?* Nashville: Broadman Press, 1971. Discusses the conflicts in the death and dying issue between Christian ethics and medical ethics.

———. "Moral Problems in Genetic Counseling," *Pastoral Psychology* 23, no. 223 (April 1972): 47–60. An investigation of 25 couples who decided to have an abortion after diagnostic amniocentesis showed that they would have deformed children.

Foley, Grover. "Reaping the Whirlwind; The Question of Faith in an Obsolete World," *Cross Currents* 23, no. 3 (Fall 1973): 279–296. Powerful critique of secular and optimistic theology. "The question today is whether the churchmen and theologians can recognize a catastrophe *before* it arrives . . ." (p. 281).

Fritsch, Albert J., S. J. *A Theology of the Earth.* 839 17th St., N.W., Washington, D.C.: CLB Publishers, 1972. Pp. 88. An attempt to help "technologic man" face the crisis of the power of uncontrolled technology and the environmental pollution resulting from this lack of control. Aims to use the teachings of faith to help transform technologic into "cosmic man." Second law of "Christodynamics": "All communicating cosmic units will proceed in Christ toward a state of maximum order." Fritsch is co-director of the Center for Science in the Public Interest. Laudatory review by Robert Lebel, S.J., *Theological Studies* 34, no. 4 (December 1973), 738–739.

Fudpucker, Wilhelm Elmer, S.J. "Die technische Möglichkeit aus der Omega Mann" [The technological possibility of Omega man]. *Zeitschrift für gläubig Forschung,* occasional research paper no. 41 (May 1, 1973). Pp. 2. Argues that genetic engineering, particularly cloning, in conjunction with the possibility that the Shroud of Turin contains drops of blood of Jesus of Nazareth, creates the conditions for a "technological new virgin birth."

Gibbs, John C. *Creation and Redemption: A Study in Pauline Theology.* Long Island City, N.Y.: Brill, 1971.

Gibson, Arthur. "Technology as Sacrament," *Ecumenist* 11, no. 6 (September–October 1973): 92–97. "For technology like every human creation stands in need of redemption; but technology like everything human *is redeemable*" (p. 92). Sees three frontier areas where technology can especially serve man's redemption—cybernetics, genetic engineering, and space research (so we can meet our friendly neighbors!).

Gill, David M. *From Here to Where? Technology, Faith and the Future.* Geneva: World Council of Churches, 1970. Pp. 111. Report from an exploratory conference in Geneva (June 28–July 4, 1970).

Graham, W. Fred. "Technology, Technique, and the Jesus Movement," *Christian Century* 90, no. 18 (May 2, 1973): 507–510. Contrasts the "instant enlightenment" mentality of the pseudo-Eastern mystics in this country with the basically ascetic and antitechnocratic attitudes of the Jesus youth.

Haughton, Rosemary. "Judgement on the Earth," *Catholic World* 214 (November 1971): 54–55.

Heelan, P. "Nature and Its Transformations," *Theological Studies* 33, no. 3 (September 1972).

Hilgers, Thomas, and Dennis J. Horan, eds. *Abortion and Social Justice.* New York: Sheed and Ward, 1973. Pp. xxv, 328. A series of essays putting forth biological, medical, psychological, sociological, legal, demographic, and ethical arguments against abortion. Perhaps the best total statement of the religious position which opposes abortion.

Hodgson, P. "Technology and Man," *Way* 13, (April 1973): 98–111.
Hogan, J. "Technology and Human Values," *Priest* 29, (July–August 1973): 24ff.
Jeffko, W. G. "Ecology and Dualism," *Religion in Life* 42 (Spring 1973): 117–127.
Lanzenstiel, Georg, ed. *Technik und kritische Weltgestaltung* [Technology and the critical world situation]. Munich: Claudius-Verlag, 1971. Pp. 51. Papers presented at the 15th meeting of the Evangelischer Philologen, September 11–15, 1970, in Tutzing, Germany.
Locher, Gottfried W. "Glauben und Wissen" [Faith and science], *Reformatio* 22 (1973): 82–92. An abstract of this paper can be found under the title "Can Technology Exist Without Belief?" *Theology Digest* 21, no. 3 (Autumn 1973): 221–223. Argues that technology is escaping the direction of faith.
McCormick, Richard A. "Moral Notes: The Abortion Dossier." *Theological Studies* 35, no. 2 (June 1974): 312–359. A bibliographical essay sorting out and commenting on the enormous abortion literature.
Miles, John H., Jr. "Wife of Onan and the Sons of Cain: To Abort, or Not to Abort," *National Review* 25, no 33 ((August 17, 1973): 891–894. The debate over abortion is being argued under the context of murder. The wider context that should alarm us even more is the danger inherent in social technology. Discussion, *Ibid.* 25, no. 39 (September 28, 1973): 1024 ff.
Mortimer, John. "The Minister Who Gives the Works to Unions and Bosses," *Engineer* 237, nos.6145–6146 (December 20/27): 32–33. Methodist minister, principal of Luton Industrial College in Britain, brings management and labor together.
"The New Icarus," *America* 130, no. 1 (January 12, 1974): 2–3. An editorial. Freed from necessity, modern man, the "new Icarus," must turn his energies toward prayer and creative leisure so that he may "soar within."
"The New Life Through Technocracy." Part VII, pp. 481–541, in Frederick J. Streng, Charles L. Lloyd, Jr., and Jay T. Allen, eds., *Ways of Being Religious* (Englewood Cliffs, N.J.: Prentice-Hall, 1973). Analyzes technology as a form of religion, with selections from R. B. Fuller, A. Harrington, A. Clarke, B. F. Skinner, V. C. Ferkiss, Sri Aurobindo, and E. Brunner. The middle transition section of this text, on "The Shift to Nontranscendent Ultimates" (pp. 333–358), is also relevant.
Parsons, Talcott. "Religion in Postindustrial America: The Problem of Secularization," *Social Research* 41, no. 2 (Summer 1974): 193–225.
Ramsey, Ian T. "The Influence of Technology on the Social Structure," *Teilhard Review* 6, no. 2 (Winter 1971–72).
Riley, Thomas J. "Is Christianity Responsible for the Excesses of Technology?" *The Pilot* (May 4, 1973): 7. Answer is No. Not profound. But indicates in brief the informed, moderate Catholic viewpoint. Along this same line see "A World in Transition: Faith and Technology," Section II of *To Teach as Jesus Did* (a pastoral message on Catholic education from the National Conference of Catholic Bishops, November 1972) (Washington, D.C.: United States Catholic Conference, 1973); and Pope John XXIII's encyclical *Pacem in Terris* which calls for the Christian transformation of science and technology from within (no. 148) and the creation of "a synthesis between scientific, technical, and professional elements on the one hand and spiritual values on the other" (no. 150).
Rosner, Fred. *Modern Medicine and Jewish Law*. New York: Yeshiva Univ. Press, Dept. of Special Publications, 1972. Pp. 216.
Ruether, Rosemary. *Liberation Theology: Human Hope Confronts Christian History and American Power*. Paramus, N.J.: Paulist Press, 1973. Pp. 194. See chapter 8 (pp. 115–126), "Mother Earth and the Megamachine: A Theology of Liberation in a Feminine, Somatic and Ecological Perspective."
Sarvis, Betty, and Hyman Rodman. *The Abortion Controversy*. New York: Columbia Univ.

Press, 1973. Pp. x, 222. A discussion of moral, legal, political, medical, and social aspects of the abortion debate.

Schmidt, Helmut, ed. *Weltveränderung durch Technik; Die Voraussetzungen der Technik-Wissenschaften und das Dilemma der Humanität* [World responsibility through technology; the assumptions of technology-science and the dilemma of humanity]. Stuttgart: Radius-Verlag, 1971. Pp. 72. Contents: Klaus Babner's "Reine Forschung — Fiktion, Möglichkeit oder Realität?", Klaus Meyer-Abich's "Weltveränderung durch Technik, Über den Zusammenhang von Naturwissenschaft, Technik und Entfremdung," Michael Drieschner's "Die Objektivität der Physik," and Gunter Rohrmoser's "Das Dilemma der verwirktlichten Humanität, Über die Dialektik des Menschsein in der Industriekultur."

Schnellmann, Guido. *Theologie und Technik; 40 Jahre Diskussion um die Technik, zugleich einer Beitrage zu einer Theologie der Technik* [Theology and technology; 40 years discussion of technology, together with a contribution to the theology of technology]. Cologne and Bonn: Hanstein, 1974. Pp. x, 390.

"Selected Issues in Medical Ethics." *Journal of Religious Ethics* 2 (Spring 1974): 3-98.

Stevens, William P. H. *Are We Ready for Leisure?* New York: Friendship Press, 1966. Pp. 64. A Sunday school pamphlet.

Stinson, Charles. "Theology and the Baron Frankenstein: Cloning and Beyond," *Christian Century* 89 (January 5, 1972): 60-63. A brief survey of the ideas of men such as J. Lederberg, P. Ramsey, K. Rahner, L. Kass.

Thomas, J. L. "Family, Sex, and Marriage in a Contraceptive Culture," *Theological Studies* 35, no. 1 (March 1974): 134-153. Calls for a reformulation of traditional values due to cultural changes and new theological insights. Part of a special issue specifically devoted to the problems of population.

Veller, Reinhard. *Theologie der Industrie- und Sozialarbeit; zur Theologie der evangelischen Industrie- und Sozialarbeit* [Theology of industrial and social work; toward a theology of Protestant industrial and social work]. Cologne: P. Hanstein, 1974. Pp. 266. Originally presented as the author's thesis, Erlangen University, 1971. Bibliography pp. 253-262.

Verghese, Paul. "Does Humanity Have a Future?" (World Council of Churches Study on Science and Technology), *Christian Century* 91, no. 29 (August 21-28, 1974): 799-800. Report on the conference "Science and Technology for Human Development—The Ambiguous Future and the Christian Hope" held in Bucharest (June 1974) and sponsored by the World Council of Churches. Speakers included Lynn White, Jr., Margaret Mead, Magnus Pyke, and Langdon Gilkey.

"West African Church Conference on Science, Technology, and the Future of Man and Society," Univ. of Ghana, March 24-30, 1972. *Ecumenical Review* 24, no. 3 (July 1972). This issue contains a "Report" on the conference (pp. 341-351), plus four papers: B. C. E. Nwosu's "Scientific Technology and the Future of Africa," Peter Sarpong's "The Search for Meaning: The Religious Impact of Technology in Africa," S. A. Aluko's "Social Prerequisites for Technological Development—An African Perspective," and G.-C. M. Mutiso's "Tools Are for People: Towards an Africanized Technology."

Wickham, Edward R. *Encounter with Modern Society*. London: Lutterworth Press, 1964. Pp. 125. Essays by the Anglican bishop of Middleton dealing with the Church's encounter with modern technology.

Wilkes, Keith. *Religion and Technology*. New York: Religious Education Press, 1972. Pp. xii, 177.

Wynne, C. "Technology Assessment and Quality of Life," *Teilhard Review* 8 (February 1973): 14-16.

IV. METAPHYSICAL AND EPISTEMOLOGICAL STUDIES

A. Primary Sources

Ahlers, Rolf. "Technologie und Wissenschaft bei Heidegger und Marcuse" [Technology and Science according to Heidegger and Marcuse], *Zeitschrift für Philosophische Forschung* 25, no. 4 (October–December 1971): 575–590. A review of the rise of the philosophy of technology with special reference to the metaphysical critiques of Heidegger and Marcuse.

Alexander, Edwin M. *Martin Heidegger's "The Question about Technic"—A Translation and Commentary.* Hamilton, Ontario: McMaster Univ., 1973. Pp. 73. A master's thesis done under the direction of George Grant.

Arbib, Michael, A. "Man-Machine Symbiosis and the Evolution of Human Freedom," *American Scholar* 43, no. 1 (Winter 1973–1974): 38–54. "Our task . . . is to try to pool our diverse ideologies, our increasing technological knowledge of computers and cybernetics, and an evolving sense of what constitutes human freedom so that humans, in symbiosis with their machines, may eventually become truly free" (p. 54). For another optimistic analysis, see also Arbib's "Complex Systems: The Case for a Marriage of Science and Intuition," *Ibid.* 42, no. 1 (Winter 1972–1973): 46–56.

Baruzzi, Arno. *Mensch und Machine: Das Denken sub specie machinae* [Man and machine: thinking sub specie machinae]. Munich: Fink, 1973. Pp. 218. Originally a dissertation. Historico-philosophical study of mind-body problem moving from the materialism of LaMettrie to cybernetics. Bibliography pp. 211–216.

Baum, R. F. "The Uniqueness of Modern Technology," *South Atlantic Quarterly* 71, no. 1 (1972): 54–61. Argues that man is not essentially a technological animal.

Borgmann, Albert. "Orientation in Technology," *Philosophy Today* 16, no. 2 (Summer 1972): 135–147. A Heideggerian-phenomenological analysis of the relationship between switches, devices, and needs. Conclusion: technology causes "the disappearance of things in their immediate presence and the emergence of a slate of terminal normalcy" (p. 143). It is this normalcy which makes the problem of orientation in (i.e. understanding, control of) technology so difficult.

Bozonis, George A. "Some Remarks on Mechanical Explanation in Biology," *Diotima* 1 (1973): 61–80. Despite similarities, living beings are not machines because of (a) the principle of wholeness and (b) the unpredictable plasticity that each living organism expresses. Furthermore, machines are always the product of mind and never of chance.

Bunge, Mario. "Les Presupposes et les produits metaphysiques de la science et de la technique contemporaines," *Dialogue* (Canada) 13 (September 1974): 443–453. Sketches the metaphysics of science and a scientific system of metaphysics. Scientific research employs a number of metaphysical principles, the study of which constitutes the metaphysics of science. Science and technology have also produced a number of theories (such as Lagrangian dynamics and automata theory) which are so general as to qualify as metaphysics. These constitute a scientific metaphysics.

———. "The Role of Forecast in Planning," *Theory and Decision* 3 (March 1973): 207–221. Planned courses of action are distinguished into their various components, and forecasting is related to two particular components in some detail. Discusses the confusion between technological forecast and prophecies of technological development. Argues that futurology is not an independent science.

Cambiano, Giuseppe. *Platone e le techniche* [Plato and techne]. Turin: Einaudi, 1971. Pp. 269.

Campbell, Richmond, and Alexander Rosenberg. "Action, Purpose, and Consciousness Among the Computers," *Philosophy of Science* 40, no. 4 (December 1973): 547–557. A discussion of K. Sayre's theory of consciousness and machines.

Carpenter, Stanley R. "Modes of Knowing and Technological Action," *Philosophy Today* 18, no. 2 (Summer 1974): 162–168. Examines the possible relationships between technology considered as a process and five types of knowledge—i.e., tacit skills, technical maxims, descriptive or empirical laws, low-level scientific theory, and high-level scientific theory.

Caturelli, Alberto. "La Corrupcion de lo natural en la civilizacion immanentista" [The corruption of the natural in immanentist civilization], *Sapientia* 28 (October–December 1973): 251–277.

Chihara, Charles S. "On Alleged Refutations of Mechanism Using Gödel's Incompleteness Results," *Journal of Philosophy* 69, no. 17 (September 1972): 507–526. A critique of the argument that uses Gödel's theorem to show either that I cannot be a Turing machine (J. R. Lucas), or that if I am I cannot know my own program (P. Benacerraf). According to Chihara both arguments are either fallacious or implausible.

Davenport, Manuel M. "Kant and Maritain on the Nature of Art," *British Journal of Aesthetics* 12, no. 4 (Autumn 1972): 359–368. Good introduction to the positions of both thinkers on the nature of art as a human activity, its relation to morality, and the concept of beauty.

Eekels, J. *Industriële doelontwikkeling; Een filosofisch-methodologische analyse* [Industrial new business development; a philosophical and methodological analysis]. Assen: Van Gorcum; 1973. Pp. xii, 353. English summary pp. 324–337. Begins with a conceptual analysis of human action and proceeds to specify the structure of the D aspect of R & D.

Esposito, Joseph L. "Synechism, Socialism, and Cybernetics," *Transactions of the Charles S. Peirce Society* 9 (Spring 1973): 63–76. After explaining Peirce's idea of synechism and its use as a theory of evolution, the principle is applied to machine evolution, cybernetics, and information theory.

Fandozzi, P. R. *The Heideggerian Perspective on Nihilism: A Critique of Modern Technology Through Its Manifestations in Literature, Philosophy and Social Thought.* Honolulu: Univ. of Hawaii, 1974. A doctoral dissertation done under the direction of F. L. Bender.

Fernandez-Lomana del Rio, Ramon. "Anotaciones Historico-Culturales al Concepto de Factibildad" [Historical-cultural notes on the concept of practicability], *Anuario Filosofico* 6 (1973): 119–143. A study of the practicable or feasible as it occurs in three successive historical periods: in the magical vision of nature, according to M. Mauss; in the "natural" conception of nature, as represented by Aristotle; and as it is found in the framework of modern science.

Franks, Dean. "An Interpretation of Technology Through the Assertorical-Problematic Distinction," *Kinesis* 4, no. 1 (Fall 1971): 22–30. In its precision and exactness technology reveals itself as assertorical, an authority of being. Yet in its drive to assert it raises the problem of the distinction between technology, as the essent, and being. This distinction discloses technology in conflict and unity with being.

Funke, Gerhard. "Technik als Herausforderung und als Aufgabe" [Technology as challenge and as problem], *Zeitschrift für philosophische Forschung* 28, no. 1 (1974): 43–67.

Gibbons, M., and C. Johnson. "Relationship Between Science and Technology," *Nature* 227, no. 5254 (July 11, 1970): 125–127. "This case study suggests that the relation between science and technology is symbiotic rather than that technology is applied science . . ." (p. 127).

Granarolo, P. "Heidegger, penseur de l'époque planétaire" [Heidegger, thinker of the planetary age], *Annales de la Faculté des Lettres et Sciences humanines de Nice,* no. 20 (1973): 41-66.

Grange, Joseph, "Magic, Technology and Being," *Religious Humanism* 8, no. 2 (Spring 1974): 88-91. Uses Heidegger's ideas about technology to give an optimistic interpretation of contemporary American interest in the occult. "As Care for the Earth and its connective link, the human body, Being Human is beginning to find the depths of wisdom that have lain forgotten in the womb of nature. In the sense of a technique that can disappear as technique, technology might very well free us to care for Being itself" (p. 90).

Grau, Nestor. "Reflexiones Sobre el Tiempo de los Objetos Tecnicos" [Reflections on the Temporality of Technical Objects], *Ensayos y Estudios* (1973), pp. 6-14. Seeks to describe a difference in entitative rhythm among three types of objects—living beings, technical objects, and spiritual layers which conform to the essence of man. Further analyzes the connection between the time of technical objects and human time through the relationship of work.

Heidegger, Martin. "The End of Philosophy and the Task of Thinking," pp. 55-73, in *On Time and Being,* trans. Joan Stanmbaugh (New York: Harper and Row, 1972). This is the English translation of a lecture from 1964 which has previously been available only in French in a collected volume entitled *Kierkegaard vivant* (Paris: Gallimard, 1966). Argues that in the modern period, "Philosophy turns into the empirical science of man, of all of what can become the experiential object of his technology for man, the technology by which he establishes himself in the world by working on it in the manifold modes of making and shaping." In this period, also, "the new fundamental science ... is called cybernetics." Cybernetics "is the theory of the steering of the possible planning and arrangement of human labor. Cybernetics transforms language into an exchange of news." Considers the "task of thinking" under such conditions.

―――. "Messkirch's Seventh Centennial," *Listening* 8, nos. 1-3 (1973): 40-54. On the occasion of its seventh centennial celebration, Heidegger meditates on the future of his home village. "Messkirch tomorrow? It will be entangled in the network of the technological era. The question will arise ... whether under the domination of modern technology and amidst the world-transformation caused by it there can still be homeland in any sense at all. Perhaps man is settling into homelessness.... Perhaps, however, a new relationship to Home is likewise being prepared in the midst of the pressing force of the alien" (p. 49). Indeed, Heidegger argues that in modern man's experience of boredom and ennui, "amidst the alienation of the modern technological world, there nevertheless still is homeland.... — but *as that for which we are searching*" (p. 51). This is a new, improved translation (in bilingual format) of "700 Jahre Messkirch," which was first rendered as "Homeland: Festival Address at a Centennial Celebration," *Listening* 6 (1971): 233-238. Translated by Thomas J. Sheehan.

Huchingson, James E. "Toward a Naturalized Technology," *Zygon* 8, nos. 3-4 (September-December 1973): 185-199. Attempts to relate technology to the natural order in ways that suggest norms for human action toward nature. Uses elements from information theory and Whiteheadian process thought to describe the process of nature and technology.

Ihde, Don. "Vision and Objectification," *Philosophy Today* 16, no. 1 (Spring 1973): 3-11.

―――. "The Experience of Technology: Human Relations," *Cultural Hermeneutics* 2, no. 3 (November 1974): 267-279. A preliminary descriptive phenomenology of certain basic human-machine experiences using the examples of ordinary tools, complex machines and scientific instrumentation.

Iseminger, Gary. "The Work of Art as Artifact," *British Journal of Aesthetics* 13, no. 1

(Winter 1973): 3-15. Defends the definition that "something is a human artifact to the extent that some person or group of persons has responsibility for its having nonintentional properties."

Isenberg, Arnold. "The Technical Factor in Art," pp. 53-69 in *Aesthetics and the Theory of Criticism*. Chicago: Univ. of Chicago Press, 1973. A careful analysis of the nature of artistic technique which applies at least in part to engineering techniques. Argues against Croce that "in the mere activity of imagination there is an essential reference to overt production" and "to have worked overtly with materials at least in the *past* is a condition of imaginative efficacy."

Kavolis, Vytautas. "Paradigms of Order: Nature, the Factory, Art," *Salmagundi* no. 26 (Spring 1974): 69-84. On the character of action in advanced technological society.

King, Magda. "Truth and Technology," *Human Context* 5, no. 1 (1973): 1-34. A careful summary and commentary upon Heidegger's lecture "Die Frage nach der Technik."

Kuypers, K. "Relation Between Knowing and Making as an Epistemological Principle," *Philosophy and Phenomenological Research* 35, no. 1 (September 1974): 60-78. "The principle of the philosophy of science that assumes a close relation between understanding and making—one does not understand something until one can make it oneself—goes a good deal further than the one of the relation between knowing and being able to predict. It is difficult to exaggerate the significance of this principle for the whole of the future life of mankind, because of the numerous fresh possibilities entailed by the technical mastery of our environment and the ability also to interfere with man and his ways of life and culture." A broad ranging, if somewhat diffuse, historico-philosophical study, with references to Plato, Descartes, Kant, Vico, Marx, cybernetics, etc.

Layton, Edwin T., Jr. "Technology as Knowledge," *Technology and Culture* 15, no. 1 (January 1974): 31-41. A historian's critical and sophisticated development of the concept of the science-technology relationship to be found in A. Koyré. "We may view technology as a spectrum, with ideas at one end and techniques and things at the other, with design as a middle term. Technological ideas must be translated into designs. These in turn must be implemented by techniques and tools to produce things. The current model of science-technology relations looks at only one end of the spectrum. It would be an equal distortion to see technology solely as thought. Both aspects are needed for a balanced view" (pp. 37-38). Footnotes provide a good set of references to alternative arguments on this topic and various definitions of technology used by historians and engineers. For earlier development of Layton's thesis in a concrete historical study see his "Mirror Image Twins: The Communities of Science and Technology in 19th Century America," *Technology and Culture* 12, no. 4 (October 1971), 562-580.

Lovitt, William. "A 'Gespraech' with Heidegger on Technology," *Man and World* 6, no. 1 (February 1973): 44-59. "We have here ostensibly two objectives. The first is to show Heidegger typically at work with words, in one of his later essays, 'The Question Concerning Technology.' The second is to move through the words themselves onto the path of thought and 'thereby prepare a free relationship' to technology."

Lubbe, Hermann. "Technik und Gesellschaft; Zur Metakritik der Kritik an der technischen Intelligenz" [Technology and society; toward a meta-critique of the critique of technical intelligence], *VDI-Zeitschrift* 116, no. 2 (February 1974): 93-98.

Lycan, William G. "Mental States and Putnam's Functional Hypothesis," *Australasian Journal of Philosophy* 52, no. 1 (May 1974): 48-62. Sets out the basic tenets of Fodor's and Putnam's functionalism, and then argues that it contains a crucial equivocation.

Magnard, Pierre. "Le discours de la machine" [The language of the machine], *Revue de Métaphysique et de Morale* 79, no. 1 (January-March 1974): 108-117. A discussion of the theories of Pascal.

Margolis, Joseph. "Ascribing Actions to Machines," *Behaviorism* 2 (Spring 1974): 85-93. A critique and modification of K. Sayre's argument that consciousness can be ascribed to machines.

———. "Works of Art as Physically Embodied and Culturally Emergent Entities," *British Journal of Aesthetics* 14, no. 3 (Summer 1974): 187–196. An ontology of works of art developed on the basis of a nonreductive materialism. Persons and works of art, it is argued, are culturally emergent entities of similar generic character.

Maser, Siegfried. "Einige Probleme zur Philosophie der Technik: Über die Aktualität ontologischer Fragestellungen" [Some problems of the philosophy of technology: on the topicality of the ontological statement of the problem], pp. 227–235, in Kurt Hübner and Albert Menne, eds., *Natur und Geschichte*, X Deutscher Kongress für Philosophie, Kiel, October 8–12, 1972. (Hamburg: Meinen, 1973.)

Mayz Vallenilla, Ernesto. *Ebozo de una crítica de la razón técnica* [Study of a criticism of technical reason]. Caracas: Ediciónes de la Universidad Simón Bolivar, 1974. Pp. 249.

Medina, Angel De L. "Ortega: Action and Ontology," *Cultural Hermeneutics* 1, no. 2 (July 1973): 177–203. A study of how for Ortega technology grows out of man's nature as free, self-creative activity.

Moser, Paulo Roberto. "Cibernética & filosofia," *Revista brasileira de Filosofia* 22 (1972): 212–221.

Moser, Simon. "Technologie und Technokratie; Zur Wissenschaftstheorie der Technik" [Technology and technocracy; on the scientific theory of technique], pp. 169–177, in H. Lenk, ed., *Neue Aspekte der Wissenschaftstheorie* (Braunschweig: Vieweg, 1971).

Murray, Michael. "Art, Technology, and the Holy: Reflections on the Work of J. M. W. Turner," *Journal of Aesthetic Education* 8 (April 1974): 79–90. Interprets Turner's "Rain, Steam, and Speed: The Great Western Railway" (1884) as depicting man's new technological relationship to the earth in a way which seeks to reveal the deceptive character of man's conquest of nature. In doing this Turner's work assigns to both human and nonhuman things a place within nature's immensity.

Olszewski, Eugeniusz. "O soderzhanii poniattiia 'tekhnika'" [On the meaning of the term 'technology'], *Voprosy istorii estestvoznaniia i tekhniki* 39 (1972): 20–26. Various definitions discussed and divided into four main categories. Final definition: "Technology is a branch of civilization and culture which is a measure of man's mastering of nature and covers actual means and tools of economic activity and the ability to use these means and tools."

Omori, Shoichi. "Art and Technology" (in Japanese), *Bigaku* (Japanese Journal of Aesthetics) 23 (December 1972): 42. H. Sedlmayr describes how the development of automation in the nineteenth and twentieth centuries has influenced art. This influence, however, does not explain the original relationship between art and technology. For an understanding of this relationship one must turn to aesthetics and the work of Heidegger.

Ostrowski, Jan, R. Pichon, and R. Durand-Auzis. *Alfred Espinas, precurseur de la praxeologie, ses antecedents et ses successeurs*. Paris: Librairie General de Droit de Jurisprudence, 1973. Pp. 394. Espinas was the author of *Les origines de la technologie* (1897); by technology he meant what is now called praxiology, the science of efficient actions. Favorably reviewed by H. Skolimowski, *Technology and Culture* 16, no. 4 (October 1975): 682–683.

Pacini, Dante. *Problemática de uma nova filosofia metafísica de communicação do entendimento humano; ensaio de nôvo discurso do método metafísico em relação lógico-dialética com o positivo, místico, moral e estético*. Rio de Janerio: Distribuidora Record, 1970. Pp. 66.

Pasqualotto, Giangiorgio. *Avanguardia e tecnologia; Walter Benjamin, Max Bense e i problemi dell'estetica tecnologica* [The avant-garde and technology; Walter Benjamin, Max

Bense, and the problem of technological aesthetics]. Roma: Officina, 1971. Pp. 244.

Paz, Octavio. "Use and Contemplation," pp. 17–24, in *In Praise of Hands: Contemporary Crafts of the World*. Greenwich, Conn.: New York Graphic Society, in association with the World Crafts Council, 1974. A thoughtful consideration of the distinctions between works of art, craft, and industrial production. Argues that whereas art is made only to be contemplated and technological objects are made only to be used, handmade things "are pleasing because they are useful *and* beautiful. This copulative conjunction defines craftwork, just as the disjunctive conjunction defines art and technology: usefulness *or* beauty" (p. 21). By an important Mexican philosopher and poet. This volume also includes a helpful preface, entitled "A World Family," by James S. Plant, secretary-general of the World Crafts Council. But most of the book consists of pictures of work from the First World Crafts Exhibition, Toronto, Canada, June 11–September 2, 1974. As such it gives a good feeling for the contemporary scope of craft in both developed and underdeveloped countries. However, the reliance on photographs, which by their very nature cannot be handcrafted, in a mass-produced book, is at odds with the stated commitments of the exhibit.

Pochtar, Ricardo. "Teoria del lenguaje, tecnica y filosofia" [Theory of language, technology and philosophy], *Cuadernos de Filosofía* 12 (July–December 1972): 297–306. A critique of Jerrold J. Katz's philosophy of language as an attempt to technologize part of philosophy.

Radnitzky, G. "Towards a 'Praxiological' Theory of Research," *Systematics* 10, no. 3 (December 1972): 129–185. Outlines a theory of research based neither on formal logic nor psychology nor sociology, but on the nature of efficient work.

Rapp, Friedrich. "Die Technik in wissenschaftstheoretischer Sicht" [Technology in theoretical scientific perspective], in H. Lenk, ed., *Neue Aspekte der Wissenschaftstheorie* (Braunschweig: Vieweg, 1971): 179–185.

Robinet, André. *Le Défi cybernétique, l'automate et la pensée* [The challenge of cybernetics, the automaton and thought]. Paris: Gallimard, 1973. Pp. 232. Cf. the commentary on the mind-body problem occasioned by this book by Paul Gochet, "Le défi cybernétique," *Revue Internationale de Philosophie* 27, no. 1 [continuous no. 103] (1973): 112–119.

Robinson, Guy. "How to Tell Your Friends from Machines," *Mind* 81, no. 324 (October 1972): 504–518. A commentary on "the problem of minds and machines (or, for that matter, its twin—the problem of 'other minds')." Attempts, first, to define machine. "My claim is that intelligence and machines are related as user and used. An intelligent being is precisely one that is capable of turning something into an instrument, whether it is a paleolith or a sophisticated weapons system. A machine is what stands on the other side of this relationship as something that has been contrived for, or given, a use" (p. 514). Thus, "to decide that something is intelligent is not just to record that it has passed certain tests or exhibited certain features but is to define one's whole human relation to it, the sort of behavior and attitudes that are appropriate or even necessary" (p. 515). Concludes with a brief discussion of how machines also use men. For a strong critical reply see Yorick Wilks, "Your Friends and Your Machines," *Mind* 83, no. 332 (October 1974): 583–585.

Rollins, P. C. "Worf Hypothesis as a Critique of Western Science and Technology," *American Quarterly* 24 (December 1972): 563–583.

Ropohl, Günter. "Prolegomena zu einem neuen Entwurf der allgemeinen Technologie" [Prolegomena to a new design for general technology], pp. 217–226, in Kurt Hübner and Albert Menne, eds., *Natur und Geschichte*, X Deutscher Kongress für Philosophie, Kiel, October 8–12, 1972. Hamburg: Meinen, 1973.

―――. "Was heisst 'Technologie'?" *VDI-Nachrichten* 26, no. 6 (1972): 11.

Rostenne, Paul. "L'Alienation rationaliste et la culture occidentale" [Rationalistic alienation and Western culture], *Giornale di Metafisica* 26 (September–December 1971): 405–425. The history of modern Western thought discloses how the rationalistic conception of reason pursues a totalitarian desire for scientific knowledge and its technical prolongation. This determines first a mode of thought, then a mode of life which has transformed Western culture into an anticulture, that is, into an instrument of dehumanization, by reducing the significance of the human to that which is revealed solely by the scientific method of cognition.

Rubinoff, Lionel. "The Contest between Faust and Prometheus: Reflections on the Crisis of the Person in a Technological Society," *Philosophy in Context* 3 (1973): 7–22. Article seeks to identify the sources of functional rationality, to describe its structure, and to assess its influence on modern technology. Functional rationality argued to be an unnecessary Faustian outgrowth of Promethean rationality.

Sachsse, Hans. "Die kybernetische Methode in der Technik, in der Biologie und in den Gesellschaftswissenschaften" [The cybernetic method in technology, in biology, and in sociology], *VDI-Zeitschrift* 115, no. 8 (June 1973): 613–620.

Schischkoff, Georgi. "Wissenschaftstheoretische Betrachtung zum Informationsbegriff" [The view of scientific theory on the concept of information], *Zeitschrift für Philosophische Forschung* 25 (1971): 60–88.

"Science." *Marxism, Communism and Western Society: A Comparative Encyclopedia.* C. D. Kernig, ed. New York: Herder and Herder, 1973. This article contains a section, pp. 280–281, on "Pure and Applied Science," which gives the standard Marxist theory of their relationship.

Skolimowski, Henryk. "Technology v. Nature," *Ecologist* 3, no. 2 (February 1973): 50–55. Modern technology has its origins not in Greek science nor in the Judeo-Christian religion, but in Renaissance metaphysics. A good historico-philosophical survey.

——. "La humanización de la mente technológica o la validez del conocimiento technológico" [The humanization of the technological mind or the validity of technical knowledge], *Folia Humanística* 11 (1973): 945–959.

——. "The Scientific World View and the Illusions of Progress," *Social Research* 41, no. 1 (Spring 1974): 52–82. Argues that modern science, in both its cognitive and pragmatic realms, undermines rather than supports the notion of progress. "I argued . . . that in its development science has undermined not only its well-established theories but also its *modus operandi* as the discoverer of ultimate truths. With the disappearance of its Platonic ethos, science began to exhibit increasing difficulties in justifying its progress in cognitive terms" (p. 68). On the practical side material progress has only been an illusion of progress when viewed in the light of our ecological and spiritual regression.

Steinbuch, Karl, and Simon Moser. *Philosophie und Kybernetik.* Munich: Nymphenburger, 1970. Pp. 198. Includes K. Steinbuch's "Grundbegriffe und Fragestellungen der Kybernetik" (pp. 13–25), G. Färber's "Kybernetik und Biologie" (pp. 26–35), S. Moser's "Philosophie an der Technischen Hochschule" (pp. 36–45), S. J. Schmidt's "Sprache und Denken; Eine Strukturskizze ihres möglichen Zusammenhangs" (pp. 46–56), A. Hoppe's "Sprache und Denken" (pp. 57–78), E. Oldemeyer's "Überlegungen zum phänomenologisch-philosophischen und kybernetischen Bewusstseinsbegriff" (pp. 79–93), H. Kilian's "Überlegungen zer Metanoetik; Ein Beitrag zur kritischen Theorie unbewusster Strukturen des bewussten Denkens" (pp. 94–121), S. Moser's "Theorie und Erfahrung" (122–135), K. Steinbuch's "Realität und Modell" (pp. 136–150), D. Bierlein's "Bemerkungen zum Thema 'Realität und Modell'" (pp. 151–155), H. M. Lipp's "Problemlösen und Erkennen aus der Sicht der Nachrichtenverabeitung" (pp. 156–162), J. Peter's "Information und Signal" (pp. 163–173), B. Hassenstein's "Abbildende Begriffe" (pp. 174-181), K. Steinbuch's "Die Zukunft im Rahmen der Kybernetik" (pp. 182-190).

Taminiaux, Jacques. "Sur Marx, l'art et la vérité" [On Marx, art and truth], *Revue Philosophique de Louvain* 71, 4th series no. 14 (May 1974): 311-327. "This study attempts to show that a certain ontology of production as access to self-presence and to self-enjoyment grounds Marx's thought on art. . . . But it is also this ontology of production which grounds, apparently in contradictory fashion, the attribution of an aesthetic connotation to the original movement of praxis."

Tavares, José Augusto T. "Cibernética e pensamento filosófico" [Cybernetics and philosophic thought], *Revista brasileira de Filosofia* 20 (1970): 175-185.

Traupel, Walter. "Ziele und Grundlagen der technischen Wissenschaften" [Ends and foundations of the technological sciences], in Helmut Holzhey, ed., *Wissenschaft/ Wissenschaften; Interdisziplinäre Arbeit und Wissenschaftstheorie* (Basel and Stuttgart: Schwabe, 1974).

Van Wyk, Jacobus Daniel. *Die tegniese wetenskap, afgebeeld in die elektronika: struktuur, wese en grense* [Technological science, as reflected in electronics: structure, nature and boundaries]. Johannesburg: Randse Afrikaanse Universiteit, 1972. Pp. 36. "Traditionally investigation of the structure, nature and boundaries of Technology has been conducted from outside by Philosophy and not from inside by itself. The question as to what extent this is still possible and desirable at present, is posed. Technical Science is intimately related to Technology itself and consequently the structure, nature and boundaries of the former are investigated. A structure based on the elements of System, Energy, Matter and Information is proposed . . ."—from the English summary. Structure exemplified by the case of electronics. Lecture by an engineer who has been influenced by H. Van Riessen. Thirty-one-item bibliography with about one fourth of these being South African publications.

Walker, Jeremy. "Philosophy in the Present Age," *Dialogue* (Canada) 13, no. 3 (September 1974): 561-576. Raises the question of whether philosophy, in the ancient sense understood as a form of wonder and worship, is possible in a technological age. "For it is clear that this modern concept of knowledge . . . that is derived from the dissection of its object is quite other than knowledge derived from loving its object" (p. 575). The problem lies in finding a knowledge which is both technological and contemplative. Excellent essay.

Young, Charles M. *The Concept of "Techne" in Plato's "Republic" and Other Dialogues.* Baltimore: Johns Hopkins University, 1974. A doctoral dissertation done under the direction of David Sachs and Stephen Barker.

Zieleniewski, J. "Remarks of a Polish Praxiologist on the Subject of a Paper by Gutiérrez," *Theory and Decision* 1 (1971): 359-368. Surveys the development of theories of (efficient) human action; defense of T. Kotarbiński.

B. Secondary Sources

Bichowsky, Francis Russell (1889-1951). *Industrial Research.* New York, Arno Press, 1972. Pp. vi, 126. "Technology and Society" series reprint of the 1942 edition of an early attempt to define industrial research by an influential scientist-engineer.

Bigelow, Jacob (1787-1879). *The Useful Arts.* Vols. 1 and 2. New York: Arno Press, 1972. Pp. 384 and 396. "Technology and Society" series reprint of the 1842 edition of an important survey of the whole field of technology, by the author of *The Elements of Technology* (1831).

Dale, Rodney. "Teaching Creative Thinking Might Solve All Your Problems," *Engineer* 236, no. 6100 (February 8, 1973): 61. Innovation through Creative Analysis combines lateral thinking, brainstorming, and synectics to give more efficient ideas generation.

Händle, Frank, and Stefan Jensen, eds. *Systemtechnik und Systemtheorie; sechzehn Aufsätze* [Systems engineering and systems theory; six essays]. Munich: Nymphenburgen, 1973. Pp. 321.

Heuer, Georg. "Industrielle Forschung und technischer Fortschritt" [Industrial research and technical development], *VDI-Zeitschrift* 116, no. 4 (March 1974): 261–265. Industrial research and development produce knowledge of the goals which a new product ought to meet as well as the means out of which and by which the product ought to be produced. Technical innovation, since it is first of all dependent on technical realization (problem solving), results from the combination of goal and means. Only then can one decide about the actual product to be produced, given a range of different product possibilities. Technical progress or development is achieved through the development of new products.

Hill, L. S. "Systems Engineering in Perspective," *IEEE Transactions* EM 17, no. 4 (November 1970): 124–131. Presents a background of the evolution of the systems engineering process, with a detailed and comprehensive resolution.

Hollingum, Jack. "Group Technology: The Red Revolution Is Changing Its Face," *Engineer* 236, no. 6102 (February 22, 1973): 49–51. An article introducing "group technology"—a concept of production which focuses on the grouping of jobs for a single machine tool. Brief history of the concept from its Russian origin in the 1920s to contemporary Western debates. Text includes bibliographic references.

———. "Total Technology Is the Spirit of Modern Engineers," *Engineer* 237, no. 6137 (October 25, 1973): 52–53. University studentships are being introduced in Britain to widen the scope of engineering study.

Hubka, Vladimir. *Theorie der Maschinensysteme; Grundlagen einen wissenschaftlichen Konstruktionslehre* [Theory of machine systems; principles of a scientific discipline of design]. Berlin, Heidelberg, and New York: Springer-Verlag, 1973. Pp. 142. Illustrated. A contemporary sequel to the work of F. Reuleaux. Borrowing from cybernetics, Hubka develops a comprehensive theory of machines which, according to a review by Wolfrom Borchert in *VDI-Zeitschrift* 115, no. 18 (December 1973), p. 1499, "far exceeds previous efforts, without losing itself in pure theory." Exposition of basic concepts followed by definition and subsequent logical classification of different machine systems according to function and structure. Includes subject index and a large number of references.

Matchett, Edward. "Fundamental Design Method—A Means of Controlled Thinking and Personal Growth," *Systematics* 11, no. 1 (June 1973): 29–53.

———. "From Fundamental Design Method to Logosynthesis," *Systematics* 11, no. 2 (September 1973): 83–96.

———. "Introduction to the Discipline of Fundamental Design Method," *Systematics* 11, no 3 (December 1973): 163–173.

———, and A.G.E. Blake. "Logosynthesis: A Meta-Controlled Design Discipline," *Systematics* 11, no. 2 (September 1973): 97–121.

Note: The four preceding articles give a somewhat eclectic synthesis of engineering, psychology, and esoteric spiritual disciplines which is nevertheless quite relevant to much discussion of engineering design methodology. Other papers by Matchett (and others) on the subject of design and the "creative process" appear regularly in *Systematics*. For an explication of the theory of systematics itself, of which the fundamental design method is a practical application, see John G. Bennett, "General Systematics," *Systematics* 1, no. 1 (June 1963): 5–19, which describes systematics as "the study of systems and their application to the problem of understanding ourselves and the world." Systematics also attempts to make connections between various traditional spiritual techniques and the techniques of design thinking; for instance, fundamental design method is presented not simply as abstract logical process but as a practice which one must be initiated into. On the relation between systematics and systems theory see "Editorial: Systematics and General Systems Theory," *Systematics* 1, no. 2 (September 1963): 105–110; and "Systematics and Systems Theory," *Systematics* 7, no. 4 (March 1970): 273–278. "General Systems Theory defines a system in terms of interaction, a nonqualitative concept, and Systematics in terms of mutual relevance, a qualitative concept."

Parsegian, V. L. *This Cybernetic World of Men, Machines and Earth Systems.* New York: Doubleday, 1972. Pp. xiv, 209. Illustrations. A readable study for the layman of the complex science of cybernetics. Cybernetics is discussed in all of its aspects—political, neurological, electronic, ecological, etc. Favorably reviewed by Stafford Beer in *Technology and Culture* 14, no. 2 (April 1973): 331–333.

Ribereau-Gayon, Jean. *Problemes de la recherche scientifique et technologique: les hommes et les groupes* [Problems of scientific and technological research: individuals and groups]. Paris: Dunod, 1972. Pp. x, 220. Preface by Jean Fourastié. Bibliography, pp. 215–220.

Robinson, Herbert W. and Douglas E. Knight. *Cybernetics, Artificial Intelligence, and Ecology.* New York and Washington: Spartan Books, 1972. Pp. 333. Proceedings of the fourth annual symposium of the American Society for Cybernetics. Articles on three topics: "General Cybernetics," "Artificial Intelligence and Robotics," "Ecological Cybernetics"; by W. R. Ashyby, G. Pask, etc.

Speiser, Ambros P. "Forschung in Hochschule und Industrie: Wechselwirkungen zwischen Wissenschaft und Technik" [Research in university and industry; interrelationships between science and technology], pp. 7–10, in *Forschungstheorie Forschungspraxis,* (Vienna and New York: Springer-Verlag, 1971). Attempts to map out the interrelationships in terms of a series of overlapping but not completely parallel activities. Not hard-core philosophy; reflects general engineering understanding.

Williams, Christopher G. *Craftsmen of Necessity.* New York: Random House, 1974. Pp. 182. "When man first began to build, he watched nature's performance and followed it. His ways rested easily in the environment because the environment was his control and reference; he lacked the power to do otherwise. . . . Machine technology drove a steel wedge between man and his home . . . [and] has given builders materials and methods that have little to do with climate or environment. . . . Most of the indigenous people of the world still practice organic technology. This is the opposite of machine thinking. It is a way, not a device, a philosophy to govern the methods of selecting action. It does not plan for an event but lets the event develop on its own and then coordinates its effort with the development. . . . The following pages examine people, town, mud, forest and stone and attempt to give substance to these perilously broad definitions." With photographs. Although not hard-core philosophy in the sense of rigorous conceptual analysis or dense metaphysical speculation, does point with a sure intuition toward an aspect of technology which tends to be overlooked.

V. APPENDIX

A. Philosophical and Historical Background

Bailes, K. E. "Politics of Technology: Stalin and Technocratic Thinking Among Soviet Engineers," *American Historical Review* 79, no. 2 (April 1974): 445–469. Very informative history of the conflict between Russian professional engineers and their attempt to develop a technical philosophy which would oppose orthodox Marxist ideology.

Boorstin, Daniel J. *The Americans; The Democratic Experience*. New York: Random House, 1974. Pp. xiv, 717. This third volume of his work on American history describes the impact of technology on America, in such aspects as transportation, the communication media, mass production, consumerism, and public education. Includes a good description of Edison and the growth of R & D industries. Excellent bibliographic references.

Cardwell, D. S. L. *Technology, Science and History: A Short Study of the Major Developments in the History of Western Mechanical Technology and Their Relationships with Science and Other Forms of Knowledge*. London: Heinemann Educational, 1972. Pp. 244. Illustrated. Bibliography.

———. "Technology," pp. 357–365 in Philip P. Wiener, ed., *Dictionary of the History of Ideas*, vol. 4 (New York: Scribners, 1973). A Good historical overview for philosophers.

Carlisle, Robert B. "The Birth of Technocracy: Science, Society, and Saint-Simonians," *Journal of the History of Ideas* 35, no. 3 (July–September 1974): 445–464. A historico-philosophical study of attempts to relate technology and society. See also F. A. Hayek, *The Counter-Revolution of Science* (Glencoe, Ill.: Free Press, 1955), whose own study of Saint-Simon is discussed in this paper. Saint-Simon's work is easily consulted in Henri de Saint-Simon, *Social Organization, the Science of Man, and Other Writings*, ed. and translated by F. Markham (New York: Harper and Row, 1964).

Cipolla, Carlo M. "The Diffusion of Innovations in Early Modern Europe," *Comparative Studies in Society and History* 14 (January 1972): 46–52. The borrowing of technical devices and knowledge is shown to be a cultural process with no economic relevance.

Davenport, William H. "Antitechnology Attitudes in Modern Literature," *Technology and Society* (Boston Univ.) 8 (April 1973): 7–14.

Gille, Bertrand. "Prolégomènes à une histoire des techniques," *Revue des mines et de la métallurgie* 4 (1972): 3–65. Discussion of the history and theory of technological progress.

Haraway, Donna Jeanne. "The Transformations of the Left in Science: Radical Associations in Britain in the 30's and the U.S.A. in the 60's," *Soundings* (Vanderbilt Univ.) 58, no. 4 (Winter 1975): 441–462. "The question of the proper role of science in modern culture is so large that I would like to restrict this paper to a glance at the opinions and organizations of dissident scientists themselves in periods of major world and national crises" (pp. 442–443). Latter part of this article concerned with the theories and influence of Marcuse, Habermas, and Roszak.

Heilbroner, Robert L. "The Paradox of Progress: Decline and Decay in 'The Wealth of Nations'," *Journal of the History of Ideas* 34, no. 2 (April–June, 1973): 243–262. Discusses the paradox of an economic growth which terminates in long-run decline according to Adam Smith's economic theory in *The Wealth of Nations* (1776).

Israel, G., and P. Negrini. "Science and the French Revolution," *Scientia* 108, nos. 3–4, 5–6 (1973): 376–392. Preceded (pp. 357–375) by the same article in Italian. This is the second part of a two-part article, the first of which has apparently not been published.

"The first part of this article deals with physical mathematical science in eighteenth-century France up to the eve of the Revolution. The second part gives an outline of the crisis of scientific institutions and the debate which took place on the role of science in the years 1789–1795. The authors show how a conception of the relationship between science and technology grew up which no longer subordinated the latter to the former."

Kranzberg, Melvin. "Engineers," *Vital Speeches* 38 (September 1, 1972): 676–683.

Kuhn, Thomas S. "The Relations between History and History of Science," *Daedalus* 100 (Spring 1971): 271–304. A stimulating article. One wishes he would have said more about the history of technology.

Kursunoğlu, Behram, and Arnold Perlmutter, eds. *Impact of Basic Research on Technology*. Studies in the Natural Sciences, vol. 1. New York: Plenum, 1973. Pp. 301. Illustrated. Good historical perspective.

Lerori-Gourham, André. *Évolution et techniques*, II: *Milieu et techniques*. 2nd revised edition. Paris: A. Michel, 1973. Pp. 475. Revision of a classic anthropological study which contains an interesting taxonomic classification of machines.

Linden, Stanton J. "Francis Bacon and Alchemy: The Reformation of Vulcan," *Journal of the History of Ideas* 35, no. 4 (October–December 1974): 547–560.

Rürup, Reinhard. "Historians and Modern Technology: Reflections on the Development and Current Problems of the History of Technology," *Technology and Culture* 15, no. 2 (April 1974): 161–193. A good overview of European thought about technology in general, although the focus is on historical research.

Simmons, H. "System Dynamics and Technocracy," *Futures* 5, no. 2 (April 1973): 212–228. Compares the technocracy movement of the 1930s to the Forrester, Meadows, *et al.* computer analyses of technical-economic development.

Smith, Cyril Stanley. "Technology in History," *Minerva* 8 (1970): 469–470. Essay review of Lynn White, Jr.'s *Machina ex Deo: Essays in the Dynamism of Culture* (Cambridge, Mass., 1968).

Treue, Wilhelm. "Zur jüngeren Technikgeschichtsschreibung in der DDR über den 'deutschen Imperialismus'" [On the recent historiography of technology in East Germany concerning German imperialism], *Technikgeschichte* 38, no. 3 (1971): 220–232.

White, Lynn, Jr. "Technology Assessment from the Stance of a Medieval Historian," *American Historical Review* 79, no. 1 (Feb. 1974): 1–13. Contemporary technology assessment needs to be concerned with more than cost-benefit calculations and to develop a greater sense of the depth of time. Argues these points by referring to the historical impact of a number of Western inventions — i.e., brandy, weapons, eyeglasses, chimney flue, buttons, knitting, spinning wheel, etc.

———. "The Flavor of Early Renaissance Technology," pp. 36–57 in Bernard S. Levy, ed., *Development in the Early Renaissance*. Albany, N.Y.: State Univ. of New York Press, 1972. Argues that "the innovative quality of Renaissance engineering is to be credited neither to application of scientific discovery nor to stimulus from antiquity" nor did it reflect "the new needs of a rapidly expanding commercial society." "Quite the contrary, the engineers seem most often to be indulging in a speculative empiricism which is almost playful in flavor" (pp. 50–51).

B. Textbooks

Campbell, A. V. *Moral Dilemmas in Medicine; A Coursebook in Ethics for Doctors and Nurses*. Baltimore: Williams and Wilkins, 1972. Pp. 214.

Cross, Nigel, David Elliott, and Robin Roy, eds. *Man-Made Futures: Readings in Society, Technology and Design*. London: Hutchinson, 1974. Pp. 365. Textbook for Open University second-level course.

DeNevers, Noel. *Technology and Society*. Reading, Mass.: Addison-Wesley, 1972. Pp. 307. A text developed for a course on technology in the College of Engineering at the University of Utah. Includes readings and discussion questions on the following topics: "The Complaints About Technology," "The History of Technological Change, the Acceleration of Technological Change, and the Transfer of Technology," "How We Respond to Technological Change," "The Predictions of Disaster," "How Safe Is Safe Enough?," "The Interrelations of Technology," "Science-Technology and Government." Selected from works by N. Calder, W. H. Ferry, H. G. Rickover, J. K. Feibleman, P. F. Drucker, G. Hardin, A. M. Weinberg, C. Starr, S. C. Gilfillan, etc.

Douglas, Jack D., ed. *The Technological Threat*. Englewood Cliffs, N.J.: Prentice-Hall, 1971. Pp. vi, 185. Contents: D. Bell's "American Society as a Technological Society: Notes on the Post-Industrial Society," J. Taviss's "The Technological Society," R. Nisbet's "The Impact of Technology on American Values: The Impact of Technology on Ethical Decision-Making," K. Mannheim's "The Crisis in Valuation," D. Riesman's "The Impact of the Electronics Revolution: Leisure and Work in Post-Industrial Society," N. Wiener's "Some Moral and Technical Consequences of Automation," D. Bell's "The Threats of Alienation and Social Technology: The Disjunction of Culture and Social Structure," R. C. Rogers and B. F. Skinner's "Some Issues Concerning the Control of Human Behavior," C. W. Mill's "The Impact of Technology on Political Values: Liberal Values in the Modern World," L. K. Frank's "The Need for a New Political Theory," E. T. Chase's "Politics and Technology."

Goran, Morris. *Science and Anti-Science*. Ann Arbor: Ann Arbor Science, 1974. Pp. 132. Deals with the factual elements in the antiscience movement. Suggested as a text in science curricula.

Lauda, Donald P., and Robert D. Ryan, eds. *Advancing Technology: Its Impact on Society*. Dubuque, Iowa: William Brown, 1971. Pp. xvii, 536. Readings for social science courses dealing with the problems of technology. Includes articles by R. Theobald, J. Ellul, M. Kranzberg, B. Seligman, S. deGrazia, E. Fromm, S. Chase, A. Westin, N. Calder, E. J. Mishan, and R. Buckminster Fuller.

Layton, Edwin T., ed. *Technology and Social Change in America; Interpretations of American History*. New York: Harper and Row, 1973. Pp. 181. Bibliography. Intended as a text.

Michalos, Alex C. *Philosophical Problems of Science and Technology*. Boston: Allyn and Bacon, 1974. Pp. 623. A text divided into ten sections—I, The Nature of Science and Technology; II, The Logic of Discovery and Growth; III, Types of Scientific Explanation; IV, The Nature and Function of Scientific Laws; V, Scientific Theories; VI, Observation and the Meaning of Scientific Terms; VII, Induction, Probability, and the Appraisal of Hypotheses; VIII, The Role of Values in Science; IX, The Social Responsibilities and Control of Science; X, The Impact of Science and Technology on Society—with essays by M. Bunge, N. R. Hanson, P. K. Feyerbend, K. R. Popper, C. G. Hempel, E. Nagel, D. Callahan, etc. A good broad-ranging anthology. As the editor notes, "the selections are intended to be *introductory*." Weaknesses: ignores cosmological questions; much more on science than on technology, which tends to be seen merely as the practical application of science; technology discussed merely as a social issue in the last section of the book.

Newton, David E., ed. *Science and Society*. Boston: Holbrook Press, 1974. Pp. x, 306.

Parkman, Ralph. *The Cybernetic Society*. New York: Pergamon, 1972. Pp. x, 396. An "interdisciplinary" textbook describing cybernetics, its history, and its social implications.

Piel, Emil J., and J. G. Truxal. *Man and His Technology; Problems and Issues*. New York: McGraw-Hill, 1973. Pp. 261. Illustrated.

Pyle, James L. *Chemistry and the Technological Backlash.* Englewood Cliffs, N.J.: Prentice-Hall, 1974. Pp. xiii, 354. Seeks to gain an understanding of technological problems by "considering them from the chemical point of view" (p. 5). The author begins with a detailed yet readable description of the carbon cycle in nature and its relation to the energy problem. Next follow chapters on separate topics such as water pollution, the pesticide problem, solid-waste disposal, chemical additives, and the chemistry of contraception involved in population control. Diagrams and chemical structures comprise a good deal of the book. At the end of each chapter is a short bibliography. Pyle concludes with the usual statements about the scientist's and the politician's ethical responsibilities in "taming technology." Technology defined as "the systematic application of knowledge to generate new materials, machines, or other developments" (p. 344). The book is designed to be used as a supplement to basic chemistry texts. It may be used alone, assuming the student has at least a one-quarter course in college chemistry. In this respect it is recommended.

Rochlin, Gene I., ed. *Scientific Technology and Social Change; Readings from Scientific American.* San Francisco: W. H. Freeman, 1974. Pp. 403. Bibliography, pp. 391–395.

Susskind, Charles. *Understanding Technology.* Baltimore: John Hopkins Univ. Press, 1973. Pp. x, 163. Courses in the humanities are required of engineers to supplement their deficiency in the liberal arts. "Yet in a 'liberal' education," Susskind warns, "technology is virtually ignored, despite its central place in contemporary culture" (p. ix). The purpose of this work is to give the technically illiterate a "first acquaintance with contemporary technology" to make up for their deficiency. Opens with a history of technology. Technology is defined as "man's efforts to satisfy his material wants by working on physical objects" (p. 1). Beginning with the Industrial Revolution, inventions and their developments are listed within the context of such categories as energy conversion (e.g., the steam engine), materials processing, transport, and communications technology. The "Second Industrial Revolution" leads to a consideration of cybernetics and computers. Following this history is a brief discussion of how technology is beneficial to medicine, the fine arts, education, and humanistic studies like archeology and linguistics. Reviews the ideologies of technology ranging from Ellul to the Marxists. Closes with a suggested "Engineer's Hippocratic Oath" and an analysis of the ethical question, "Technologist: Benefactor or Monster?" Susskind exhausts the depths of his profundity by concluding: "The truth doubtless lies somewhere in between" (p. 103). Peter Drucker, in a review in *Technology and Culture* 15, no. 1 (Winter 1974), pp. 80–82, strongly criticizes Susskind's historical overview as superficial.

Wertz, Richard W., ed. *Readings on Ethical and Social Issues in Biomedicine.* Englewood Cliffs, N.J.: Prentice-Hall, 1973. Pp. 306. Inexpensive paperback. Articles by important scholars in the biomedical field. Includes three articles on women and medicine and four controversial articles on health-care delivery.

RESEARCH IN PHILOSOPHY AND TECHNOLOGY

Volume 2. December 1978 Cloth 540 pages (Tent.)   Institutions: $30.00
ISBN NUMBER 0-89232-101-6   Individuals: $15.00

TENTATIVE CONTENTS:

Part I: Historico-Philosophical Studies

On the Nature of Nature, Jacob Klein, St. John's College, Annapolis. **Philosophy of Technology: Origins and Issues,** Carl Mitcham, St. Catharine's College. **The Concepts of "Nature" and "Techniques" According to the Greeks,** Wolfgang Schadwaldt, Universitat Tubingen. **Simone Weil (1901–1943), Two Unpublished Letters on Machines. Heidegger and Marcuse: Technology as Ideology,** Michael Zimmerman, Newcomb College, Tulane University.

Part II: Conference and Non-Conference Papers

Euthyphronics and the Problems of Adapting Technical Progress to Man, Jozef Banka, Katowice, Poland. **An Engineering Critique of Philosophy of Technology,** George Bugliarello, New York Polytechnic. **The Normative Side of Technology,** Edmund Byrne, Indiana-Purdue University. **Art in a Technological Society,** Phillip Frandozzi, University of Montana. **Marx, Machinery, and Alienation,** Bernard Gendron and Nancy Holmstrom, University of Wisconsin, Milwaukee. **The Physical and Social Foundation of Technology,** Friedrich Rapp, Technical University, Berlin.

Part III: Reviews and Bibliography

Philosophy of Technology in France, C. Beaune, Universite de Clermont. **The Philosophy of Technology of Jean Brun,** Daniel Cerezuelle, Universite de Grenoble. **Systems Theory: A Critical Analysis and Bibliography,** William O'Neil, University of Illinois, Chicago Circle. **Philosophy of Technology in Italy,** Lucia Palmer, University of Delaware. **Technology Assessment: A Critical Analysis and Bibliography,** F. A. Rossini, Georgia Institute of Technology. **Philosophy of Technology in the Netherlands,** Egbert Schuurman, Vrije University, Amsterdam.

A 10 percent discount will be granted on all institutional standing orders placed directly with the publisher. Standing orders will be filled automatically upon publication and will continue until cancelled. Please indicate which volume Standing Order is to begin with.

RESEARCH IN SOCIOLOGY OF KNOWLEDGE, SCIENCES AND ART
An Annual Compilation of Research
Series Editor: Robert Alun Jones, Department of Sociology, University of Illinois.

The essays in this annual series consist of original research done in the fields of the sociology of knowledge, science, and art. As the contents of the first volume suggest, the focus of the series will be explicitly interdisciplinary, including contributions from philosophers and historians as well as sociologists, and extending to the philosophy of science and history of ideas as well as sociology. A number of theoretical and methodological perspectives, as well as nationalities, will be represented. The series will also serve as the vehicle for essays of a length or content inappropriate to more conventional scientific journals.

Volume 1. March 1978. Cloth 350 pages
ISBN NUMBER: 0-89232-026-5

Institutions $25.00
Individuals $12.50

CONTENTS:

Explanation and Understanding in the Social Sciences, Mihailo Markovic, University of Belgrade, with comments by Murray Murphey, University of Pennsylvania, G. H. von Wright, University of Helsinki, Quentin Skinner, Cambridge University, Ernest Nagel, Columbia University. **Remarks Concerning Epistemological Problems of Objectivity in the Social Sciences,** Joachim Israel, University of Lund. **Friedrich Buchholz as a Sociologist of Ideas,** Hans H. Gerth, Glashutten, West Germany. **The Sins of the Fathers: British Anthropology and African Colonial Administration,** Henrika Kuklick, University of Pennsylvania. **Social Movements: The Anatomy and Indirection of a Sociological Field,** Joseph R. Gusfield, University of California—San Diego. **The Sociology of Nonknowledge: A Paradigm,** Deena Weinstein, DePaul University, and Michael Weinstein, Purdue University. **Scientific Reward Systems: A Comparative Analysis,** Stephen Cole, State University of New York—Stony Brook. **The Sociometric Structure of a Scientific Discipline,** Judith R. Blau, Baruch College, City University of New York. **On Integrating Behavioral and Philosophical Systems: Toward a Unified Theory of Problem Solving,** Ian I. Mitroff and Ralph H. Kilmann, University of Pittsburgh. **The Sociology of Literature: An Historical Introduction,** Priscilla P. Clark, University of Illinois—Chicago Circle. **Aesthetic Theory and the Sociology of Art: The Social Foundations of Classicism and Romanticism,** Don Martindale, University of Minnesota. **Toward a General Sociology of the Folk, Popular and Elite Arts,** Marcello Truzzi, Eastern Michigan University. **Index.**

A 10 percent discount will be granted on all institutional standing orders placed directly with the publisher. Standing orders will be filled automatically upon publication and will continue until cancelled.

STUDIES IN SYMBOLIC INTERACTION
An Annual Compilation of Research
Series Editor: Norman K. Denzin, Department of Sociology, University of Illinois.

The essays in this annual series consist of original research and theory within the general sociological perspective known as Symbolic Interactionism. Longer than conventional journal-length articles, the essays wed micro and macro sociological concerns within a qualitative, field-method empirical orientation. International in scope, this series draws upon the work of urban ethnographers, phenomenologists, ethnomethodologists, critical theorists, humanistic sociologists, as well as symbolic interactionists and conflict theorists. The emphasis is on new thought and research which bridge links to an emergent theory of self, socialization, interaction, social relationships, social organization and society.

Volume 1. April 1978 Cloth Approx. 350 pages
ISBN NUMBER: 0-89232-065-6

Institutions $25.00
Individuals $12.50

CONTENTS:

Foreword, Norman K. Denzin, University of Illinois. **Social Unrest and Collective Protest,** Herbert Blumer, University of California—Berkeley. **Markets as Social Organization,** Elihu Gerson, Pragmatica Systems, Inc. **Organizational Crime: A Theoretical Perspective,** Edward Gross, University of Washington. **Crime and the American Liquor Industry,** Norman K. Denzin, University of Illinois. **Symbolic Interaction and Social Worlds,** Anselm L. Strauss, University of California—San Francisco. **Toward a Ritual Theory of Self: Durkheim and Goffman Reconsidered,** Norbert Wiley, University of Illinois. **Ritual as a Game: Playing to Become a Sanema,** Gregory P. Stone, University of Minnesota, and Gladys I. Stone, University of Wisconsin. **Toward a Symbolic Interactionist Theory of Learning: A Reapproachment with Behaviorism,** Edward Gross, University of Washington. **The Social Psychology of Sexual Arousal: A Symbolic Interactionist Interpretation,** Jeffrey S. Victor, Jamestown Community College. **The Social Construction and Reconstruction of Physiological Events: Acquiring the Pregnant Identity,** Rita Sieden Miller, City University of New York—Brooklyn. **Tyranny,** Dan Miller, University of Manitoba, Marion W. Welland, Wichita State University, and Carl J. Couch, University of Iowa. **The Meditation Movement: Symbolic Interactionism and Synchronicity,** Richard Bibee and Julian B. Roebuck, Mississippi State University. **Bodies and Selves: Notes on a Dilemma in Demography,** David R. Maines, Rockefeller University. **Deviance and Choice,** Regan G. Smith, Sangamon State University. **Cooley's Economic Sociology: The Theory of Pecuniary Evaluation,** Glenn Jacobs, University of Massachusetts—Boston. **The Use of Improvisation and Modulation in Natural Talk: An Alternative Approach to Conversational Analysis,** Reyes Ramos, University of California—La Jolla. **Chicago Sociology: A Metatheoretical Analysis,** Harvey A. Farberman, State University of New York—Stony Brook. **Index.**

A 10 percent discount will be granted on all institutional standing orders placed directly with the publisher. Standing orders will be filled automatically upon publication and will continue until cancelled.